Circuit and Interconnect Design for RF and High Bit-rate Applications

ANALOG CIRCUITS AND SIGNAL PROCESSING SERIES

Consulting Editor: Mohammed Ismail. *Ohio State University*

Titles in Series:

THE GM/ID DESIGN METHODOLOGY FOR CMOS ANALOG LOW POWER INTEGRATED CIRCUITS
Jespers, Paul G.A.
ISBN-10: 0-387-47100-6
CIRCUIT AND INTERCONNECT DESIGN FOR RF AND HIGH BIT-RATE APPLICATIONS
Veenstra, Hugo, Long, John R.
ISBN: 978-1-4020-6882-9
HIGH-RESOLUTION IF-TO-BASEBAND SIGMADELTA ADC FOR CAR RADIOS
Silva, Paulo G.R., Huijsing, Johan H.
ISBN: 978-1-4020-8163-7
SILICON-BASED RF FRONT-ENDS FOR ULTRA WIDEBAND RADIOS
Safarian, Aminghasem, Heydari, Payam
ISBN: 978-1-4020-6721-1
HIGH-LEVEL MODELING AND SYNTHESIS OF ANALOG INTEGRATED SYSTEMS
Martens, Ewout S.J., Gielen, Georges
ISBN: 978-1-4020-6801-0
MULTI-BAND RF FRONT-ENDS WITH ADAPTIVE IMAGE REJECTION
A DECT/BLUETOOTH CASE STUDY
Vidojkovic, V., van der Tang, J., Leeuwenburgh, A., van Roermund, A.H.M.
ISBN: 978-1-4020-6533-0
BASEBAND ANALOG CIRCUITS FOR SOFTWARE DEFINED RADIO
Giannini, Vito, Craninckx, Jan, Baschirotto, Andrea
ISBN: 978-1-4020-6537-8
DESIGN OF HIGH VOLTAGE XDSL LINE DRIVERS IN STANDARD CMOS
Serneels, Bert, Steyaert, Michiel
ISBN: 978-1-4020-6789-1
CMOS MULTI-CHANNEL SINGLE-CHIP RECEIVERS FOR MULTI-GIGABIT OPT...
Muller, P., Leblebici, Y.
ISBN 978-1-4020-5911-7
ANALOG-BASEBAND ARCHITECTURES AND CIRCUITS
FOR MULTISTANDARD AND LOW-VOLTAGE WIRELESS TRANSCEIVERS
Mak, Pui In, U, Seng-Pan, Martins, Rui Paulo
ISBN: 978-1-4020-6432-6
FULL-CHIP NANOMETER ROUTING TECHNIQUES
Ho, Tsung-Yi, Chang, Yao-Wen, Chen, Sao-Jie
ISBN: 978-1-4020-6194-3
ANALOG CIRCUIT DESIGN TECHNIQUES AT 0.5V
Chatterjee, S., Kinget, P., Tsividis, Y., Pun, K.P.
ISBN-10: 0-387-69953-8
LOW-FREQUENCY NOISE IN ADVANCED MOS DEVICES
von Haartman, M., Östling, M.
ISBN 978-1-4020-5909-4
SWITCHED-CAPACITOR TECHNIQUES FOR HIGH-ACCURACY FILTER AND ADC...
Quinn, P.J., Roermund, A.H.M.v.
ISBN 978-1-4020-6257-5
ULTRA LOW POWER CAPACITIVE SENSOR INTERFACES
Bracke, W., Puers, R. (et al.)
ISBN 978-1-4020-6231-5
BROADBAND OPTO-ELECTRICAL RECEIVERS IN STANDARD CMOS
Hermans, C., Steyaert, M.
ISBN 978-1-4020-6221-6
CMOS SINGLE CHIP FAST FREQUENCY HOPPING SYNTHESIZERS FOR WIRELESS
MULTI-GIGAHERTZ APPLICATIONS
Bourdi, Taoufik, Kale, Izzet
ISBN: 978-1-4020-5927-8
CMOS CURRENT-MODE CIRCUITS FOR DATA COMMUNICATIONS
Yuan, Fei
ISBN: 0-387-29758-8
ADAPTIVE LOW-POWER CIRCUITS FOR WIRELESS COMMUNICATIONS
Tasic, Aleksandar, Serdijn, Wouter A., Long, John R.
ISBN: 978-1-4020-5249-1
PRECISION TEMPERATURE SENSORS IN CMOS TECHNOLOGY
Pertijs, Michiel A.P., Huijsing, Johan H.
ISBN-10: 1-4020-5257-X

Hugo Veenstra • John R. Long

Circuit and Interconnect Design for RF and High Bit-rate Applications

Springer

Dr. Hugo Veenstra
Philips Research
High Tech Campus 37
Building WY 1.004
5656 AE Eindhoven
Netherlands

Prof. Dr. John R. Long
Delft University of Technology
Electronics Research Lab.
Mekelweg 4
2628 CD Delft
ET Bldg.
Netherlands

ISBN 978-1-4020-6882-9 e-ISBN 978-1-4020-6884-3

Library of Congress Control Number: 2008926523

Printed on acid-free paper

9 8 7 6 5 4 3 2 1

springer.com

Contents

Preface

Circuit and interconnect design techniques that tackle many of the greatest difficulties and uncertainties in the development of ICs for RF and high bit-rate applications are the subject of this book. At bit-rates above 10 Gb/s, the impact of on-chip interconnect on circuit performance can be detrimental to the performance of the IC. The bottlenecks in interconnect design, circuit design and on-chip signal distribution for high bit-rate applications are analysed, and solutions for circumventing them presented. These methodologies can be applied to analyse whether target bit-rates and frequencies can be reached in a given IC technology, or to provide guidelines for further IC process optimisation in support of today's and tomorrow's high bit-rate circuit design. It should be noted that specific amplifier requirements such as low noise and intermodulation distortion are not discussed in this book.

The main topics addressed in this book and how they relate to each another are illustrated in Fig. 1. High bit-rate circuit design using advanced SiGe and InP HBT IC technologies is the core subject of this monograph.

The bottlenecks in IC design for high bit-rate applications tackled in this book are:

1. Interconnect design and modelling
2. IC process technology: transistor performance and optimisation; relevant metrics
3. On-chip signal distribution; joint optimisation of circuit and interconnect
4. Reduced breakdown voltage of transistors in next-generation IC processes
5. LC-VCO design at microwave frequencies
6. IC design flow

Several key innovations are presented in this book. For on-chip transmission lines, configurations are proposed that are minimally sensitive to their surroundings (see Fig. 2.20). These configurations enable low loss, low crosstalk and well-controlled line characteristics (e.g., characteristic impedance and delay). Improved performance margins provide flexibility in the layout floorplan of an IC. The floorplan is usually ill defined in the initial phase of an IC design. However, a design library of interconnects that are desensitized to their surroundings enables accurate predictions of the impact of interconnects on circuit performance, simplifying the floorplanning

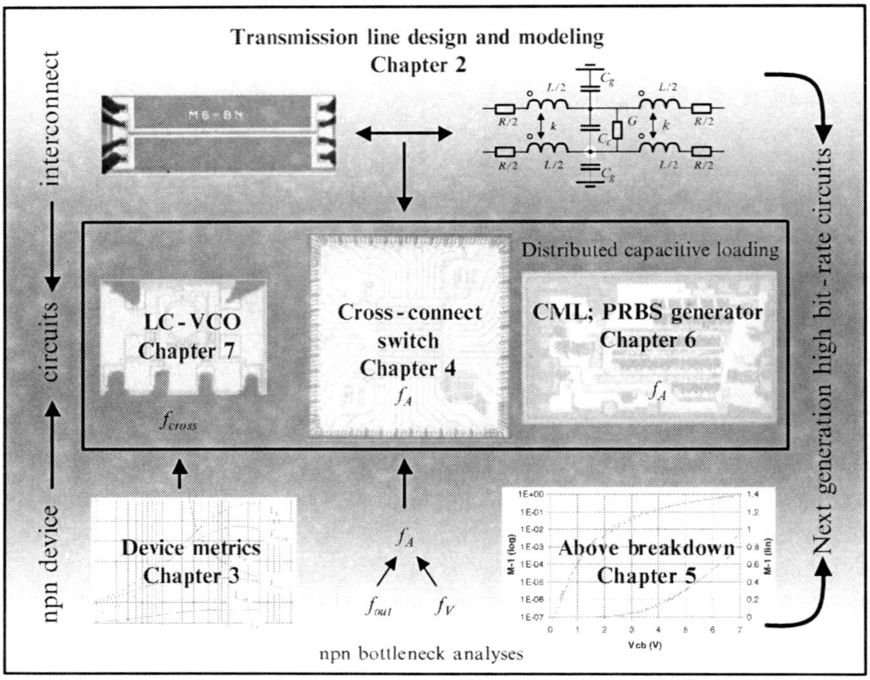

Fig. 1 Structure of the book, showing how the various topics relate to each other

phase. The proposed configurations for single-ended and differential applications can be applied in any IC technology, provided that at least two metal layers are available for interconnections.

Lumped-element transmission line models that capture the characteristic impedance, delay and loss of the lines, and the relationships between element values and the characteristic impedance and time delay are presented. The equivalent circuit model for a differential transmission line (see Fig. 2.6) provides an accurate representation of the differential and common-mode characteristics of the line. As a result, these models can be used during the initial design stages, *before* the physical layout of the transmission lines.

IC technology requirements for various RF and high bit-rate functions are similar. The figure of merit f_A, introduced in Chapter 3, is a valuable parameter when analysing the capabilities of the npn transistor in an IC process for broadband applications. Circuit design supporting bit-rates up to f_A (for highly complex circuits such as a cross-connect switch), and up to $2 \cdot f_A$ (for CML circuits of average complexity such as a PRBS generator) is feasible. Available bandwidth f_A can be further sub-divided in two parameters: the input bandwidth f_V and output bandwidth f_{out}. Analysis of f_V and f_{out} as a function of bias directly reveals which dominates when biasing the transistor at peak-f_T. This information is invaluable for IC process optimisation.

The design of a cross-connect switch IC for optical networking described in Chapter 4 is an excellent vehicle to highlight the topics addressed in this book, since it integrates many of the disciplines that are important to high bit-rate design. The cross-connect demonstrates the feasibility of a complex circuit operating at an aggregated bandwidth of up to 250 Gb/s in a SiGe process with a 12 GHz f_A. However, it also shows that significant improvements in circuit design and IC technology are needed in order to realise a similar function at 40 Gb/s per input.

When higher supply voltages are used, circuit topologies that may enable improved circuit bandwidths become feasible. Moreover, the trend towards lower breakdown voltages in modern IC processes is driven by the fact that a lower breakdown voltage BV_{CEO} usually allows a higher unity-gain frequency, f_T. Chapter 5 deals with the consequences of circuit design at supply voltages above the breakdown voltage of the transistor. When a transistor is operated at a collector-emitter voltage above its breakdown voltage BV_{CEO}, current flows out of the base terminal due to avalanche multiplication from the collector-base junction. The impact of avalanche current on bias circuits is analyzed, and avalanche current compensation techniques are introduced to improve the accuracy of current mirrors when operated above breakdown.

Chapter 6 discusses the design of high bit-rate current-mode logic circuits (e.g., 40 Gb/s and higher). A PRBS generator designed in an InP HBT technology which achieves record performance in terms of output bit-rate is used as a test and demonstration vehicle.

Many high-frequency oscillators use a cross-coupled differential pair to synthesise a negative resistance. The maximum attainable oscillation frequency for such a topology is given by metric f_{cross}, as shown in Chapter 7. Thus, f_{cross} can be used for IC process optimization, in a similar way to the unity-power-gain frequency, f_{max}.

A capacitively loaded emitter follower provides a negative shunt input resistance up to f_{LIMIT}. For practical values of the load capacitance, f_{LIMIT} is considerably higher than f_{cross}, thus enabling oscillator design at frequencies that cannot be reached with topologies based on the cross-coupled differential pair. Moreover, a capacitively loaded emitter follower can be implemented as a double emitter follower with a resistive load. The negative resistance and output buffer functions of an LC-VCO can be combined in this way, as described in Chapter 7.

The work presented in this book has been carried out at the Philips Research Laboratories, Eindhoven, the Netherlands, as part of the Philips Research program. The authors would like to thank the Philips Research management for their support and encouragement. Many colleagues and students contributed to the results presented. We wish to thank all of them, especially the valuable contributions from Fred Hurkx, Edwin van der Heijden, Hans Brekelmans, Dave van Goor, Wei Liat Chan, Jeroen Paasschens, Peter Deixler, Dennis Jeurissen, Johan Klootwijk, Jos Bergervoet, Cicero Vaucher and the Optical Networking team of Philips Semiconductors in Caen, France. We wish to acknowledge Domine Leenaerts for his constructive feedback and proofreading. This work has also been reviewed by Prof. Dr. ir. J.W. Slotboom, Prof. Dr. ir. B. Nauta, Prof. Dr. ir. A.H.M. van Roermund,

Prof. Dr. M.J.S. Steyaert, Prof. Dr. H.-M. Rein and Prof. Dr. ir. R.J. v.d. Plassche. We thank them for their constructive feedback and suggestions for improvements.

Philips Research *Dr. Ir. Hugo Veenstra*
High Tech Campus 37
5656 AE Eindhoven
The Netherlands

Delft University of Technology *Prof. Dr. John. R. Long*
Mekelweg 4
2628 CD Delft
The Netherlands

Chapter 1
The Challenge

The advance of modern IC processes has supported increasing bit-rates in many consumer and professional applications, such as hard disk drives and optical networking. Achieving a higher bit-rate by applying a new generation of an IC process for analog circuits and systems is not a simple matter of scaling existing solutions. The reduced feature size of new generations of IC technology drives the improvement of high-frequency performance of transistors and passive elements, but at the same time requires a reduction of supply voltages. This poses significant challenges to the design of high-frequency building blocks. Example applications that highlight these challenges are transceivers and cross-connect switch ICs for optical networking.

In optical networks, bit-rates in the physical layer have increased over the past two decades from 155 Mb/s to approximately 40 Gb/s (see Fig. 1.1).

Network capacity is being increased by two technologies simultaneously. One is higher data processing speeds and electronic time division multiplexing (ETDM), which drives the increase of bit-rates. The second is wavelength division multiplexing (WDM), which allows the use of multiple independent data streams per fibre, each assigned a different colour and thereby multiplying the data transmission capacity per fibre by the number of colours used. The WDM technique will not be further discussed.

Due to its high bit-rate, optical networking has been the driving force behind several generations of bipolar IC technologies. For example, IBM targets >100 Gb/s communication systems for their 0.12 μm silicon-germanium:carbon (SiGe:C), $f_T = 207\,\text{GHz}$, $f_{max} = 285\,\text{GHz}$ technology [4].

A block diagram of the physical layer of a typical optical networking system is shown in Fig. 1.2. The operation of this example implementation can be briefly explained as follows.

The transmit path of the physical layer includes a clock multiplier unit (CMU) and a multiplex (MUX) function, usually combined in a single IC. The incoming N parallel data bits are multiplexed into a high bit-rate serial stream. Usually, N equals 4 or 16, due to the hierarchical nature of the format with binary data. The voltage controlled oscillator (VCO), with oscillation frequency f_0 in this example equal to the bit frequency f_{bit}, is locked to the incoming f_{bit}/N-clock using a phase-locked

H. Veenstra and J.R. Long, *Circuit and Interconnect Design for RF and High Bit-Rate Applications*,
doi: 10.1007/978-1-4020-6884-3_1,

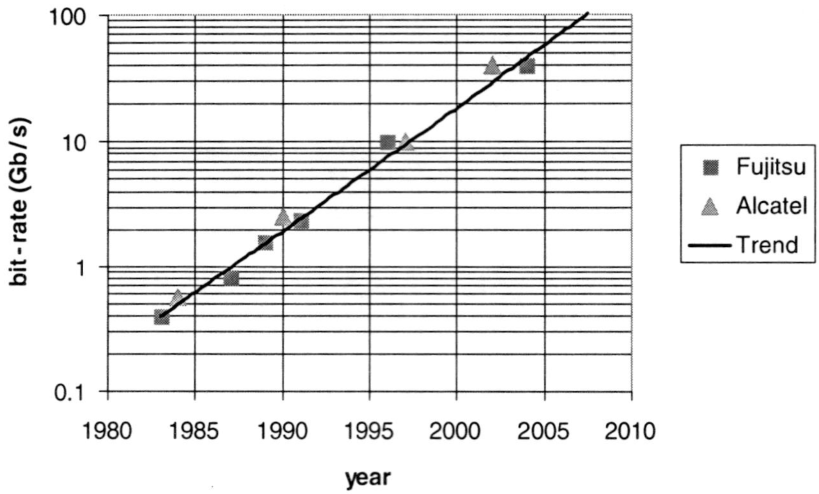

Fig. 1.1 Evolution and extrapolation of the bit-rate in optical networks [1–3]

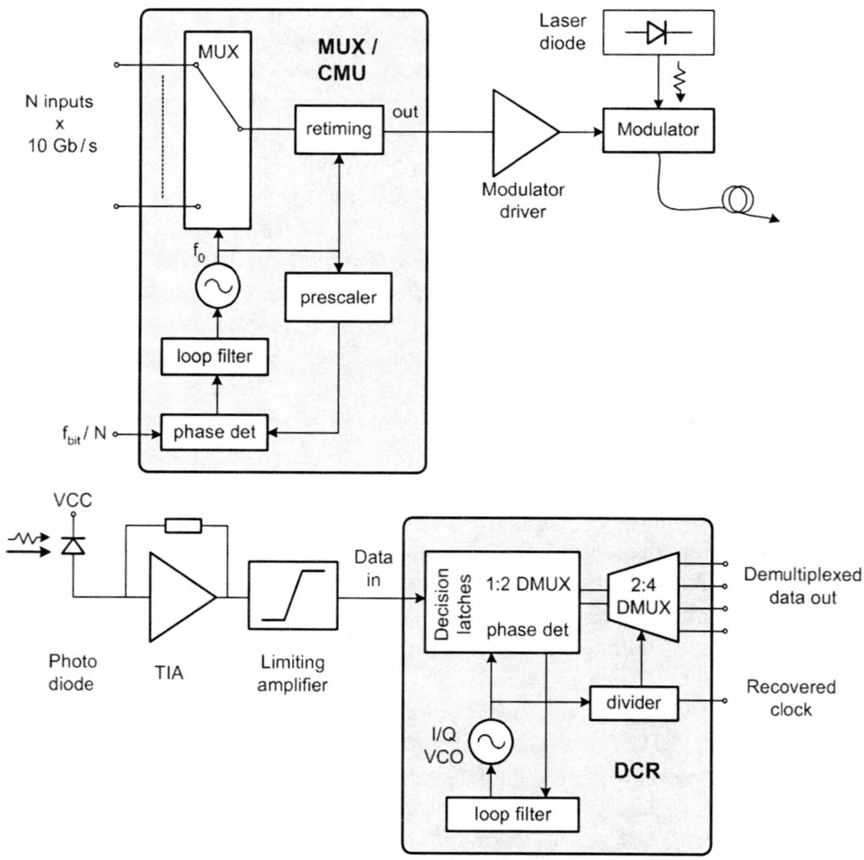

Fig. 1.2 Typical block diagram of the physical layer of an optical networking system. This example shows a full-rate MUX/CMU and half-rate DCR implementation

loop (PLL). The serialised data at the output of the multiplexer is retimed, typically using a data flip-flop (DFF) clocked at f_{bit}. This retiming, important for low jitter in the serial data, requires a full-rate transmit architecture: $f_0 = f_{bit}$. The serial data output stream is amplified by the modulator driver, driving an external modulator of the laser diode light output. This modulates the light coupled to the fibre, thereby performing electrical to optical conversion of the transmit data.

In the receive path, a photodiode converts the incoming light from the fibre into an electrical current. This current is amplified by the transimpedance amplifier (TIA). Usually, the output amplitude of the TIA is further amplified to a fixed amplitude by a limiting amplifier, driving the data and clock recovery (DCR) function. Inside the DCR, the data and the clock are recovered, and the demultiplexing function (DMUX) is usually implemented, too. The VCO inside the DCR unit needs to lock to the incoming bit-rate. Usually, a PLL performs this function.

In some high bit-rate receivers, a high-Q bandpass filter such as a dielectric resonator is used to recover the clock. This avoids the need for a VCO, but results in a receiver that operates at only a fixed bit-rate. The use of high-Q filtering is typically seen only in very high bit-rate circuits [5]. Using a PLL has the advantage of achieving a higher degree of monolithic integration, and enables operation over a wider range of input bit-rates.

The multiplexer of the transmit path is often implemented using cascaded 2-to-1 multiplexer building blocks, clocked at binary scaled frequencies ranging from f_{bit}/N for the input multiplexers to $f_{bit}/2$ for the final multiplexer. A cascade of frequency divide-by-2 circuits generates the required clock frequencies from the VCO frequency. The design of the on-chip clock distribution network is critical to the performance of the IC. The timing alignment between the multiplexers in relation to the data needs to be carefully analysed and optimised. Each of the multiplexers is usually built from current-mode logic (CML), using latches and selectors. The DCR function also uses latches for data recovery, demultiplexing and a (bang-bang) phase detector, as in for example [6, 7]. This makes the design of high-speed CML circuits an important element of high bit-rate circuit design.

For MUX/CMU output bit-rates of 10 Gb/s and beyond, the design of the fully integrated oscillator is a challenging task. The VCO needs to achieve a low phase noise, since phase noise translates into jitter at the data output. Typically, LC-type VCOs are used to meet the phase noise specification.

A half-rate CMU relaxes the required oscillator frequency by a factor of two. In return, however, the duty cycle of the VCO output signal becomes important. In a half-rate DCR system with quadrature VCO, the phase accuracy between in-phase (I) and quadrature (Q) outputs is also an important specification. Also, a large tuning range may be required for DCRs that need to support several transmission standards, operating at different bit-rates.

The DCR is sometimes implemented as full-rate, but often as half-rate, with both the I- and the Q-signals driving a DCR function operating at half the incoming bit-rate. For 40 Gb/s, systems have been published in various IC technologies such as indium-phosphide (InP), silicon-germanium (SiGe) and recently the first CMOS implementation at quarter-rate [8]. Implementing the DCR at half-rate halves the required oscillation frequency f_0, but requires the availability of in-phase and quadrature oscillator output signals for phase detection. Similarly, the quarter-rate implementation of [8] needs a 10 GHz four-phase VCO output.

The design of the on-chip clock distribution network is critical to the performance of the IC. Distribution is needed to a multitude of latches, implementing the phase detector.

To conclude, there are several critical elements for DCR and MUX/CMU performance including the VCO, CML latch and gates, clock distribution, and input/output signal amplifiers (to operate always at full-rate). The latch performance plays a highly critical role in the DCR decision function and the MUX/CMU output retiming function. In addition, the clock distribution in the transmit and receive functions is critical to the performance of the ICs.

The problems encountered in the design of DCR and MUX/CMU ICs are also involved in the design of many other ICs for high bit-rate applications. High-speed digital functions and GHz VCO circuits are for example part of ICs for high bit-rate optical networking functions with a built-in self-test feature. This book discusses the design of circuits that can be used for high bit-rate applications, for example in a cross-connect switch. In Chapter 4, the design of a cross-connect switch IC with built-in self-test will be described. This cross-connect function will be introduced below.

Optical cross-connect switches (OXCs) are widely used for routing data in optical networks. The basic topology for optical backbone networks is a ring structure with optical add drop multiplexers (OADM) and optical cross-connect switches, as in Fig. 1.3 [3].

Each ring uses multiple fibres to provide protection in the case of cable cuts. Different categories of switches exist [1]. Three example implementations of OXCs are shown in Fig. 1.4. These optical switching solutions are referred to as: electrically

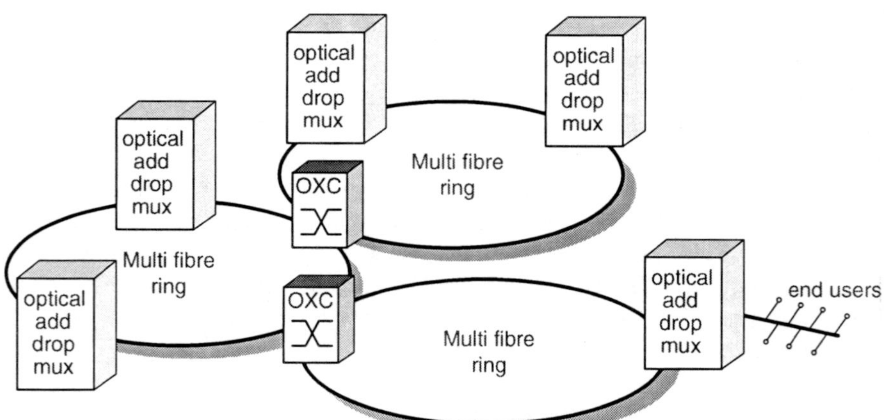

Fig. 1.3 Basic structure of an optical backbone network

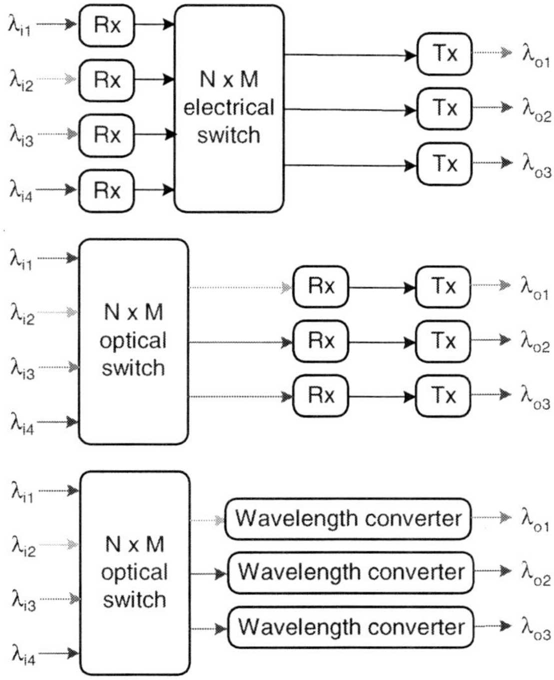

Fig. 1.4 Different solutions for OXCs

switched router/transponder (top), optically switched router/transponder (middle), and all-optical wavelength router (bottom).

Note that an *electrically* switched router/transponder is still referred to as an *optical* cross-connect switch. In all cases, wavelength routing is performed by tuning the wavelength of the output ports.

The electrically switched router/transponder is usually combined with an electrically implemented retiming function [3]. This type of switch dominates the market today. The all-optical switch solution is an interesting vehicle for research, since it allows independent bit-rate and modulation formats for the switches, but makes retiming significantly more difficult. In the following, only the electrically switched router/transponder will be considered.

The bandwidth of a switch is often expressed as aggregated bandwidth, defined as the maximum bit-rate per input multiplied by the total number of inputs. To route the data in the backbone of the network, achieving the highest possible aggregated bandwidth per switch is needed to lower cost, number of components in the switching network and thereby increase reliability. Achieving the highest aggregated bandwidth per switch IC means both achieving the largest possible number of inputs and outputs, and achieving the highest possible bit-rate per input. For many practical applications, the input bit-rate needs to support standard SDH/SONET rates such as 2.5–3.125 Gb/s or 10–12.5 Gb/s [9].

The following challenges need to be addressed for the design of high-speed switch ICs: the design of high-bandwidth input and output buffer circuits, the design of high-bandwidth matrix circuits, and distribution of all input signals through the IC with minimum jitter generation and crosstalk. This includes the design and modelling of RF interconnect.

The two high bit-rate example applications described – transceivers and cross-connect switches for optical networks – involve similar challenges for the design of the ICs, which can be summarised as:

The design of circuits and interconnect for high bit-rate and RF applications, and their combined optimisation.

This is the subject of this book. The following sections introduce the fundamental issues of this subject, relating to interconnect, IC technology, RF building blocks and design techniques.

1.1 Interconnect

In the case of nearly all high bit-rate and RF circuits, the interconnections between circuits require detailed analysis and modelling. This includes routing on printed circuit boards and assessing the effect of bondwires and on-chip interconnect. However, not all on-chip interconnect is of equal importance to the performance of the IC.

A first class of interconnect lines requiring accurate analysis and modelling are the RF signal lines. Several transmission line configurations can be used for RF interconnect. Some widely used examples are shown in Fig. 1.5.

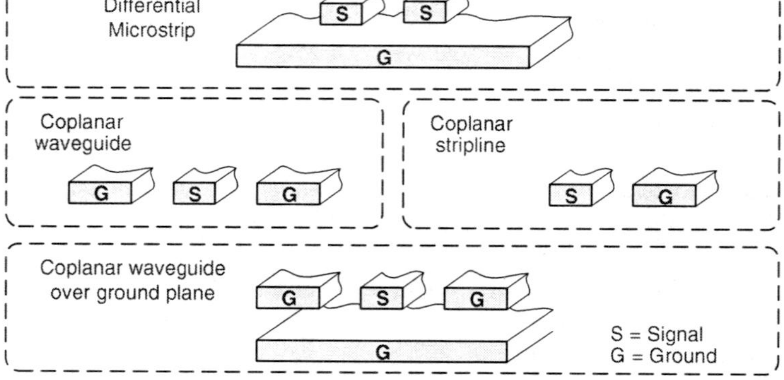

Fig. 1.5 Some widely used transmission line configurations

Transmission line models are required for computer-aided design. The term 'transmission line' in this respect means that the time delay along the line is important, meaning that the line inductance is included in the model, as for example for 10 Gb/s applications in [10] where it is recommended to use such models for all interconnects of >1.5 mm length. At 10 Gb/s, the wavelength λ of the $f = 5$ GHz fundamental of a ...0101010... pattern, assuming a relative permittivity $\varepsilon_r = 4$, equals $\lambda = c/(f \cdot \sqrt{(\varepsilon_r)}) = 3$ cm, while the on-chip physical distance between 2 bits equals 1.5 cm. Thus, the suggested 1.5 mm corresponds to 5% of the wavelength or 10% of the distance between two consecutive bits. It is common practice to use transmission line models for interconnects of length $l > 0.05 \cdot \lambda$, as suggested in [10].

In [16], a cross-connect switch implemented in gallium-arsenide (GaAs) technology is described, in which substrate losses may be ignored due to the high resistivity of the GaAs substrate. The models themselves are lumped element RLC models, describing a single-ended coplanar transmission line. The use of *differential* transmission lines is not considered, although these are widely used in differential circuit design.

In [17], the use of microstrip lines for longer RF interconnects is proposed. Lines are classified as 'critical', 'less critical' or 'non-critical', and the lengths of the 'critical' lines are minimised at the expense of increase in length of the 'non-critical' lines. This approach can also be applied to cross-connect switch ICs, but it needs to be understood which lines are critical and which lines are less critical in such an application. In cross-connect switch ICs, the chip size will readily exceed $0.05 \cdot \lambda$ in two directions and because each signal needs to travel across the complete IC, many signal lines are electrically long. The electrically long lines (including the supply lines) must be considered as transmission lines. Furthermore, other options besides microstrip interconnect are possible (see for example Fig. 1.5) that may be more attractive for cross-connect switch applications, where the transmission line density plays an important role in the chip area.

In [18], interconnect is analysed for digital (microprocessor) applications. Important parameters considered are line inductance, loss and delay. A lumped element model is presented that captures the frequency dependence of series resistance and inductance by using a parallel-network of resistors and inductors per section. Given the application, only single-ended interconnect configurations are considered.

The use of interconnect models that include both differential and common modes is mentioned in combination with RF circuit design in [20]. Here, a pseudo-random binary sequence (PRBS) generator generating a 10 Gb/s output signal is described. Post-layout simulation was done with interconnect models generated from finite element software for RF lines longer than 100 μm. Although this approach is correct, it does not provide a-priori knowledge on how to predict the influence of the RF interconnect on the signal integrity. A more structured approach for the interconnect design and modelling is needed.

Early in the design phase of high bit-rate and RF ICs, accurate (but simple) interconnect models are needed. For most applications, time domain analyses for studying (for example) jitter need to be supported. Lumped element models fulfil this requirement.

The interconnect models described in Chapter 2 will be used in combination with high bit-rate and RF circuit design in the rest of the book. Lumped element models are used for modelling selected single-ended and differential interconnect configurations. These models are only valid for interconnects shielded from the substrate. This shielding is important for minimisation of crosstalk coupled via the substrate, in order to achieve low loss and to obtain a well-controlled line impedance, independent of other interconnect and circuitry near the interconnect under study.

Another class of interconnect that deserves equal attention is the supply routing, a subject that is not very often discussed in high bit-rate and RF circuit literature. For wafer probing this is less important than for wire-bonded ICs, since the supply line inductance is typically lower. Still, supply line inductance in combination with on-chip high-Q decoupling capacitors can cause severe ringing in supply networks. Such ringing typically has a dramatic impact on all the signals in the IC. Even in differential circuits, in which signal energy at the supply line is suppressed by the common-mode rejection of the circuits, it is common practice to *evaluate* the output signals using single-ended measurements.

The decoupling strategy requires a more structured analysis for RF ICs, in order to avoid resonance while applying the best possible on-chip decoupling. The supply network needs to be analysed for potential resonance. If such resonance exists, damping may be applied to avoid ringing of the supply voltage of the circuits. For fully differential circuits, the supply decoupling strategy may differ from the strategy for single-ended circuits [17]. Several supply domains can be used on-chip; each domain requires individual supply decoupling analysis and design.

Transmission line interconnect modelling can also be applied to supply lines, in order to better understand and predict the effect of supply line impedance on circuit performance. The supply line modelling and decoupling strategy should be an integral part of the design of all microwave ICs, and will be discussed for several IC implementations described in this book.

1.2 Device Metrics

The performance constraints of transistors play an important role in fundamental circuit limitations. For example, relating circuit performance to widely accepted technology parameters allows one to predict the impact of a new technology. The most commonly used device metric is the unity current gain bandwidth, or f_T of the transistors, defined as the (extrapolated) frequency where the magnitude of the current gain, $|h_{21}|$, equals 1, as shown in Fig. 1.6.

The curve shows a typical $|h_{21}|$ as a function of frequency, together with the asymptotic response. The f_T value is derived from the asymptotic response, with the extrapolation frequency chosen in the frequency range between f_T/β_0 and f_T, at a frequency where the slope is -20 dB per decade.

Circuit performance is often benchmarked against the peak-f_T of the process. To judge whether an IC process will perform adequately in a certain application,

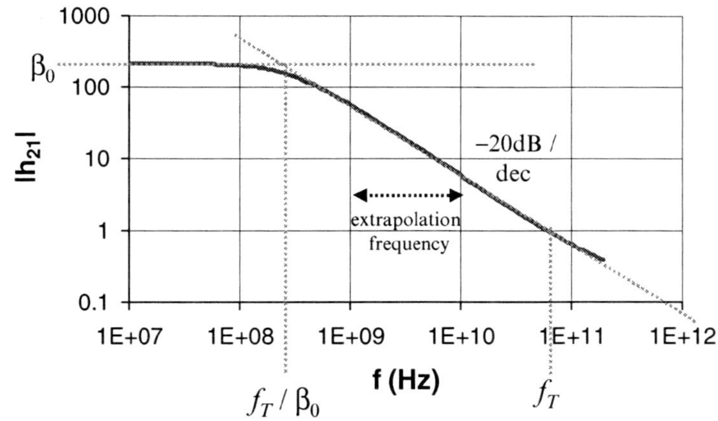

Fig. 1.6 Definition of f_T

representative building blocks such as ring oscillators and frequency dividers can be designed and characterised. CMOS or CML ring oscillators with a large number of inverters are often implemented to demonstrate the capabilities of an IC process because gate delay is derived from simple low-frequency measurements. This gate delay is an indication of the propagation delay that can be expected from more complex (digital) functions.

A more accurate performance indicator is the maximum toggle frequency of a static CML divide-by-2 circuit, because the basic cell of the static divider, the latch, also forms the basic element of many building blocks inside a high bit-rate optical networking system. The maximum toggle frequency of a bipolar CML static divide-by-2 circuit is usually related to the peak-f_T of the process. An often-used benchmark for the maximum toggle frequency of the frequency divider is $f_T/2$, although the $f_T/2$ value is an oversimplified relation [11] and therefore not simply obtained. For example, the static frequency divider described in [11] is realised in a InP bipolar technology with $f_T = 198\,\text{GHz}$ and reached a speed of 72.8 GHz. The f_T is indicative of, but not definitive for the maximum toggle rate of a frequency divider since it does not take into account all the delay contributions in circuits. To be more specific, the input bandwidth of the transistor when driving the base with a voltage source hardly affects the f_T but is important for the maximum speed of CML circuits. The f_T is hence a poor metric for CML circuits. Consequently, f_T is an important, but not the only, performance indicator for RF circuits.

The fundamental maximum frequency of oscillation that can be obtained for a single transistor is by definition equal to f_{max}, defined as the frequency at which the power gain of the transistor equals 1, assuming a conjugate match for input and output ports of the transistor. Since such a power match cannot be assumed for most oscillator circuits, the practical maximum oscillation frequency remains well below f_{max}. The maximum oscillation frequency f_{max} can be approximated by [12]

$$f_{\max} \approx \sqrt{\frac{f_T}{8\pi R_b C_{bc}}} \quad (1.1)$$

where R_b is the base series resistance and C_{bc} the base-collector capacitance. In contrast to f_T, metric f_{max} is a function of the base resistance R_b, and consequently a function of the input bandwidth of the transistor, important for the performance of many RF circuits.

In Chapter 3, the device metrics important for RF applications will be briefly reviewed. This will cover f_T and f_{max} as well as the less frequently used metrics f_A, f_V and f_{out}. In addition, a new metric f_{cross} will be introduced that relates the maximum oscillation frequency for oscillators using a cross-coupled differential pair to technology. Trends in recently published bipolar and BiCMOS IC processes targeting RF and microwave applications will be summarised. The overview of this chapter is important for high bit-rate and RF circuit design, because in this book a link is made between these device metrics and the performance of several high speed/high frequency circuits.

1.3 Cross-Connect Switches

In 1974, a monolithic 4-input, 4-output (e.g., 4×4) cross-connect switch based on CML was presented, intended for use in a space-division network for digitised video distribution [13]. Later, cross-connect switches were applied to couple high-speed processors, sharing data in a wideband communication network, as in [14].

Recent high bit-rate switches for optical networking applications are implemented in GaAs or InP technologies [15, 16, 19]. Bit-rates up to 25 Gb/s have been published in InP technology, supporting 2 inputs, achieving an aggregated bandwidth of 50 Gb/s. An aggregated bandwidth of 160 Gb/s, implemented as 16 inputs, each supporting up to 10 Gb/s, has been achieved in GaAs technology with bipolar junction transistors, the highest throughput reported up to the year 2003. These ICs do not include in situ test functionality, such as a boundary scan test or a built-in random data generator and error detector.

Some switch ICs use an architecture in which a demultiplexer is used per input, demultiplexing each input signal to M outputs. A multiplexer is used per output, selecting one out of N possible input signals. A block diagram of such an architecture is shown in Fig. 1.7 [15]. This architecture does not support multicast nor broadcast functionality, since the inputs cannot be connected to multiple outputs simultaneously. Moreover, there are a large number of wires between demultiplexer outputs and multiplexer inputs: ($N \times M$) signal paths (of which M carry an RF signal). Multicast functionality is desired, since it allows transmission of (for example) advertisements to multiple users simultaneously.

A more favourable switch IC implementation, supporting multicast and broadcast functions, requires distribution of each input signal to the inputs of all MUX circuits, leading to the architecture of Fig. 1.8. In the literature, this switch architecture has

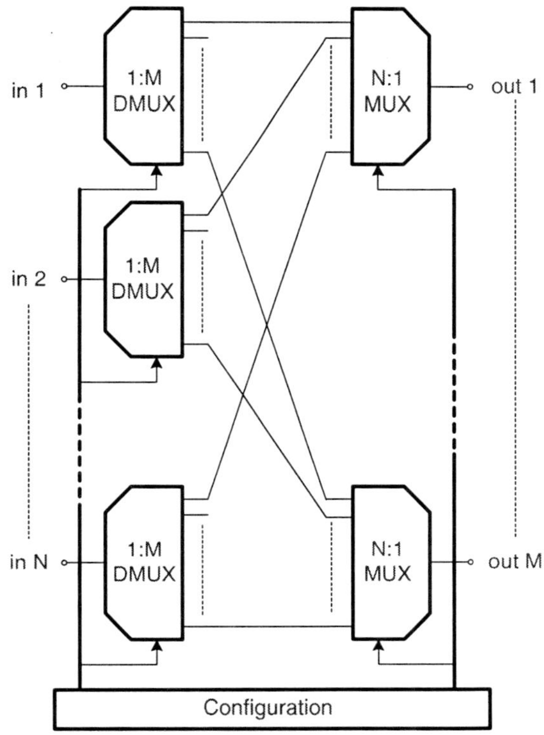

Fig. 1.7 Block diagram of an N × M cross-connect switch based on DMUX–MUX architecture [15]

been referred to as broadcast-and-select architecture [16]. Similar functionality can be achieved with a matrix architecture.

Recently, a 20-input 20-output cross-connect switch supporting up to 12.5 Gb/s per input was presented [21]. A block diagram of this IC, achieving the highest reported aggregated bandwidth to date of 250 Gb/s, is shown in Fig. 1.9. The IC is implemented in a 0.25 μm SiGe process with 70 GHz f_T [22].

The design of a cross-connect switch IC with high aggregated bandwidth poses several challenges, covering many of the subjects described in this book. The circuits of the RF path such as the input buffer, output buffer, and matrix, must be designed with sufficient bandwidth. The RF interconnect from bondpads up to input/output circuits needs to be designed and accurately modelled. The matrix circuits and RF interconnect inside the matrix need to be jointly optimised. Issues requiring attention in this context are (among others): losses in interconnect, characteristic impedance of the interconnect, interconnect configuration for low crosstalk, input/output impedance of circuits connected to the interconnect, signal transfer across loaded interconnect, power supply routing and supply decoupling. The complete RF signal path needs to be verified and optimised. For testability of the IC, a PRBS generator and error detector are included. The design of the on-chip PRBS

Fig. 1.8 Block diagram of an N × M cross-connect switch based on a distribute-MUX architecture

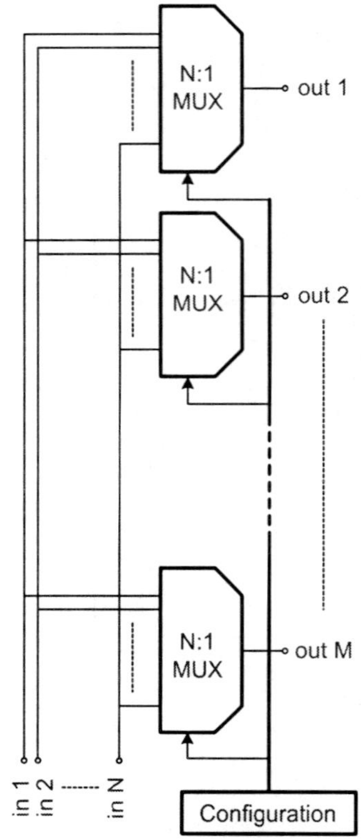

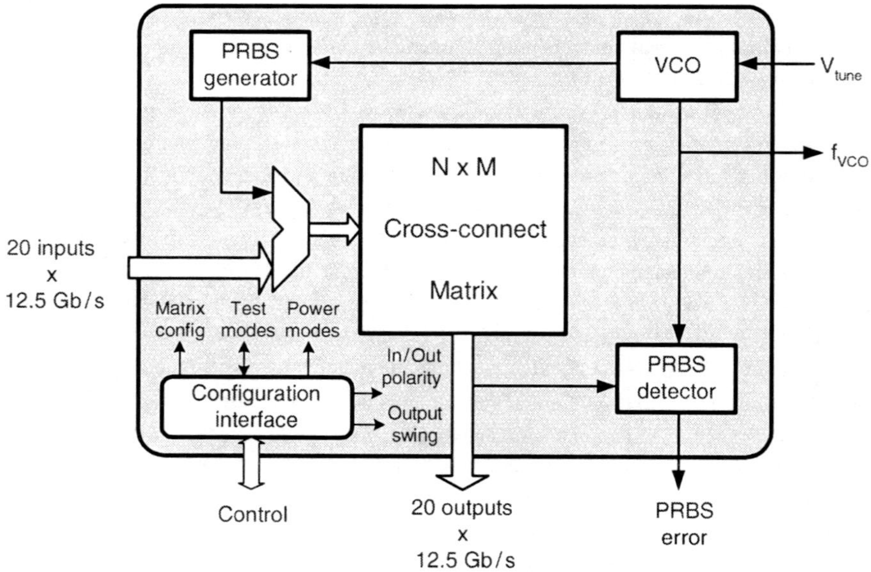

Fig. 1.9 Cross-connect switch based on a matrix architecture

generator and distribution of the PRBS signal to all inputs requires analysis of clock and PRBS data timing and distribution. The IC includes a 12.5 GHz VCO, to drive the on-chip PRBS generator and error detector.

Thus, the 12.5 Gb/s cross-connect switch IC described in Chapter 4 is an example realization of high bit-rate signal distribution and circuit design. It builds on the interconnect design and modelling techniques described in Chapter 2 and the transistor analyses based on device metrics described in Chapter 3.

To implement a similar cross-connect switch function operating at up to 40 Gb/s per input is a major challenge, and forms the framework for the building blocks employed in the rest of this book. A factor of almost 4 in speed improvement is needed in relation to the cross-connect switch described in Chapter 4. This speed improvement will come only partially from IC technology improvements (e.g., increase of f_T). Consequently, improved circuit techniques are needed to achieve 40 Gb/s.

While the cross-connect switch described in Chapter 4 operates from a supply voltage VCC $\approx$ BV_{CEO}, achieving the highest possible bit-rate for a given IC technology requires typical supply voltages well above BV_{CEO}. The problems relating to circuit operation at VCC $>$ BV_{CEO} will be addressed in Chapter 5. The challenges relating to the design of high bit-rate digital functions will be discussed within the context of a PRBS generator targeting 40 Gb/s operation in Chapter 6. The challenges relating to the design of a 40 GHz VCO will be addressed in Chapter 7.

1.4 Transistor Operation above BV_{CEO}

Another critical device parameter for RF circuit performance is BV_{CEO}, defined as the collector-emitter breakdown voltage in the open base configuration. This configuration does not occur frequently, since a relatively low impedance is typically seen from the base terminal to ground in high-speed circuits. Depending on the circuit topology, collector-emitter voltages above BV_{CEO} may be tolerated. Still, BV_{CEO} is an important parameter for the design of such circuits since it is related to the maximum useable collector-emitter voltage and thereby the possible circuit topologies.

Bipolar circuits with a supply voltage VCC above BV_{CEO} are common today. The trend towards lower breakdown voltages of modern IC processes is driven by the fact that a lower breakdown voltage BV_{CEO} usually allows a higher f_T. For a given IC technology and transistor structure, a trade-off between f_T and BV_{CEO} can be realised via the emitter to collector distance L. By approximation, the breakdown voltage scales via $BV_{CEO} \sim L$, while the transition frequency f_T scales via $f_T \sim 1/L$. The theoretical maximum attainable product $f_T \cdot BV_{CEO}$ is for silicon (Si) processes limited to $\approx$200 GHz $\cdot$ V, often referred to as the Johnson limit [38]. Although modern SiGe:C processes surpass the Johnson limit, the trade-off for a given IC process generation remains valid. The Johnson limit has recently been re-evaluated and is now believed to be $\approx$500 GHz $\cdot$ V [39].

The trend towards lower BV_{CEO} of modern SiGe and SiGe:C IC technologies, down to 1.4 V [29] combined with a typical V_{be} of 0.9 V, requires the

Fig. 1.10 NPN with collector-base avalanche current source I_{avl}

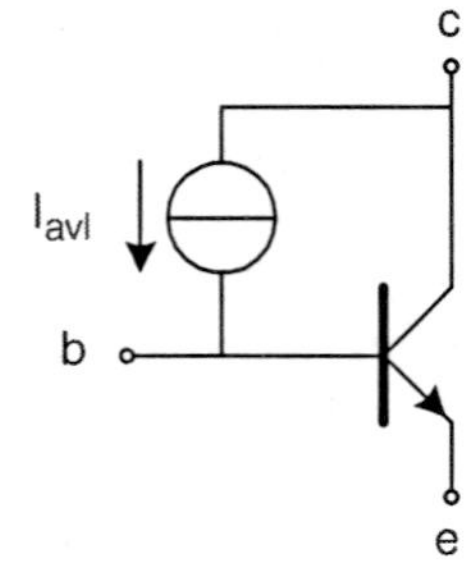

use of VCC > BV_{CEO} for many applications. Limiting the supply voltage to VCC < BV_{CEO} results in unconditional circuit safety against breakdown, but limits possible circuit topologies and thereby the maximum attainable speed. High-speed broadband circuits make extensive use of (dc-coupled) emitter followers, and thus require a supply voltage of several Volts.

When a transistor is operated at a collector-emitter voltage $V_{ce} > BV_{CEO}$ (and the base terminal is not open-circuited), the base terminal current flows out of the base terminal. This is due to the avalanche multiplication current from the base-collector junction, indicated as I_{avl} in Fig. 1.10. This avalanche multiplication current is generated due to impact ionisation [30].

From the circuit point of view, base current resulting from avalanche multiplication must be analysed and managed in the design. For example, high-speed current-mode logic such as emitter-coupled logic and double emitter-coupled logic (ECL and EECL) require supply voltages of 3–5 V, depending on common-mode biasing and the number of stacked logic inputs.

In ECL circuits, the (current-mode) logic functions, implemented using stacked and/or cascaded differential pairs, are coupled via single emitter followers. In EECL circuits, the logic functions are coupled via two cascaded emitter followers. Both ECL and EECL circuits are examples of current-mode logic implementations. Due to the low current gain $\beta(f) = i_c/i_b$ of transistors operating close to their f_T, cascading two emitter followers can increase the impedance transformation ratio and thereby reduce the input capacitance of the buffering/coupling function, making the EECL style preferable to ECL for high-speed logic [17].

The EECL current-mode logic buffer circuit shown in Fig. 1.11 demonstrates that some transistors in CML circuits may operate at $V_{ce} > BV_{CEO}$ under certain operating conditions.

In this circuit, $I \cdot R$ equals the logic swing (which is typically 0.2 V), V_{be} equals a base-emitter voltage and V_{deg} equals the degeneration voltage of current mirror Q_7/Q_8.

This circuit can be operated at supply voltages exceeding BV_{CEO}. For transistor Q_1, V_{ce} will not exceed $V_{be} + I \cdot R \approx 1.1$ V. Similarly, for Q_2 and Q_3, V_{ce} will not exceed ≈ 2.0 V and ≈ 2.9 V, respectively. These V_{ce} may exceed BV_{CEO}. To avoid this, diodes may be added in series with the collectors.

The bias current of differential pair Q_3/Q_6 defines the logic swing, and is generated using a bias current source Q_7. The collector-emitter voltage of the bias current transistor Q_7 in Fig. 1.11 equals

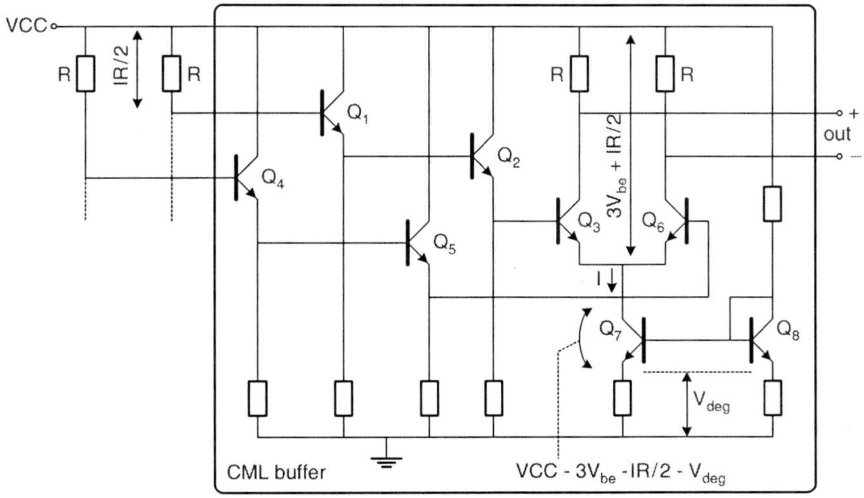

Fig. 1.11 Example EECL current-mode logic buffer circuit

$$V_{ce,Q7} = VCC - 3 \cdot V_{be} - \frac{IR}{2} - V_{\text{deg}} \tag{1.2}$$

Transistor Q_7 has to cope with a large operating range in V_{ce}, caused mainly by temperature and supply voltage variations. In the example circuit of Fig. 1.11, a typical supply voltage specification is $VCC = 5\,V +/- 10\%$, resulting in a potential 1 V variation in the V_{ce} of transistor Q_7. In addition, the collector-emitter voltage V_{ce} of transistor Q_7 may vary as much as 0.5 V due to temperature variation of the base-emitter voltages (V_{be}) of $Q_1 - Q_6$, assuming a $-40\,°C$ to $120\,°C$ operating range and $dV_{be}/dT = -1.1\,mV/°C$ for a typical SiGe process. This leads to a total 1.5 V required operating range in V_{ce} of transistor Q_7, added on top of the minimum required V_{ce}. Consequently, a V_{ce} close to 2 V may occur, which exceeds the BV_{CEO} of modern SiGe:C bipolar IC processes. In contrast to the solution proposed for transistors $Q_1 - Q_6$, addition of level shifts in the collector of Q_7 does not alleviate this problem. It is therefore of interest to study the behaviour of current sources for operation at output voltages beyond BV_{CEO}. The circuit shown in Fig. 1.11 is also representative as output buffer circuit. In such an application, the required output signal swing may force the output transistors Q_3/Q_6 to operate at collector-emitter voltages V_{ce} well above BV_{CEO}. A further example of transistor operation above BV_{CEO} is related to power amplifiers, as shown in Fig. 1.12.

The signal swing at the collector terminal may force the transistor to operate at collector-emitter voltages V_{ce} well above BV_{CEO}. How this affects circuit behaviour depends on the output impedance of the bias circuit and its capability to handle negative base currents from the power amplifier transistor.

Depending on the function, circuits for 40 Gb/s in SiGe and SiGe:C technologies operate at a typical supply voltage VCC in the range from 1 to 3 times BV_{CEO}. The

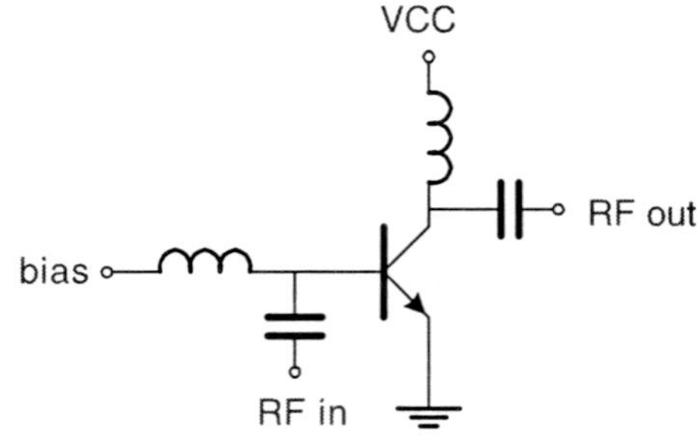

Fig. 1.12 Simplified power amplifier circuit

highest ratio is found for output driver circuits, for which output swings of several Volts are often required. Commercial ICs are available with supply voltages as high as $VCC/BV_{CEO} = 2.9$ [31]. The modulator driver for optical networking mentioned in that paper delivers an output swing of up to 3.5 V_{pp}, at a supply voltage of 5.2 V. IC technologies often provide different transistor styles. In addition to the standard high-speed transistor, a type with increased breakdown voltage BV_{CEO} (and reduced f_T) is often available. Such an increased breakdown type is particularly suitable for implementing output driver circuits.

It is common practice to operate the output transistors of biasing and driver circuits above BV_{CEO}, but below BV_{CBO}. This can be accomplished by driving the base of the output transistor by a voltage source (i.e., a source with a low output resistance) rather than a current source (or high-ohmic driving impedance). The exact limit for the output voltage as a function of circuit topology is not widely known. This problem will be addressed in Chapter 5 of this book, in which several bias circuit topologies and their behaviour at output voltages beyond BV_{CEO} will be analysed. The goal of this study is to find improved circuit implementations for bias circuits operating at output voltages continuously above BV_{CEO}.

1.5 CML Circuits, PRBS Generator

CML circuits are essential elements in many high bit-rate circuits such as static frequency dividers, phase detectors, multiplexers, demultiplexers, etc. The performance of complex digital functions can often be related to the propagation delay of the CML latch. For high bit-rate CML circuits, however, the performance is not only related to gate delays. The data and clock signal distribution also play a role in the performance, since the propagation delay across the interconnect may become a significant portion of a bit period. A pseudo-random binary sequence (PRBS) generator is an excellent example of a function with which performance is substantially improved when both CML gate delays and signal distribution are optimised.

A PRBS generator can be used to implement a built-in self-test (BIST) function in a high bit-rate application. To guarantee that such an IC meets all specifications, functions need to be tested at full speed. Testing is preferably done at several stages

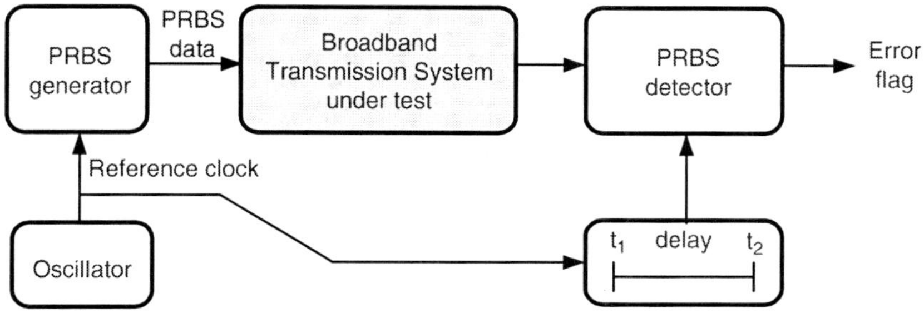

Fig. 1.13 Testing a communication system using pseudo-random data

during production. Wafer testing is performed in order to package/assemble only the samples which meet specifications. The high-speed requirements of such RF tests are now beyond the capabilities of even the most advanced test equipment for 40 Gb/s applications. A solution to this problem is to either test the fully assembled product or to provide the IC with a BIST feature.

A suitable test configuration for broadband communication systems involves applying pseudo-random data to the communication system under test, and measuring the bit-error rate at the output. This configuration is shown in Fig. 1.13. This setup can be used to test communication systems in a laboratory environment. When applying pseudo-random data to the input, eye diagrams can be generated and analysed. For example, the jitter generation from a cross-connect switch is measured by comparing jitter from the input signal with jitter from the output signal.

PRBS sequences can be generated with various lengths, but sequence lengths of $2^7 - 1$ or $2^{31} - 1$ bits are often used. PRBS data at rates up to 40 Gb/s can be generated using commercially available equipment. For example, such equipment is used for testing the DCR/DEMUX IC in [32]. The PRBS generator and bit-error rate tester (BERT) can also be included on-chip, implementing a BIST system [33]. Such systems are already being used for testing (large-scale) digital ICs. Implementing such a BIST system on a high-speed cross-connect switch IC has been demonstrated up to 3 Gb/s (e.g., the CX20462 developed by Conexant with 68 inputs and 68 outputs). Implementing a BIST system (consisting of a high-speed PRBS generator and error detector) with Gb/s bit-rates poses significant challenges in the following domains. The PRBS signal must be distributed to all inputs of the cross-connect switch. Also, the PRBS generator needs to be driven by an on-chip VCO; the design of this VCO is a challenge because of the high frequency of operation. The clock distribution inside the PRBS generator plays an important role in the maximum output bit-rate of the PRBS generator. Finally, the design of the multiplexer circuit needed to operate the cross-connect IC in the BIST mode is not straightforward, because of the high number of inputs involved in combination with the high-speed requirement.

Table 1.1 gives an overview of previously published single-chip PRBS generators. Note that all designs use two clock inputs of identical frequency, of which the phase relationship requires accurate external alignment to obtain the reported

Table 1.1 Benchmarking previously published PRBS generators

Reference	Kromat [20]	Chen [35]	Schumann [36]	Knapp [37]
Year	1998	2000	1997	2002
Max. bit-rate	11.5 Gb/s	21 Gb/s	25 Gb/s	40 Gb/s
Core bit-rate	2.875 Gb/s	10.5 Gb/s	12.5 Gb/s	20 Gb/s
Sequence	$2^{15}-1$, $2^{23}-1$	2^7-1	2^7-1	2^7-1
Auto start	Yes	No	No	Yes
Trigger out	Yes	Yes	No	Yes
# clock inputs	2	2	2	2
Technology	Si	GaAs HBT	Si	SiGe:C
f_T	25 GHz	40 GHz	50 GHz	106 GHz
Bit-rate/f_T	0.46	0.53	0.50	0.38
Size (mm^2)	4×8	3.2×3.2	1.1×0.86	0.86×0.7
Power	6.2 W	1.1 W	2.3 W	1.2 W

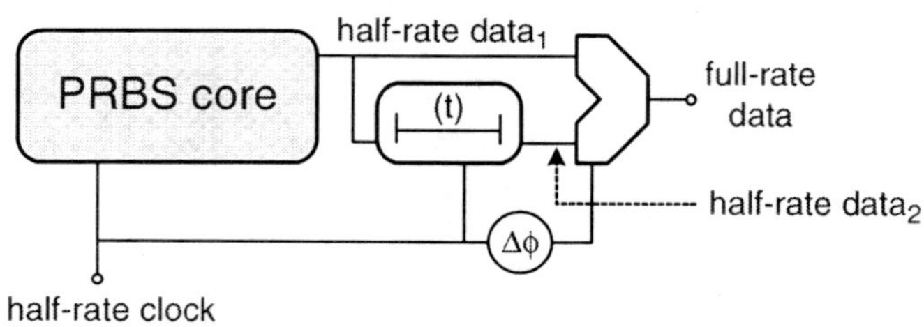

Fig. 1.14 PRBS generator requiring phase alignment of the PRBS and multiplexer clock signals

maximum bit-rate. One clock is used for driving the PRBS generator core at half of the desired bit-rate, while the other clock is used for the 2:1 multiplexer which interleaves two bit streams to realize the serial Gb/s data output, as shown in Fig. 1.14.

Having two clock inputs requiring external phase alignment makes the circuits unsuitable for BIST applications, and therefore the need for two clock inputs must be eliminated. This requires accurate modelling and analysis of the on-chip clock distribution so that correct phase alignment of the multiplexer and PRBS clocks is realized. Signal integrity across the clock lines and the effect of loading the clock lines with latches needs to be analysed and optimised in the design. One of the goals of this book is to investigate the possibility of integrating a PRBS generator for 40 Gb/s requiring only a single clock input. This is the subject of Chapter 6.

1.6 Oscillators

Many publications deal with voltage controlled oscillator (VCO) design for operation at frequencies above 1 GHz. In the clock conversion function of optical networking systems, LC oscillators are preferred due to their low jitter generation, or low phase noise when viewed in the frequency domain. In the data and clock recovery function, RC oscillators are often used since they can provide a large tuning range. In this book, only LC oscillators are considered. Many tuneable LC oscillators apply the cross-coupled differential pair to undamp the LC-tank circuit, using the basic configuration shown in Fig. 1.15.

While transistor performance dominates the circuit performance in frequency dividers, passive elements (L and C) also play an important role in LC-VCOs. Here, performance is often expressed via a more complicated figure of merit (FOM), accounting for power dissipation and phase noise [28]:

$$FOM = 10\log\left(\left(\frac{f_0}{\Delta f}\right)^2 \cdot \frac{10^{-3}}{P_d} \cdot 10^{\frac{-L}{10}}\right) \tag{1.3}$$

where f_0 is the oscillation frequency, Δf the distance from the carrier at which the phase noise L is obtained with L in dBc/Hz, and P_d is the power dissipation in mW. This FOM is widely accepted for comparing oscillator performance. It is however not the only FOM in use for VCOs. As an alternative, the tuning range may be included in the FOM [28]. The FOMs have to be used with care because different features are included in different publications (for example, for power dissipation: VCO core only, VCO core plus biasing, or VCO core plus biasing and output signal buffering). Also, values are sometimes extrapolated (for example, the frequency tuning linearised per Volt and multiplied by the supply voltage).

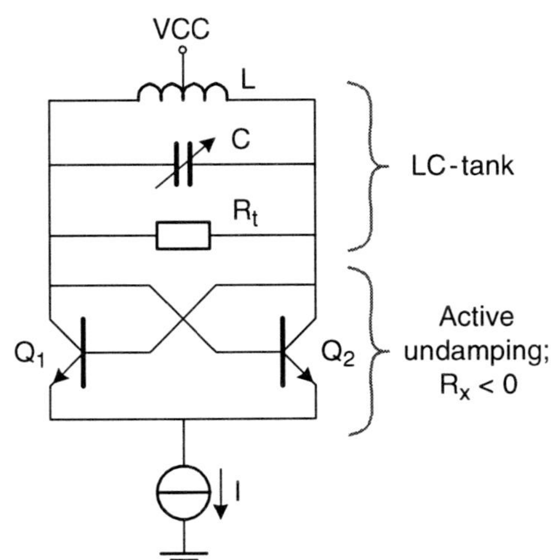

Fig. 1.15 LC oscillator using a cross-coupled differential pair (Q_1, Q_2) to compensate the losses of the tank. LC-tank losses are represented by the parallel resistance R_t

To stress the difficulties of implementing a high oscillation frequency for a given IC technology, the f_0/f_T-ratio is sometimes mentioned in addition to the FOM. Whether an IC technology provides adequate performance for reaching a certain target oscillation frequency is not addressed. It is important to understand what IC technology requirements are relevant to the implementation of (for example) a 40 GHz VCO, needed for a full-rate 40 Gb/s CMU.

To reach 40 GHz, several implementations are demonstrated in the literature. The use of frequency doublers allows an oscillator core operating at lower frequencies. In a similar way, push–push oscillators combine signal generation and frequency doubling, thereby enabling higher frequency ranges for a given technology [23]. A limitation of the use of frequency doublers or push–push oscillators is that these typically generate single-ended output signals. The first fully integrated monolithic VCO operating at 40 GHz with wide tuning range was implemented in an InP bipolar technology with $f_T = 185\text{GHz}$ [24]. This VCO is based on the circuit shown in Fig. 1.15. Since the bipolar cross-coupled pair limits the maximum swing across the tank, some variations on the topology of Fig. 1.15 exist, such as ac-coupling of the cross-coupled pair to the tank or including a level-shift between tank and cross-coupled pair with emitter follower buffers, as applied in [25] for example. Other implementations ac-couple the varactor to the tank, thereby allowing a larger voltage range for the tuning input for increased tuning range.

LC oscillators operating at up to 50 GHz in SiGe have been published in the literature [26]. The oscillators in [26] are not based on a cross-coupled differential pair. Instead, a capacitively loaded emitter follower is used to implement a negative resistance in parallel to the LC-tank. Again, the maximum attainable oscillation frequency for such a topology in a given IC technology has not yet been analysed.

In a DCR, the oscillator signal needs to drive multiple latches and/or demultiplexers. Therefore, the VCO should be able to drive an on-chip transmission line, with typical impedance levels of 40–100 Ω (single-ended). An impedance of 50 Ω is often required if the VCO signal has to be driven off-chip. Thus, buffering of the VCO signal (or signals in the case of multiple outputs) is needed to increase the output voltage swing and also to reduce loading effects on the oscillator (e.g., frequency pulling or de-Qing of the tank). Usually, an oscillator output buffer is designed as a separate building block. The input impedance of the output buffer loads the tank, however, and should be taken into account during the design of the oscillator.

I/Q oscillators are widely used for half-rate DCR functions and quadrature demodulators. In such systems the oscillator needs to provide a frequency (f_0) at half the bit-rate, with in-phase (I) and quadrature (Q) outputs. The highest oscillation frequency published for an I/Q LC-VCO so far equals 28 GHz [27]. This VCO is considered as a technology demonstrator.

It is possible to implement even more oscillator outputs at equally spaced phase differences, using multiple identical cores in a ring structure. This principle was applied in the first 40 Gb/s CMOS DCR IC [8], in which a quarter-rate DCR was implemented using four differential VCO outputs at 45° phase difference. Such systems have not yet found commercial use. One of the reasons for this is that, in sub-rate systems, the input circuit still requires full-rate bandwidth.

Often, VCO circuits are implemented with a wide tuning range. For example, a digital tuning mechanism may be added, implementing a programmable tank capacitance. This programming can be used for frequency trimming, to compensate for possible process variations [40]. In addition, digital tuning can be applied to reduce the sensitivity of the analog tuning input, df_0/dV_{tune}, important in many PLL designs for lowering the jitter. Moreover, the supply pushing, defined as $df_0/dVCC$, and generation of spurious tones may be reduced by applying a digital tuning mechanism. In reality, several iterations are often required before the on-chip LC-VCO performs according to its specifications, due to the difficulties in predicting oscillation frequency and spectral purity.

In Chapter 7, the maximum attainable oscillation frequency for the widely used VCO topology (given in Fig. 1.15) will be analysed. Furthermore, the analysis will be extended to include the oscillator topology with a capacitively loaded emitter follower. Circuit implementations will be demonstrated for both topologies, achieving an oscillation frequency approaching the theoretical limit in a given IC technology. This requires detailed analysis of the active part of the oscillator (which provides the means to undamp the LC-tank) and of the LC resonator. The degree of correspondence between predicted and measured oscillation frequencies and tuning ranges will be analysed for possible discrepancies. In all cases, 50 Ω output drivers will be included in the design.

When a capacitively loaded emitter follower is used to synthesize a negative resistance, it becomes possible to combine the 50 Ω output buffer function with the negative resistance function, as will be shown. The resulting new oscillator topology can also be used as part of an I/Q oscillator, as will be demonstrated.

1.7 Outline of the Book

In Chapter 2, theory and models for on-chip interconnect will be reviewed. First, a review of transmission line theory will be presented in such a format that it will provide easy to use rules of thumb for line impedance and delay. Equivalent lumped element models that allow usage in time domain simulators will be described. Both single-ended and differential transmission lines will be discussed. Equations will be provided, explaining how to fit the models to measured transmission line data. Experimental results showing measurement data and equivalent models for transmission lines in a modern IC technology will be discussed.

In Chapter 3, a brief review of transistor device metrics important for RF applications will be presented, such as f_T, f_{max} and f_A. Also, a new metric f_{cross} will be introduced. Trends in recently published bipolar and BiCMOS IC processes targeting RF and microwave applications will be summarised.

In Chapter 4, the design of the RF path of a 20-input, 20-output, 12.5 Gb/s per input, cross-connect switch IC for optical networking applications will be described. This will provide an excellent example of combined optimisation of RF circuits and signal distribution across long on-chip interconnect. First, the design and realization of a test IC, studying the signal transfer across unloaded and loaded transmission lines, will be described. This will form the basis for the RF path of the cross-connect IC, which will also be described.

To implement a similar cross-connect switch function operating up to 40 Gb/s per input is a major challenge, which will form the framework for the building blocks addressed in the rest of this book. A factor of 3 to 4 speed improvement is needed relative to the cross-connect switch described in Chapter 4. This speed improvement will only partially come from IC technology improvements (e.g., increase of f_T and f_{max}). Consequently, improved circuit techniques are needed to achieve 40 Gb/s.

While the cross-connect switch described in Chapter 4 operates from a supply voltage VCC $\approx$ BV_{CEO}, achieving the highest possible bit-rate for a given IC technology requires typical supply voltages well above BV_{CEO}. Thus, the BV_{CEO} of a transistor is becoming increasingly relevant for high bit-rate circuits. There is a clear trend towards lower breakdown voltages in modern IC processes, since a lower breakdown voltage BV_{CEO} usually allows a higher f_T. For a given IC technology and transistor structure, a trade-off between f_T and BV_{CEO} can be realised via the emitter to collector distance. Although a high f_T is important for high-speed circuits, a low supply voltage is a disadvantage. Therefore, it is a challenge to design circuits tolerating a supply voltage VCC $> BV_{CEO}$. When VCC $> BV_{CEO}$, there will usually be only a small number of transistors per circuit operating at $V_{ce} > BV_{CEO}$. These transistors will often be found as output transistors of biasing circuits and output driver circuits. Chapter 5 will discuss important consequences of operating biasing circuits at output voltages continuously above BV_{CEO}.

It is important to understand the consequences of operating at $V_{ce} > BV_{CEO}$. The effect for bias current sources has not yet been published. Several often-used bias circuit implementations will be analysed to assess their behaviour at high output voltages. Also, the goal is to find improved circuit implementations for the bias circuits with respect to operation at high output voltage.

Digital circuits are used in many front-end functions. Current-mode logic is usually applied for high bit-rate circuits. The pseudo-random data generator, which is the subject of Chapter 6, is interesting as a technology demonstrator, since it makes extensive use of high-speed digital circuits. The data generator can be used for self-testing high bit-rate transmission ICs. The design and realization of such a data generator targeting 40 Gb/s will be described. Clock distribution is a major issue requiring attention, since it deals with distribution of the high-frequency clock signal across relatively long distances on-chip to a multitude of latches. The optimisation of the clock signal distribution and latch design will also be described.

The VCO can be considered a general-purpose microwave systems building block. Voltage controlled oscillator (VCO) circuits using LC resonators are the subject of Chapter 7. The maximum attainable oscillation frequency for a given IC technology will be analysed. A study of the maximum attainable oscillation frequency for the classical LC-VCO with undamping via a cross-coupled differential pair will be presented. The goal is to relate this maximum frequency of oscillation to IC technology parameters. The target is to design LC-VCOs operating at an oscillation frequency close to the theoretical maximum, and to find alternative circuit proposals to implement oscillators beyond the maximum frequency when using a cross-coupled differential pair. The results of this study could be applied to the design of a 40 GHz VCO for a full-rate 40 Gb/s CMU, for example.

References

1. Y. Mochida, N. Yamaguchi, G. Ishikawa, "Technology-oriented review and vision of 40-Gb/s-based optical transport networks," *J. Lightwave Technol.*, vol. 20, No. 12, December 2002.
2. M. Kuznetsov, N.M Froberg, et al., "A next-generation optical regional access network," *IEEE Commun. Magazine*, pp. 66–72, January 2000.
3. T. Brenner, H. Preisach, B. Wedding, "Wired data communication; evolution and impact on semiconductor technologies," in *Proc. IEEE BCTM*, 2000, pp. 150–156.
4. B. Jagannathan, M. Khater, et al., "Self-aligned SiGe NPN transistors with 285 GHz f_{MAX} and 207 GHz f_T in a manufacturable technology," *IEEE Electron Device Lett.*, vol. 23, No. 5, May 2002, pp. 258–260.
5. R. Takeyari, K. Watanabe, et al., "Fully monolithically integrated 40-Gbit/s transmitter and receiver," in *Proc. OFC*, 2001, pp. WO-1–WO-3.
6. J. Hauenschild, C. Dorschky, T. Winkler bon Mohrenfels, R. Seitz, "A plastic packaged 10 Gb/s BiCMOS clock and data recovering 1:4-demultiplexer with external VCO," *IEEE J. Solid-State Circuits*, vol. 31, No. 12, December 1996, pp. 2056–2059.
7. B. Lai, R. Walker, "A Monolithic 622 Mb/s clock extraction data retiming circuit," *ISSCC Dig. Tech. Papers*, February 1991, pp. 144–145.
8. J. Lee, B. Razavi, "A 40Gb/s clock and data recovery circuit in 0.18 μm CMOS technology," *ISSCC Dig. Tech. Papers*, 2003, pp. 242–244.
9. [Online]. Available: http://www.tektronix.com/Measurement/App_Notes/SONET
10. K.S. Lowe, "Bufferless broadcasting: a low power distributed circuit technique for broadcasting 10-Gb/s chip input signals," *IEEE J. Solid-State Circuits*, vol. 32, No. 10, October 1997, pp. 1551–1555.
11. M. Sokolich, C.H. Fields, et al., "A Low-Power 72.8-GHz static frequency divider in AlInAs/InGaAs HBT technology," *IEEE J. Solid-State Circuits*, vol. 36, No. 9, September 2001, pp. 1328–1334.
12. P.A.H. Hart (ed.), *Bipolar and bipolar-MOS integration*, Elsevier, 1994.
13. M. Sunazawa, T. Hani, "Low-power crosspoint switch matrix for space-division digital-switching network," *ISSCC Dig. Tech. Papers*, 1974, pp. 206–207.
14. H. Shin, J. Warnock et al., "A 5Gb/s 16 × 16 Si-bipolar crosspoint switch," *ISSCC Dig. Tech. Papers*, 1992, pp. 128–129.
15. A.G. Metzger, C.E. Chang, et al., "A 10Gb/s 12 × 12 cross-point switch implemented with AlGaAs/GaAs heterojunction bipolar transistors," in *Proc. GaAs IC Symp.*, October 1997, pp. 109–112.
16. K.S. Lowe, "A GaAs HBT 16 × 16 10-Gb/s/channel crosspoint switch," *IEEE J. Solid-State Circuits*, vol. 32, No. 8, August 1997, pp. 1263–1268.
17. H.-M. Rein, M. Moller, "Design considerations for very-high-speed Si-bipolar IC's operating up to 50 Gb/s, *IEEE J. Solid State Circuits*, vol.17, No.8, August 1996, pp. 1076–1090.
18. B. Kleveland, X. Qi, et al., "High-frequency characterisation of on-chip digital interconnects," *IEEE J. Solid-State Circuits*, vol. 37, No. 6, June 2002, pp. 716–725.
19. M. Mokhtari, B. Kerzar, et al., "A 2V 120mA 25Gb/s 2 × 2 crosspoint switch in inP-HBT technology," *ISSCC Dig. Tech. Papers*, February 1998, pp. 204–205.
20. O. Kromat, U Langmann, G. Hanke, W.J. Hillery, "A 10-Gb/s silicon bipolar IC for PRBS testing," *IEEE J. Solid State Circuits*, vol. 33, No. 1, January 1998, pp. 76–85.
21. H. Veenstra, P. Barré, et al., "A 20-Input 20-Output 12.5Gb/s SiGe cross-point switch with less than 2ps RMS jitter," *ISSCC Dig. Tech. Papers*, 2003, pp. 174–175.

22. P. Deixler, R. Colclaser, et al., "QUBiC4G: a $f_T/f_{max} = 70/100$GHz 0.25 μm low power SiGe-BiCMOS production technology with high quality passives for 12.5Gb/s optical networking and emerging wireless applications up to 20GHz," in *Proc. IEEE BCTM*, 2002, pp. 201–204.
23. R. Wanner, G.R. Olbrich, "A hybrid fabricated 40 GHz low phase noise SiGe push–push oscillator," in *Proc. Silicon Monolithic Integrated Circuits in RF Systems*, 2003, pp. 72–75.
24. A. Kurdoghlian, M. Mokhtari, et al., "40 GHz fully integrated and differential monolithic VCO with wide tuning range in AlInAs/InGaAs HBT," in *Proc. GaAs IC Symp*, 2001, pp. 129–132.
25. P. Baltus, A. Wagemans, R. Dekker, A. Hoogstrate, H. Maas, A. Tombeur, J. van Sinderen, "A 3.5–mW, 2.5-GHz diversity receiver and a 1.2-mW 3.6-GHz VCO in silicon on anything," *IEEE J. Solid-State Circuits*, vol. 33, No. 12, December 1998, pp. 2074–2079.
26. H. Li, H.-M. Rein, "Millimeter-wave VCOs with wide tuning range and low phase noise, fully integrated in a SiGe bipolar production technology," *IEEE J. Solid-State Circuits*, vol. 38, No. 2, February 2003, pp. 184–191.
27. S. Hackl, J. Bock, G. Ritzberger, M. Wurzer, A.L. Scholtz, "A 28-GHz monolithic integrated quadrature oscillator in SiGe bipolar technology," *IEEE J. Solid-State Circuits*, vol. 38, No. 1, January 2003.
28. W. De Cock, M.J.S. Steyaert, A 2.5 V, "10 GHz fully integrated LC-VCO with integrated high-Q inductor and 30% tuning range," *Analog Integrated Circuits and Signal Processing*, vol. 33, No. 2, November 2002, pp. 137–144.
29. J.-S. Rieh, B. Jagannathan, et al., "SiGe HBTs with cut-off frequency of 350 GHz," in *Proc. IEDM*, 2002, pp. 771–774.
30. R.D. Thornton, D. de Witt, P.E. Grae, E.R. Chenette, *Characteristics and limitations of transistors*, Section 1.6, Wiley, New York, 1966.
31. G. Freeman, M. Meghelli, "40-Gb/s Circuits built from a 120-GHz f_T SiGe technology," *IEEE J. Solid-State Circuits*, vol. 37, No. 9, September 2002, pp. 1106–1114.
32. A. Ong, S. Benyamin, et al., "A 40–43Gb/s clock and data recovery IC with integrated SFI-5 1:16 demultiplexer in SiGe technology," *ISSCC Dig. Tech. Papers*, 2003, pp. 234–235.
33. H. Troy Nagle, S.C. Roy et al., "Design for testability and built-in self test: a review," *IEEE Trans. Ind. Electron.*, vol. 36, No. 2, May 1989, pp. 129–140.
34. [Online]. Available: http://www.mindspeed.com/web/products/index.jsp?catalog_id=16&cookietrail=0,1
35. M.G. Chen, J.K. Notthoff, "A 3.3-V 21-Gb/s PRBS generator in AlGaAs/GaAs HBT technology, *IEEE J. Solid State Circuits*, vol. 35, No. 9, September 2000, pp. 1266–1270.
36. F. Schumann, J. Bock, "Silicon bipolar IC for PRBS testing generates adjustable bit rates up to 25 Gbit/s," *Electronics Letters*, November 1997, pp. 2022–2023.
37. H. Knapp, M. Wurzer, T. Meister, J. Bock, K. Aufinger, "40 Gbit/s $2^7 - 1$ PRBS generator IC in SiGe bipolar technology," in *Proc. IEEE BCTM*, 2002, pp. 124–127.
38. E.O. Johnson, "Physical limitations on frequency and power parameters of transistors," *RCA Rev.*, vol. 26, 1965, p. 163.
39. K.K. Ng, M.R. Frei, C.A. King, "Reevaluation of the ftBVceo limit on Si bipolar transistors," *IEEE Trans. Electron Devices*, vol. 45, No. 8, August 1998, pp. 1854–1855.
40. A. Maxim, "A 10 GHz SiGe OC192 frequency synthesizer using a passive feed-forward loop filter and a half rate oscillator," in *Proc. ESSCIRC 2004*, pp. 363–366.

Chapter 2
Interconnect Modelling, Analysis and Design

2.1 Introduction

Circuits for high bit-rate applications should be designed efficiently with a thorough understanding of the interconnect in order to maximize their performance and efficiency. Predicting the impact of the interconnect on the circuit performance is essential for the joint optimisation of circuits and interconnect. This chapter provides an overview of interconnect modelling, analysis and design strategies. The results of this chapter will be used for high bit-rate circuit design described in the rest of this book.

Interconnect is defined as the wiring used to connect the electrical components of a circuit, such as transistors, resistors and capacitors. Since every current loop includes a return path, at least two wires are involved in every interconnection. A signal wire for the signal v, and a ground for the return path are needed for interconnect transporting a single-ended signal. The lowest possible impedance is required for the ground path. Different implementations for the ground path exist, such as a wire, a set of wires connected in parallel, a mesh, a plane, or a combination of these.

In differential circuits, two signal lines transport signals v^+ and v^-. In such configurations, the ground reference plays a role in the transport of common-mode signals.

Different models exist for interconnect transporting of a single-ended signal. The most widely used models are shown in Fig. 2.1.

These models are applicable to any line configuration intended for transport of a single-ended signal. The ground in all of the models in Fig. 2.1 is assumed to be ideal. The models in the Fig. 2.1a to e are lumped element models; those in Fig. 2.1f and g are multi-section models. The multi-section RLC model is an approximation to a transmission line model. The approximation is accurate only up to a certain frequency, depending on the number of sections per wavelength, as will be explained in Section 2.3. Moreover, the electromagnetic field lines are assumed to be contained near the interconnect geometry, leading to a transversal electromagnetic (TEM) or quasi-TEM propagating mode. Radiation losses are assumed to be absent.

H. Veenstra and J.R. Long, *Circuit and Interconnect Design for RF and High Bit-Rate Applications*, doi: 10.1007/978-1-4020-6884-3_2, © Springer Science + Business Media B.V. 2008

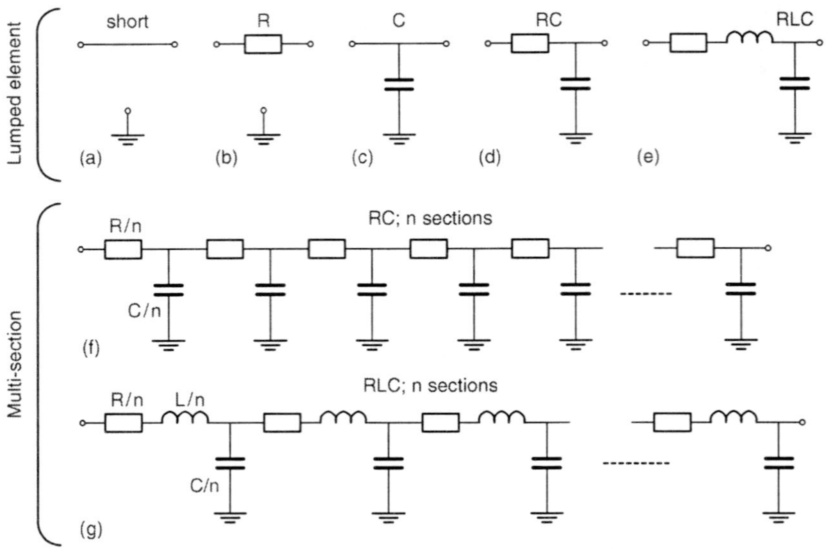

Fig. 2.1 Circuit models for interconnect transporting a single-ended signal

In this book, the multi-section RLC model is also referred to as a transmission line model. To improve accuracy, interconnect models evolve from the simple short, via lumped element models to the transmission line model. A transmission line model is required if a model for the correct line impedance and delay is desired over a wide range of frequencies. In this chapter, transmission line interconnect models are explained and used for the analysis and design of interconnect configurations. Both single-ended and differential configurations will be discussed.

Modern literature related to on on-chip interconnect analysis can be divided into two major application areas: digital and microwave. Interconnect density requirements for digital applications are usually more stringent than those for microwave applications, and more loss may be tolerated, leading to different interconnect configurations. Also, in digital applications the signals are typically driven onto the interconnect via the relatively high-impedance outputs of logic gates, and the interconnect is loaded by gate capacitances. Many different drive and load impedances may be used for microwave applications. In this chapter, the main focus is on interconnect for radio frequency (RF) and microwave applications. An overview of interconnect modelling and its behaviour for digital applications can be found for example in references [1] and [2]. A brief discussion will be presented in Section 2.11.

The definition of the terms RF and microwave requires some attention. A distinction can be made between RF and microwave on the basis of frequency range, bandwidth or the application area. The definitions of both RF and microwave change with time due to the advancement of technology. Applications that once used to be RF may now be considered analog applications. Likewise, the microwave ovens found in most households operate at 2.5 GHz – the same frequency as today's Bluetooth wireless communication ICs which are regarded as RF ICs. In this book, a differentiation is made between RF and microwave on the basis of the IC design flow.

RF ICs in the 1–5 GHz range are nowadays highly integrated functions, such as the front-end ICs for DECT, GSM and Bluetooth systems, some of which comprise more than 10,000 components per IC. Such complexity has become feasible due to the high-frequency capabilities of modern SiGe and CMOS IC processes. Such ICs are designed with a traditional analog/RF design flow, which supports these complexities in terms of number of components but with little attention to interconnect design, analysis and modelling. Other IC technologies, such as GaAs and InP, are traditionally applied in the microwave domain. High gain-bandwidth products can be obtained at the cost of a relatively high power dissipation, limiting the number of components per IC to approximately a few hundred. In microwave IC design flows, the focus is not on high complexity in terms of number of components. Moreover, microwave design flows use distributed models for interconnect and passives, and microwave circuit design is focussed on impedance matching between circuit elements (e.g., transistor and load, source and filter, etc.). Interconnect design and modelling is supported via electromagnetic simulation tools.

In this book, the traditional analog/RF IC design flow is adopted as a starting point for circuit design. Different design phases, typically described by the flow diagram of Fig. 2.2 occur within this flow. In the initial design stage, no layout is available and interconnect models are usually represented by shorts, as in Fig. 2.1a. When feasibility is demonstrated by circuit simulations, estimates for interconnect effects based on educated guesses may be included in the next design iteration. Often, only interconnect capacitance will be included for lines that are anticipated to be critical using a single lumped capacitance model. When physical layouts are generated, extraction software can be used to improve the circuit simulation accuracy. Typically, the lumped line capacitance is included for lines with a lumped capacitance exceeding a certain threshold value (usually 1 fF).

The analog/RF IC design flow does not provide sufficient accuracy for many of the critical design aspects of RF circuits, and often results in many design-processing-evaluation iterations before the ICs meet their specifications. Here, 'critical' can be defined in different ways for different applications. The first is critical with respect to signal timing. This is relevant for the phase relationship between in-phase and quadrature signals, for clock distribution and for routing two signal lines of a differential signal. In such cases the signal delay and reflections need to be accurately modelled. The second definition is critical with respect to signal amplitude, as in I/Q matching and routing of differential signals. The third is critical with respect to bandwidth and gain peaking, for example, in the case of interconnect connected to the output of an emitter follower. Here, the line impedance plays an important role in the input impedance and voltage gain of the emitter follower. The line inductance can play an important role for distribution of supply and ground paths to the circuits and supply decoupling networks. Ringing of the supply voltage critically depends on the supply and ground path inductance. Finally, the capacitance to the substrate and to other nets (e.g., other signal nets or the supply or ground nets) is an important parameter for crosstalk.

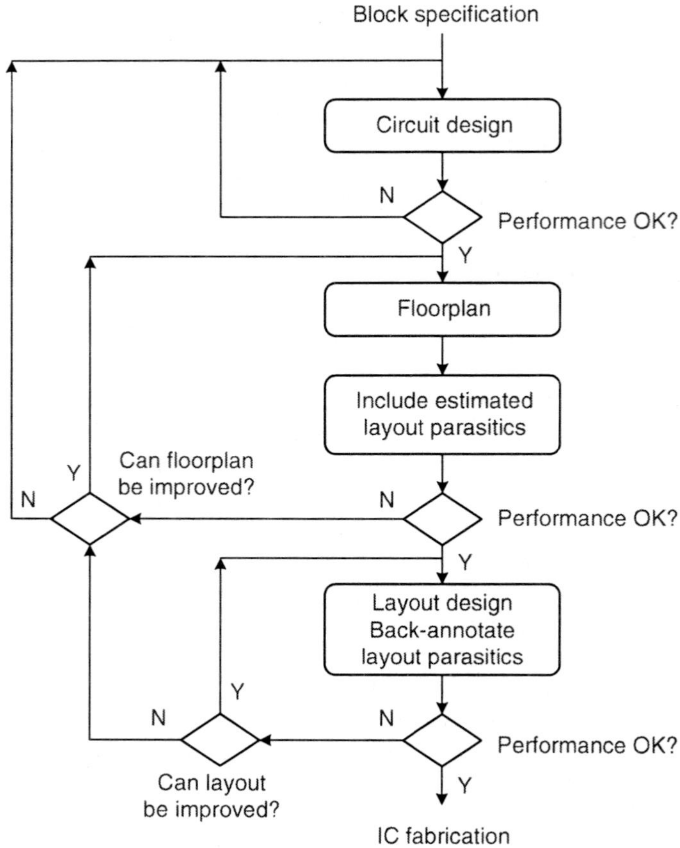

Fig. 2.2 Traditional analog/RF IC design flow

To cope with all the above-mentioned effects, the following strategy for interconnect analysis and design is adopted. As a starting point, an interconnect topology is chosen that is expected to meet the critical design aspects. An interconnect model that is appropriate for this topology is included in the circuit simulations. Depending on the length of the interconnect with respect to the wavelength of the signals on the interconnect, a lumped element or a transmission line model may be applied. On the basis of the circuit plus interconnect simulation results, the interconnect configuration and/or circuit design may be modified to optimise the overall performance.

Note that this strategy does not guarantee a first-time-right design. There are several additional aspects, such as supply decoupling, substrate connection, power supply distribution (sharing of supply pins between circuits or use of separate supply pins), etc., that also have an impact on the final IC performance. In addition, in complex system-on-chips, interactions that are not evident in sub-system test circuits may occur between blocks. Therefore, appropriate interconnect modelling and design are necessary, but they do not guarantee first-pass success with increasing chip complexity.

Building on the traditional analog/RF IC design flow shown in Fig. 2.2, this chapter considers interconnect-related aspects of the flow. Firstly, it must be understood when a simple lumped RC interconnect model is sufficiently accurate and when transmission line effects should be included. Secondary effects such as the influence of substrate and passivation layers and the skin effect on interconnect behaviour will be described. Interconnect topologies will be selected that best meet the largest possible subset of criteria for high bit-rate applications. The proposed models for lines with RF or microwave signals will include differential and common-mode behaviour where appropriate. Finally, a brief discussion of digital interconnect will be given. Since the transfer of signals over an interconnect line is a linear operation, all small-signal analyses and results presented in this chapter are also valid for large-signal operation.

2.2 Transmission Line Theory

Under certain conditions, interconnections behave as transmission lines, which have unique properties that can be described by electromagnetic theory. The important parameters of a transmission line, and their relationship to equivalent models are considered in the following sections. In all analyses, radiation losses are assumed to be absent. Also, the formulas presented are valid at frequencies where the lines do not behave dispersive so that the TEM or quasi-TEM propagation mode occurs.

2.2.1 *Single-Ended Lines*

Any two parallel conductors, one conveying the signal and the other being the reference or ground line, may be used to transport an electrical signal. The line can be regarded as a transmission line with a characteristic impedance Z_0 and a delay t_d. The ground conductor may be a wire, a number of wires, a mesh or a ground plane.

Using Maxwell's equations, the electric and magnetic fields around the conductors can be calculated and the propagation constant γ and characteristic impedance Z_0 can be found. These parameters can be related to the characteristic line parameters per unit of length R (in Ω/m), L (in H/m), C (in F/m) and G (in $1/\Omega \cdot \mathrm{m}$) with which an equivalent transmission line model can be built, which is valid per unit of length (see Fig. 2.3). Note that the model does not include radiation loss.

In this 1-dimensional multi-segment distributed model, R, L, C and G are the distributed line parameters per metre. This model is often referred to as the RLCG or RLC line model. By inspection, the Telegrapher's equations describing the voltage and current relationships can be derived:

$$\frac{i(y,t) - i(y+\Delta y,t)}{\Delta y} = G \cdot v(y^*,t) + C\frac{\partial}{dt}v(y^*,t) \tag{2.1}$$

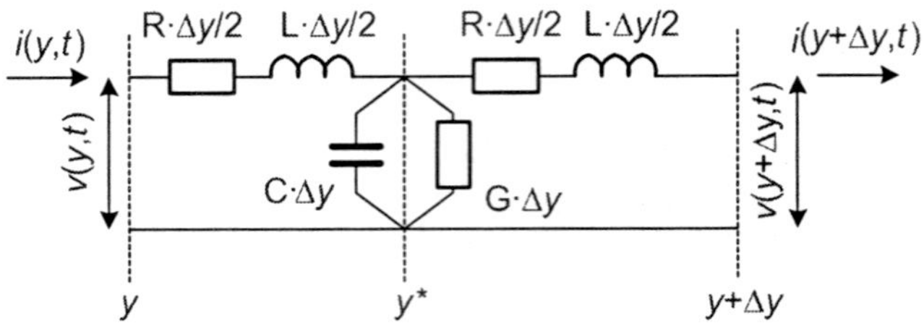

Fig. 2.3 Equivalent transmission line model representing one unit of length Δy

$$\frac{v(y,t) - v(y+\Delta y,t)}{\Delta y} = \frac{R}{2}(i(y,t) + i(y+\Delta y,t)) + \frac{L}{2}\frac{\partial}{dt}(i(y,t) + i(y+\Delta y,t)) \quad (2.2)$$

For $\Delta y \rightarrow 0$, equations (2.1) and (2.2) reduce to

$$\frac{\partial i}{\partial y} = -G \cdot v - C \cdot \frac{\partial v}{\partial t} \quad (2.3)$$

$$\frac{\partial v}{\partial y} = -R \cdot i - L \cdot \frac{\partial i}{\partial t} \quad (2.4)$$

where both v and i are a function of time t and location y, thus $v = v(y,\ t)$ and $i = i(y,\ t)$. The derivatives of the Telegrapher's equations with respect to location y are:

$$\frac{\partial^2 i}{\partial y^2} = -G\frac{\partial v}{\partial y} - C\frac{\partial^2 v}{\partial y \partial t} \quad (2.5)$$

$$\frac{\partial^2 v}{\partial y^2} = -R\frac{\partial i}{\partial y} - L\frac{\partial^2 i}{\partial y \partial t} \quad (2.6)$$

The derivatives of the Telegrapher's equations with respect to time t are:

$$\frac{\partial^2 i}{\partial y \partial t} = -G\frac{\partial v}{\partial t} - C\frac{\partial^2 v}{\partial t^2} \quad (2.7)$$

$$\frac{\partial^2 v}{\partial y \partial t} = -R\frac{\partial i}{\partial t} - L\frac{\partial^2 i}{\partial t^2} \quad (2.8)$$

Substituting (2.3) and (2.7) into (2.6) gives, after re-arranging,

$$\frac{\partial^2 v}{\partial y^2} = RGv + (RC + LG)\frac{\partial v}{\partial t} + LC\frac{\partial^2 v}{\partial t^2} \quad (2.9)$$

In a similar way, substituting (2.4) and (2.8) into (2.5) gives

$$\frac{\partial^2 i}{\partial y^2} = RGi + (RC + LG)\frac{\partial i}{\partial t} + LC\frac{\partial^2 i}{\partial t^2} \quad (2.10)$$

The general solution to both (2.9) and (2.10) is a complex exponential function. In the case of a sinusoidal excitation, the resulting voltages and currents along the line will also be sinusoidal functions of time. Therefore, the dependence on position and time of the voltages and current may be written as:

$$v = v(y,t) = v(y)e^{j(\omega t + \varphi(y))} \tag{2.11}$$

$$i = i(y,t) = i(y)e^{j(\omega t + \psi(y))} \tag{2.12}$$

Here, $v(y)$, $i(y)$, $\varphi(y)$ and $\psi(y)$ are functions of the location y only. Using the solutions of (2.11) and (2.12) in equations (2.9) and (2.10) gives

$$\frac{\partial^2 v}{\partial y^2} = v\left[RG + j\omega(RC + LG) - \omega^2 LC\right] = v(R + j\omega L)(G + j\omega C) \tag{2.13}$$

$$\frac{\partial^2 i}{\partial y^2} = i\left[RG + j\omega(RC + LG) - \omega^2 LC\right] = i(R + j\omega L)(G + j\omega C) \tag{2.14}$$

Equation (2.13) can be mapped onto the second order differential equation according to the general form

$$\frac{\partial^2 v}{\partial y^2} - \gamma^2 v = 0 \tag{2.15}$$

The general solution for equation (2.15) is

$$v(y) = K_0 e^{-\gamma y} + K_1 e^{\gamma y} \tag{2.16}$$

To map equation (2.13) onto (2.15), the complex propagation constant γ is introduced, defined as

$$\gamma = \sqrt{(R + j\omega L)(G + j\omega C)} \tag{2.17}$$

The two terms in (2.16) represent sinusoidal waves in the positive and negative y-directions. With this solution for the voltage, the current follows via equation (2.4):

$$\frac{\partial v}{\partial y} = -i(R + j\omega L) = -\gamma K_0 e^{-\gamma y} + \gamma K_1 e^{\gamma y} \tag{2.18}$$

$$i(y) = \frac{\gamma K_0 e^{-\gamma y} - \gamma K_1 e^{\gamma y}}{R + j\omega L} \tag{2.19}$$

The characteristic impedance Z_0 of the line is defined as

$$Z_0 = \frac{R + j\omega L}{\gamma} = \sqrt{\frac{R + j\omega L}{G + j\omega C}} \tag{2.20}$$

With the definition of Z_0, the current along the line from equation (2.19) can be written as

$$i(y) = \frac{K_0}{Z_0} e^{-\gamma y} - \frac{K_1}{Z_0} e^{\gamma y} \tag{2.21}$$

Note that the propagation constant γ is in general a complex value,

$$\gamma = \sqrt{(R + j\omega L)(G + j\omega C)} \mathrel{\hat{=}} \alpha + j\beta \tag{2.22}$$

Combining equations (2.21) and (2.22) gives

$$i(y) = \frac{K_0}{Z_0} e^{-\alpha y} e^{-j\beta y} - \frac{K_1}{Z_0} e^{\alpha y} e^{j\beta y} \tag{2.23}$$

From equation (2.23) it follows that the current along the line is also a sum of two sinusoidal waves, one in the positive y-direction and one in the negative y-direction. The phase and frequency of each sinusoidal wave follows from the phase constant $\mathrm{Im}(\gamma) = \beta$; the amplitudes follow from the attenuation constant $\mathrm{Re}(\gamma) = \alpha$.

All the parameters R, L, G and C are expressed per unit of length. Thus, α represents the losses in Np/(unit of length). The losses can be converted to dB/(unit of length) by multiplying with a constant factor 20/ln(10). The phase constant β represents the phase shift across the line in rad/(unit of length). A lossless line has $\mathrm{R} = \mathrm{G} = 0$, resulting in

$$\gamma = j\omega\sqrt{LC} \Rightarrow \alpha = 0; \beta = \omega\sqrt{LC} \tag{2.24}$$

The delay per unit of length can be found by taking the derivative of the phase constant. For a lossless line this gives

$$t_d = \frac{\partial \beta}{\partial \omega} = \sqrt{LC} \tag{2.25}$$

Depending on the line configuration, the simplified equations $Z_0 = \sqrt{(\mathrm{L}/\mathrm{C})}$ and $t_d = \sqrt{(\mathrm{LC})}$ often give an accurate approximation of the line characteristics for broadband applications. In the case of on-chip interconnect, this holds mostly for unloaded interconnect on a low-ohmic ground plane, shielded from the substrate and nearby unrelated wires and circuits.

The characteristic impedance is in general a complex, frequency-dependent value. For a lossless line, $\mathrm{G} = 0$ and $\mathrm{R} = 0$ and the characteristic impedance becomes a real value $Z_0 = \sqrt{(\mathrm{L}/\mathrm{C})}$. For an ideal ground plane and lossless dielectric, the main losses are due to the series resistance R in the signal line. Then, the losses from a transmission line can be judged from the phase of the characteristic impedance. At very low frequencies, e.g., at $\omega\mathrm{L} \ll \mathrm{R}$, the phase of Z_0 approaches $-45°$, while at high frequencies the characteristic impedance approaches a real value $Z_0 = \sqrt{(\mathrm{L}/\mathrm{C})}$. The interconnect between the read and write heads and the preamplifier IC in hard disk drives is a good example of interconnect on a near-ideal ground plane with $\mathrm{G} = 0$, as shown in [3]. The high-frequency characteristic impedance can still be frequency-dependent, mainly due to the frequency dependence of the inductance L via the skin effect.

The input impedance $Z_i = v(l)/i(l)$ of a uniform transmission line with length l terminated by a load impedance Z_l is analysed using the definitions of Fig. 2.4. The load is connected at the end of the line ($y = 0$) while the input of the line is at $y = l$. The choice of $y = 0$ at the end of the line where the load impedance is connected and $y > 0$ at a distance from the load leads to convenient calculations, but implies a sign reversal (e.g., $y := l - y$) for the ordinate in equations (2.16) and (2.21).

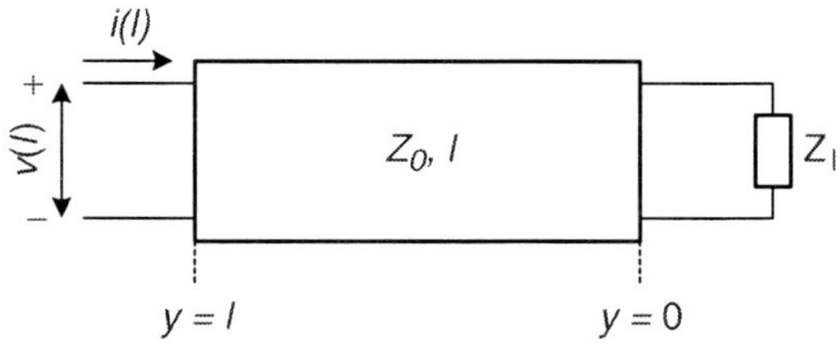

Fig. 2.4 Transmission line with length l, terminated with a load impedance Z_l

Using equations (2.16) for $v(y)$ and (2.21) for $i(y)$, the input impedance at any point y along the line can be calculated:

$$Z_i(y) = \frac{v(y)}{i(y)} = Z_0 \frac{K_2 e^{\gamma y} + K_3 e^{-\gamma y}}{K_2 e^{\gamma y} - K_3 e^{-\gamma y}} \tag{2.26}$$

The boundary condition at $y = 0$ requires $Z_i(0) = Z_l$; thus

$$Z_i(0) = Z_l = Z_0 \frac{K_2 + K_3}{K_2 - K_3} \tag{2.27}$$

The input impedance at $y = l$ follows from equations (2.26) and (2.27):

$$Z_i(l) = \frac{v(l)}{i(l)} = Z_0 \frac{K_2 e^{\gamma l} + K_3 e^{-\gamma l}}{K_2 e^{\gamma l} - K_3 e^{-\gamma l}} = Z_0 \frac{(Z_l + Z_0)e^{\gamma l} + (Z_l - Z_0)e^{-\gamma l}}{(Z_l + Z_0)e^{\gamma l} - (Z_l - Z_0)e^{-\gamma l}} \tag{2.28}$$

Using the definitions $\cosh(x) = \frac{1}{2}(e^x + e^{-x})$ and $\sinh(x) = \frac{1}{2}(e^x - e^{-x})$, the input impedance can be written as

$$Z_i(l) = Z_0 \frac{Z_l \cosh(\gamma l) + Z_0 \sinh(\gamma l)}{Z_0 \cosh(\gamma l) + Z_l \sinh(\gamma l)} = Z_0 \frac{Z_l + Z_0 \tanh(\gamma l)}{Z_0 + Z_l \tanh(\gamma l)} \tag{2.29}$$

Since $\tanh(jx) = j \cdot \tan(x)$, for a lossless line equation (2.59) reduces to

$$Z_i(l) = Z_0 \frac{Z_l + jZ_0 \tan(\beta l)}{Z_0 + jZ_l \tan(\beta l)} \tag{2.30}$$

The line input impedance is of particular interest for lines of lengths that are an integer multiple of $\lambda/4$, where λ is the wavelength of the signal on the line. The relationship between the wavelength λ and the phase constant β follows from substituting (2.22) into (2.16):

$$v(y) = K_0 e^{-\alpha y} e^{-j\beta y} + K_1 e^{\alpha y} e^{j\beta y} \tag{2.31}$$

Thus, the voltage as a function of position is a sum of two sinusoidal waves with wavelength $\beta \cdot \lambda = 2\pi$, or

$$\lambda = \frac{2\pi}{\beta} \tag{2.32}$$

Consequently, at frequencies at which the line length l corresponds to $\lambda/4$, the input impedance becomes

$$Z_i(l = \lambda/4) = Z_0 \frac{Z_l + jZ_0 \tan(\pi/2)}{Z_0 + jZ_l \tan(\pi/2)} = \frac{Z_0^2}{Z_l} \tag{2.33}$$

In the case of a uniform lossless transmission line with length $l = (\lambda/4 + \mathrm{n} \cdot \lambda/2)$, n integer, terminated with Z_l, the line input impedance will be most sensitive to the termination impedance at the end of the line. Note that when the output of the line is left open, the input impedance will behave as a short, and when the end of the line is shorted, the input will behave as an open.

In the case of a uniform lossless transmission line with length $l = (\lambda/2 + \mathrm{n} \cdot \lambda/2)$, n integer, the line input impedance will be equal to the termination impedance Z_l at the end of the line, independent of the characteristic impedance:

$$Z_i(l = \lambda/2) = Z_0 \frac{Z_l + jZ_0 \tan(\pi)}{Z_0 + jZ_l \tan(\pi)} = Z_l \tag{2.34}$$

In the case of a line with losses, $\mathrm{Re}(\gamma) = \alpha \neq 0$, the input impedance will also depend on the line length. In the case of lengths so that $\alpha \cdot l \gg 1$, the input impedance follows from equation (2.59), using $\tanh(x) \approx 1$ for $\mathrm{Re}(x) \gg 1$:

$$Z_i(l, \alpha l \gg 1) = Z_0 \frac{Z_l + Z_0 \tanh(\gamma l)}{Z_0 + Z_l \tanh(\gamma l)} \approx Z_0 \frac{Z_l + Z_0}{Z_0 + Z_l} = Z_0 \tag{2.35}$$

Thus, very long uniform lossy lines have as input impedance their characteristic impedance. Therefore, the effect of mis-termination at the output of the line will have less impact on the input impedance if this mis-termination occurs further away from the input. Obviously, the signal attenuation increases also with the line length.

2.2.2 Differential Lines

In the case of differential line configurations, capacitive and inductive coupling between the two signal lines may occur. These couplings need to be included in the equivalent circuit model. The capacitive coupling can be included by a parallel capacitance between the signal lines. The inductive coupling can be included using a coupling factor k, related to the mutual inductance M between the two signal lines via

$$k = \frac{M}{\sqrt{L_1 L_2}} \tag{2.36}$$

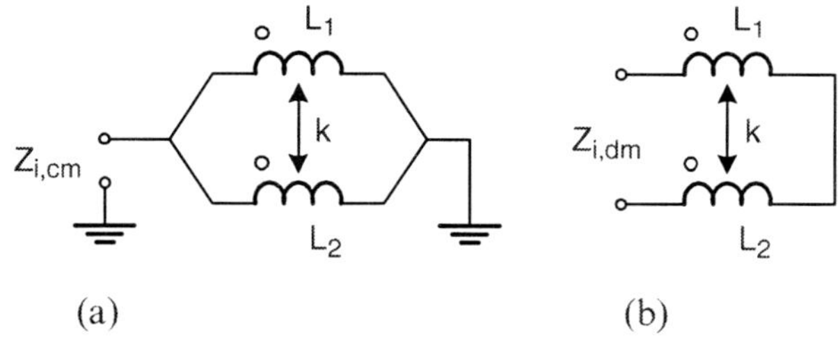

Fig. 2.5 Analysing the common-mode (a) and differential-mode (b) inductance of two coupled inductors

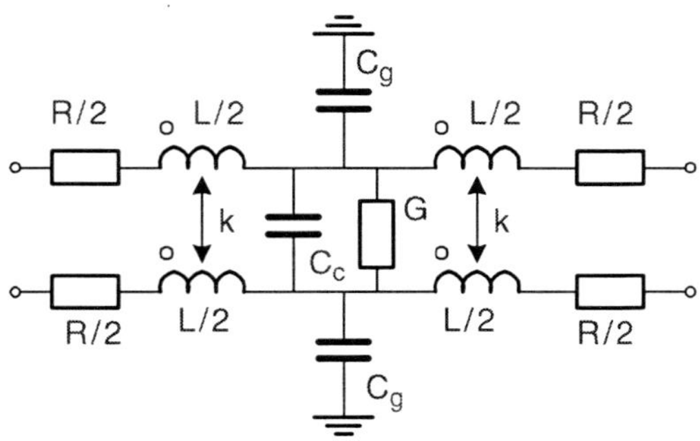

Fig. 2.6 Equivalent model for a differential transmission line, representing one unit of length

Assuming two identical signal lines that are implemented symmetrically with respect to their environment, it follows that $L_l = L_2 = L$ and thus $k = M/L$. The coupling factor k may lie in the range $k \in [-1, 1]$ and will depend on the line geometry.

In general, the differential-mode inductance L_{dm} and common-mode inductance L_{cm} for a pair of coupled inductors are defined using the analyses of Fig. 2.5.

For $L_l = L_2 = L$, this approach results in the following:

$$L_{dm} = 2L(1-k) \tag{2.37}$$

$$L_{cm} = \frac{L}{2}(1+k) \tag{2.38}$$

When the capacitive and inductive coupling between the signal lines is introduced, the equivalent circuit for a section of a differential transmission line will become as shown in Fig. 2.6.

This model can be referred to as an RLMCG or RLMC model. The model shown in Fig. 2.6 assumes that both signal lines are identical and symmetric with respect to the ideal ground. Note that when the dot convention for the inductor coupling is used as in Fig. 2.6, the coupling factor k is limited to the range $0 < k < 1$. Capacitor C_c represents the capacitance between the signal lines; capacitors C_g are the capacitance from each signal line to ground. The dielectric losses, represented by the parallel conductance G, may need to be sub-divided into a part $\alpha \cdot G$ in parallel to the capacitance C_c between the signal lines (with $0 < \alpha < 1$) and two parts

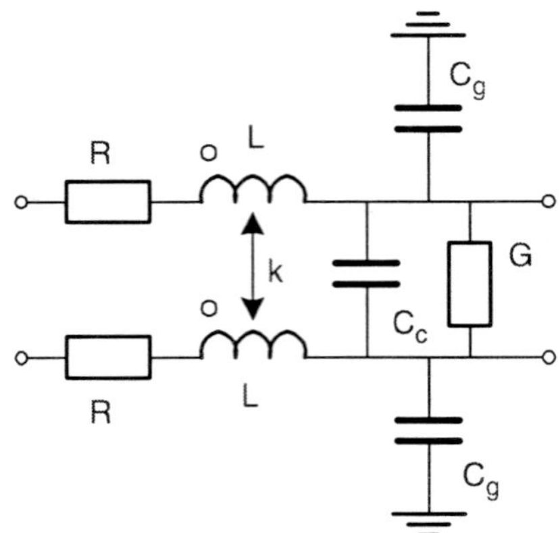

Fig. 2.7 Asymmetric equivalent model for a differential transmission line, representing one unit of length

$2(1-\alpha)\cdot G$ in parallel to each capacitance C_g to ground, to correctly divide the dielectric losses between common and differential modes. In practice, the dielectric losses will often be of minor significance, and modelling them using a single component G as in Fig. 2.6 is therefore widely accepted. Note also that the model shown in Fig. 2.6 is symmetrical with respect to the left and right sides. In the high bit-rate circuit design literature, the asymmetrical model shown in Fig. 2.7 is often used, see for example [5, 19]. When the unit of length is short with respect to the wavelength of the signals on the line, which corresponds to $\Delta y \to 0$, the difference between these approaches disappears. However, for a given number of sections, a symmetrical model provides better accuracy at high frequencies and is therefore preferred.

For a differential line there are four basic transmission line parameters: the differential-mode characteristic impedance Z_{0dm}, the differential-mode delay t_{dm}, the common-mode characteristic impedance Z_{0cm} and the common-mode delay t_{cm}. These four parameters are linked to the equivalent model via the following relationships:

$$t_{dm} = \sqrt{L_{dm}C_{dm}} = \sqrt{2L(1-k)(C_c+C_g/2)} \tag{2.39}$$

$$Z_{0dm} = \sqrt{\frac{L_{dm}}{C_{dm}}} = \sqrt{\frac{2L(1-k)}{C_c+C_g/2}} \tag{2.40}$$

$$t_{cm} = \sqrt{L_{cm}C_{cm}} = \sqrt{L(1+k)C_g} \tag{2.41}$$

$$Z_{0cm} = \sqrt{\frac{L_{cm}}{C_{cm}}} = \sqrt{\frac{L(1+k)}{4C_g}} \tag{2.42}$$

As shown by equations (2.39) to (2.42), the common-mode and differential-mode inductances (L_{cm} and L_{dm}) and capacitances (C_{cm} and C_{dm}) can be used to evaluate the differential-mode and common-mode transmission line parameters. In the literature, the terms odd mode and even-mode impedance are often used. Odd-mode impedance, Z_{odd}, is the impedance of one conductor to (virtual) ground when the pair is driven differentially. Even-mode impedance, Z_{even}, is the impedance of one conductor to ground when the pair is driven with equal signals. This leads to the following relationships [4]:

$$Z_{odd} = Z_{0dm}/2 \tag{2.43}$$

$$Z_{even} = 2 \cdot Z_{0cm} \tag{2.44}$$

In this book, the common-mode and differential-mode impedance definitions will be used, since they can be used intuitively in differential circuit design.

2.3 When to Include Transmission Line Effects

In this section, only interconnect on which RF or microwave signals are transported will be considered. In the case of digital lines, the associated RC-timeconstants will usually dominate line delays, and this results in other requirements for line modelling, as explained in [1], whereas line inductance plays a crucial role in the case of supply lines.

In [5] it is recommended to use transmission line models whenever the associated LC-delay t_d across the interconnect is equal to or larger than $t_d \geq t_r/2.5$. Here, t_r refers to the minimum rise (and/or fall) time of the signals to be transported along the interconnect. In this book, a safety factor of 4 on top of this proposal is introduced, because in the case of on-chip interconnect it is not always known a-priori how much the lumped line capacitance will increase due to crossings of other lines. Moreover, capacitive loading from circuits will increase the delay of the loaded line relative to the unloaded line.

Thus, on-chip interconnect should be modelled as a transmission line whenever the associated LC delay across the interconnect t_d is equal to (or larger than)

$$t_d \geq t_r/10 \tag{2.45}$$

In the case of single-ended applications, t_r refers to single-ended rise-time and t_d to delay as in equation (2.25); in the case of differential applications, t_r refers to the rise-time of the differential signal while t_d should be replaced by the t_{dm} of equation (2.39). A typical rise time for 10 Gb/s applications is 30 ps (20–80%). Therefore, the interconnect needs to be modelled as a transmission line if the delay t_d exceeds approximately 3 ps, scaling to 0.75 ps for 40 Gb/s applications.

For interconnect configurations with a homogeneous dielectric, the speed v of the electrical signals across unloaded interconnect will be related to the speed of light according to:

$$v \leq \frac{c}{\sqrt{\varepsilon_r \mu_r}} \tag{2.46}$$

where $c = 3 \cdot 10^8\,\mathrm{m/s}$ the speed of light, ε_r the relative permittivity and μ_r the relative permeability. This speed v is the highest speed achievable for on-chip signals. For interconnect configurations with an inhomogeneous dielectric, the relative permittivity must be replaced by the effective relative permittivity $\varepsilon_{r,eff}$. Typical values for interconnect configurations shielded from the substrate layer are $\varepsilon_r = 3.9$ (typical value for SiO_2) and $\mu_r = 1$, so $v \leq 1.5 \cdot 10^8\,\mathrm{m/s}$. As a result, in 3 ps the on-chip electrical signal travels across at most 0.45 mm distance.

The signal speed typically approaches the upper limit predicted by equation (2.46) for transmission lines shielded from the silicon substrate when the signal lines are implemented in the thick top metal layer of the RF IC process. For dense interconnect in large digital ICs, the signal speed will be well below the upper limit predicted by equation (2.46) due to additional RC-delay and/or capacitive loading of the line.

The actual value of ε_r will depend on the IC technology, the line configuration and to some extent also the package material properties. In modern IC technologies, low-k dielectrics with $\varepsilon_r \approx 3$ or less are sometimes used. These low-k dielectrics are primarily intended for minimising RC-delays in dense digital interconnect [1].

Unshielded interconnect layers close to the substrate will have electric field lines partially passing through the silicon substrate. This silicon substrate has a high $\varepsilon_r = 11.9$, lowering the speed of the electrical signal. Barrier layers for chemical mechanical polishing (CMP) are also incorporated between interconnect layers, and have a typical ε_r between 4.0–4.2. Field lines for interconnect implemented in the top-metal layer may penetrate the passivation layer plus the overlying material (e.g., plastic packaging or air).

To conclude, each line configuration will have a specific value for $\varepsilon_{r,eff}$. Once this value is known, the delay t_d across a piece of unloaded interconnect with length l will follow. This delay should be compared with the minimum rise-time of the signal to be transported. Using equation (2.45), it can be verified whether or not a transmission line model is required. For example, interconnect models for on-chip lines with a total length of up to 0.45 mm intended for 10 Gb/s communication ICs may be as simple as the associated parasitic capacitance (to substrate plus to neighbouring wires) and the series resistance. It is then not necessary to include inductive effects. If the line length exceeds 0.45 mm, considerable reflections may occur and the interconnect must be modelled as transmission line. For 40 Gb/s, this length limit scales by a factor of $^1/_4$ to a maximum line length of 0.11 mm.

2.4 Secondary Effects

In this section, secondary effects on interconnect behaviour will be analysed. The influence of the passivation layer, the substrate and the skin effect on the transmission line impedance, loss and delay will be discussed.

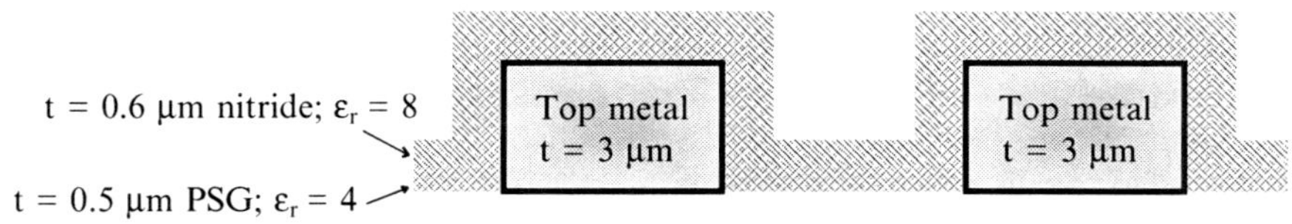

Fig. 2.8 Example configuration for two top-metal lines with a passivation layer

2.4.1 Effect of the Passivation Layer

Modern IC processes involving more than three metal interconnect layers usually include chemical-mechanical polishing to planarise the wafer before each metal layer is deposited. After the top metal layer has been deposited, the wafer no longer needs to be planarised. A nitride layer is deposited on top of the wafer as scratch protection. As a consequence, the poly-silicate glass (PSG) and nitride layers typically occur partly between the top metal lines, as visualised in Fig. 2.8.

When the thickness and material properties of the different layers are known, the exact value of $\varepsilon_{r,eff}$ can be determined, often via computer simulations. The value of $\varepsilon_{r,eff}$ for differential mode may differ from that for common mode, since the field line patterns are concentrated in different layers and/or directions. This may result in different signal delays for the differential mode and the common mode in differential interconnect configurations.

Layers with a relatively high value for ε_r, such as passivation layers, reduce the characteristic impedance and increase the delay of the line. This is most relevant to the differential-mode impedance for coplanar line configurations implemented in the top metal layer, because then the field lines are concentrated laterally between the two signal lines. If a ground layer is present underneath the signal lines, the common-mode field lines will be largely oriented vertically with respect to the ground layer and will therefore be less affected by the passivation layer above the metal. In contrast, the common-mode line parameters will depend more heavily on the ground layer properties (possibly the substrate) than the differential-mode line parameters.

2.4.2 Effect of the Substrate; Slow-Wave Effects

According to equation (2.46), the speed v of the on-chip electrical signal is related to the speed of light c via $v = c/\sqrt{(\varepsilon_{r,eff})}$, assuming $\mu_r = 1$. The factor $\sqrt{(\varepsilon_{r,eff})}$ is also referred to as the slowing factor. A typical slowing factor for a Si-based IC process for interconnect configurations shielded from the substrate is $\sqrt{(\varepsilon_{r,SiO2})} \approx 2$. Note that the slowing factor is in general frequency-dependent.

The substrate resistivity may play an important role in the RF signal transfer properties of transmission lines. Most modern SiGe BiCMOS and RF-CMOS IC processes use a substrate resistivity ρ_{sub} of 10–20 $\Omega \cdot$cm. In the case of such resistivity values, depending on the interconnect configuration and frequency, the substrate may attenuate and slow the transmitted signal.

The effect of the substrate is most pronounced in line configurations above the substrate layer, with the substrate acting as the ground layer, possibly with grounded backside metallization. For example, in microstrip configurations built from a signal line above the substrate layer, a slow-wave mode may occur for the signal transport. This can be explained as follows. The substrate provides a low-ohmic path for the electric field, thereby preventing the electric field from penetrating it. The magnetic field, however, easily penetrates the substrate due to the relatively large skin depth. Thus, the capacitance is proportional to the wire height above the substrate, while the inductance is proportional to the distance to the nearest low-ohmic ground path. The separation of electric and magnetic fields results in a slow-wave mode.

The frequency dependencies associated with multiple modes (e.g., a TEM mode and a slow-wave mode) may introduce significant timing jitter in broadband systems. Therefore, slow-wave modes are usually unwanted effects. Insertion of a metal ground shield below the signal line and above the substrate, effectively provides a boundary for both the electric and the magnetic fields, thereby avoiding slow-wave modes.

The substrate impedance may also play an important role in the high-frequency loss of a transmission line. By way of example, the coplanar line configuration shown in Fig. 2.9a behaves as a microstrip configuration above the substrate if $d \gg h$, with h being the height of the signal line above the substrate and d the lateral spacing between the signal and ground lines.

In Fig. 2.9, R_{sub} and C_{sub} represent the substrate impedance between the signal line and one ground line, C_p is the capacitance between the signal line and the substrate layer, C_l is the lateral capacitance between the signal line and one ground line and R_g represents the ground line series resistance (of one line); all expressed per unit of length. For this configuration, capacitance C_l is considerably smaller than C_p (e.g., $C_l \ll C_p$) and the impedance to ground consequently depends heavily

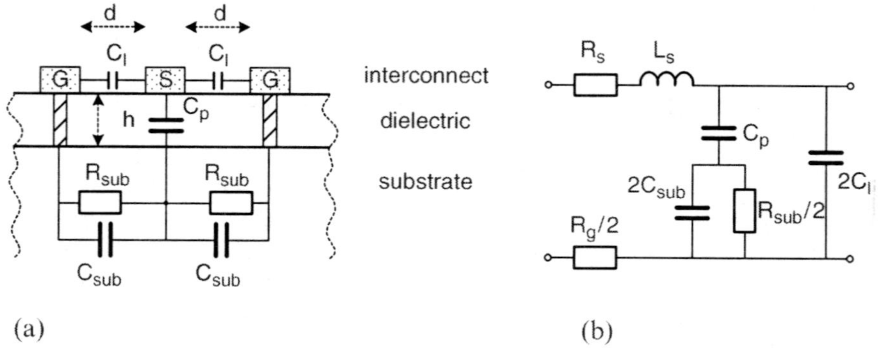

Fig. 2.9 Ground-Signal-Ground interconnect configuration above a semiconductor substrate (a) and equivalent model for a section of this line (b)

on the substrate impedance. For $d \leq h$, the configuration is referred to as coplanar waveguide (CPW). For a CPW, capacitance C_l is larger than C_p (e.g., $C_l \geq C_p$). Reducing the spacing d between the signal and ground lines increases C_l, and hence also increases the quality factor Q of the parallel impedance and reduces the high-frequency loss. The characteristic impedance of the line will also be reduced. Slow-wave effects therefore can occur in microstrip configurations that are not shielded from the substrate layer.

In an unshielded microstrip configuration there will be no nearby low-ohmic ground paths, and electric field lines will have to penetrate the substrate layer. This may result in slow-wave effects over a certain frequency range. In [6] the slowing factor for such configurations was shown to be as high as 3–4 at 10 GHz, with losses of 2.7–3.8 dB/mm at 20 GHz. The fairly substantial losses are mainly due to the low quality factor Q of the parallel impedance from the signal line to ground, e.g., the losses due to the substrate layer. The interconnect model for such slow-wave interconnect configurations on the substrate layer has to include the ground line series resistance. In fact, the model for the substrate itself in [6] is not sufficiently accurate because it ignores the substrate capacitance. A more accurate substrate model will shunt the substrate series resistance R_{sub} by a capacitance C_{sub}, as shown in Fig. 2.9. The value of the capacitance C_{sub} is independent of the substrate doping level; the substrate series resistance R_{sub} depends on the doping via $R_{sub} \sim \rho_{sub}$. The corner frequency f_ε of the substrate network, also referred to as dielectric relaxation frequency, equals

$$f_\varepsilon = \frac{1}{2\pi R_{sub} C_{sub}} = \frac{1}{2\pi \rho_{sub} \varepsilon_r} \tag{2.47}$$

Consequently, modelling the substrate as a resistance is only valid for $f \ll f_\varepsilon$, as also discussed in [7]. As an example, for a silicon substrate with $\rho_{sub} = 20\ \Omega \cdot \text{cm}$ and $\varepsilon_r = 11.9$ this yields a cut-off frequency $f_\varepsilon = 7.6\,\text{GHz}$, while f_ε will drop at higher substrate resistivities. Thus, for 10–40 Gb/s applications and $\rho_{sub} > 10\ \Omega \cdot \text{cm}$, the substrate model needs to include both resistance and capacitance. Moreover, on-chip transmission lines that are not shielded from the substrate will show a change in capacitance to ground, and hence a change in characteristic impedance and delay around f_ε.

Highly doped layers should also be modelled as a resistor in parallel to a capacitor. However, the cut-off frequency for high-doped layers is extremely high, and therefore the shunt capacitance may be ignored. For example, a $t = 1\,\mu\text{m}$ thick layer with a sheet resistance $200\ \Omega/\square$ has a resistivity $\rho = R_\square \cdot t = 0.02\ \Omega \cdot \text{cm}$, yielding a cut-off frequency $f_\varepsilon = 7.6\,\text{THz}$.

An alternative way of implementing and exploiting on-chip slow-wave structures is by using narrow metal strips placed underneath a CPW, orthogonal to the signal line [8] (see Fig. 2.10).

With this approach the substrate losses are eliminated since the stripes shield the signal line from the substrate. Like the microstrip configurations, these slow-wave interconnect configurations require a considerable chip area due to their large line widths and lateral signal to ground spacing. A slowing factor of approximately 3 has been achieved using a signal line width of 16 μm and lateral spacing to the ground lines of 20 μm.

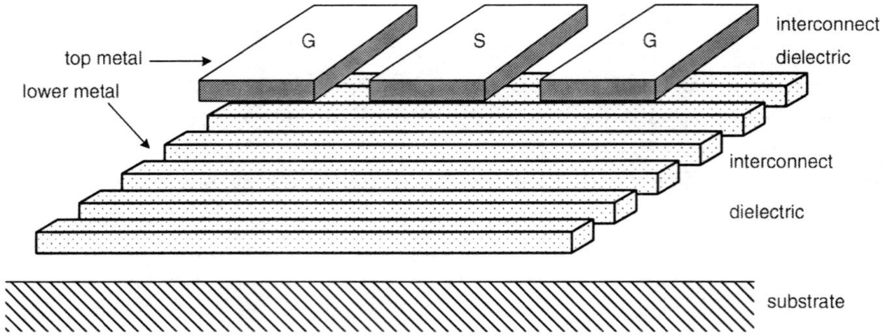

Fig. 2.10 Slow-wave CPW as presented in [8]

This approach can be useful for implementing, for example, low-loss $\lambda/4$-lines in narrowband applications needed to decouple the dc power supply domain from the RF signal domain. In broadband applications, such a low-loss slow-wave configuration can be interesting in distributed amplifier circuits, in which a transmission line connects the different amplifier stages. In most broadband applications, slow-wave effects will however be unwanted. For example, in clock distribution interconnects of digital functions such as PRBS generators and DCR circuits, the target is to minimise the clock delay between the different latches. The slowing factor of slow-wave interconnects is usually frequency-dependent, resulting in jitter in transmission of broadband data signals. Note that in the case of slow-wave interconnects, relatively short lines already require transmission line modelling and impedance matching.

2.4.3 Skin Effect

The skin effect causes the series resistance to increase and the inductance to decrease as a function of frequency. Figure 2.11 shows a visual interpretation of the skin effect for two situations: a microstrip line above a grounded substrate (Fig. 2.11a), and a differential microstrip line (Fig. 2.11b for differential mode, Fig. 2.11c for common mode) with $h \gg d$. In the differential microstrip transmission line shown in Fig. 2.11 the skin effect occurs in different directions in the common mode and the differential mode. Usually, only the differential mode is considered when analysing the skin effect in differential transmission lines [9].

To minimise the high-frequency resistance of the line, it is necessary to maximise the area of the conductor contributing to the conduction. The parts of the lines that contribute most to the conductance are oriented differently in the two cases. For a microstrip with minimum series resistance, it is best to choose a wide line width w. In the case of a differential transmission line, to minimise the series resistance for differential mode, it is best to use the thickest available interconnect. Usually, the

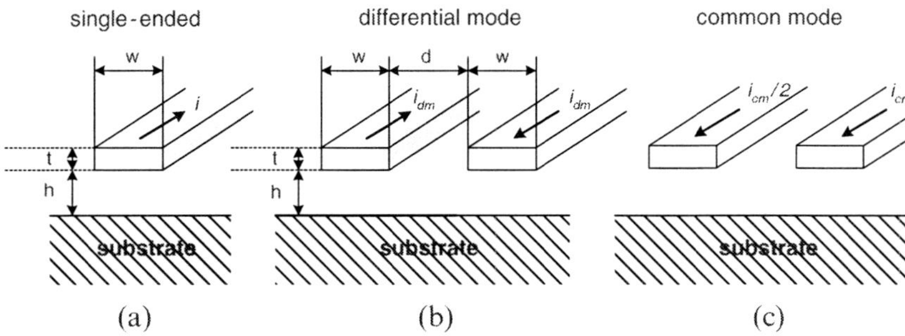

Fig. 2.11 Skin effect in a microstrip (a) and a differential microstrip line (b) and (c) above a substrate. The grey areas represent the effective skin depth δ at a certain frequency

top metal layer will have the greatest thickness t. In the case of the differential mode of the differential transmission line, there will be relatively few field lines through the substrate. As a consequence, imperfections due to the finite substrate resistivity, such as substrate losses and frequency-dependent characteristic impedances, are of less significance for the differential mode of the differential transmission line.

The effective skin depth, representing the depth of a conductor that effectively contributes to the conductance, can be calculated using the following equations (see also [11]).

For an infinitely thick conductor at a given frequency f, the skin depth δ equals

$$\delta = \frac{1}{\sqrt{\pi f \mu \sigma}} \tag{2.48}$$

with μ being the permeability of the conductor, usually $\mu = \mu_0$, and $\sigma = 1/\rho$ the conductivity. The corresponding current distribution $j(x)$ in the line equals

$$j(x) = j(0) \cdot e^{-\frac{x}{\delta}} \tag{2.49}$$

The x-direction is defined orthogonally to the surface of the signal wire, and is also referred to as the skin effect direction. At $x = \delta$, the current density equals 1/e times the current density at the surface (where $x = 0$). The current distribution for an infinitely thick conductor is visualised in Fig. 2.12.

For a wire with a finite dimension w in the skin effect direction x, the total current i has to distribute along the wire between $0 \leq x \leq$ w:

$$i = \int_0^w j(0) \cdot e^{\frac{-x}{\delta}} dx \tag{2.50}$$

Note that (2.50) holds for every frequency f, and δ can be evaluated from equation (2.48). The effective skin depth in the x-direction, δ_x, for this wire with finite width follows from solving (2.50):

$$i = j(0) \cdot \delta (1 - e^{\frac{-w}{\delta}}) \equiv j(0) \cdot \delta_x \tag{2.51}$$

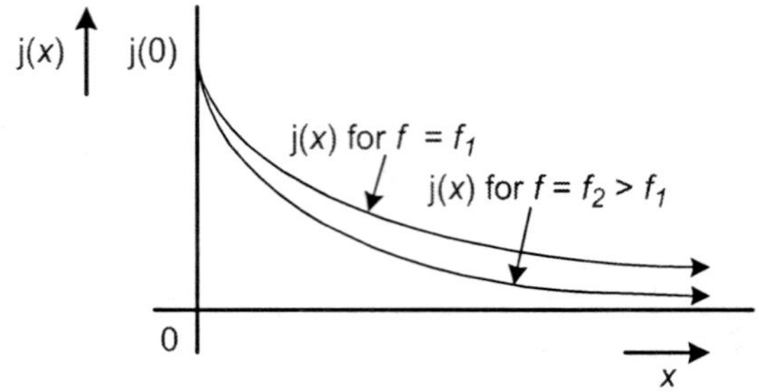

Fig. 2.12 Current distribution at frequencies f_1 and $f_2 > f_1$ in an infinitely thick conductor

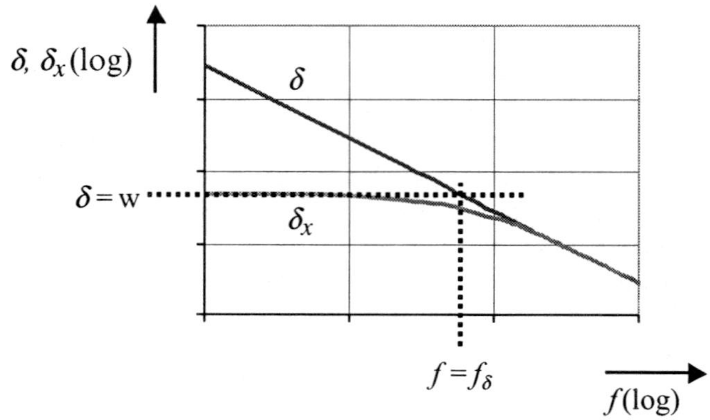

Fig. 2.13 Skin depth δ, effective skin depth δ_x and definition of the skin effect corner frequency f_δ

The relationship between δ and δ_x for a line with a finite width w in the skin effect direction x is visualised in Fig. 2.13.

The series resistance of the line can now be calculated for all frequencies, using (for the differential mode of the differential line configuration of Fig. 2.11)

$$R = \frac{\rho l}{A} = \frac{\rho l}{\delta_x t} \tag{2.52}$$

with l being the line length. Thus, the series resistance R remains close to R_{dc} for $f < f_\delta$, and increases proportionally to the frequency, $R \sim \sqrt{(f)}$, for $f > f_\delta$. From equation (2.48) it follows that for a given geometry of the interconnect, a different resistivity $\rho = 1/\sigma$ of the material will also result in a different skin depth. This is visualised in Fig. 2.14.

For example, when the interconnect material is changed from aluminium ($\rho_{Al} = 27 \cdot 10^{-9}\ \Omega \cdot \mathrm{m}$) to copper ($\rho_{Cu} = 17 \cdot 10^{-9}\ \Omega \cdot \mathrm{m}$), the resistance will decrease for all frequencies, even though the skin effect corner frequency f_δ will decrease. At $f = f_\delta$, the skin depth equals the wire size (thickness or width) in the skin effect direction. The advantage of copper is most pronounced at low

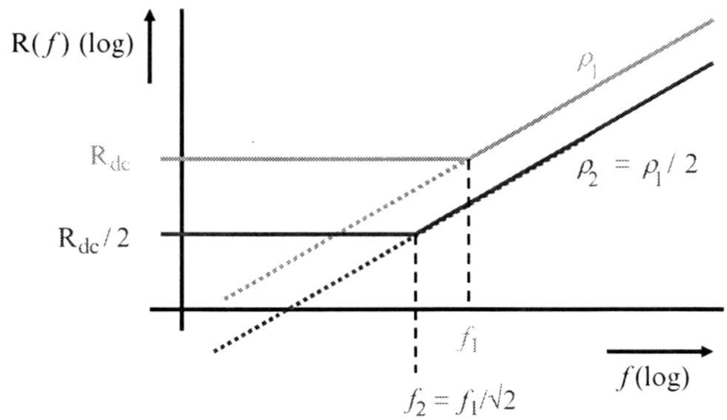

Fig. 2.14 Resistance R(f) for a given geometry, for resistivities ρ_1 and $\rho_2 = \rho_1/2$

frequencies at which the resistance scales by the ratio ρ_{Cu}/ρ_{Al}. Beyond the skin effect corner frequency f_δ, the resistance scales by a ratio $\sqrt{(\rho_{Cu}/\rho_{Al})}$.

To find the ground path series resistance for line configurations above the substrate, as for microstrip configurations and for the common-mode behaviour of a differential stripline above the substrate, it is necessary to calculate the skin depth of the substrate using equation (2.48). The resistivity $\rho = \rho_{sub}$ follows directly from the electron and hole concentration of the substrate, according to

$$\rho_{sub} = \frac{1}{q(n\mu_n + p\mu_p)} \tag{2.53}$$

Here, μ_n and μ_p are the mobility of electrons and holes, respectively; n and p are the electron and hole densities. Usually, a p-type doped substrate is used.

In the case of line configurations with a current return path through the substrate, the skin depth should also be calculated for the substrate layer. Since the resistivity of the substrate layer will be several orders of magnitude higher than that of the metal layers, the skin depth will also be significantly greater, as follows from equation (2.48). For example, at a substrate resistivity of $\rho_{sub} = 20\ \Omega \cdot \text{cm}$ at $f = 5\,\text{GHz}$, the skin depth equals 3.2 mm. This is more than the typical thickness of the substrate layer, and consequently the substrate is nearly transparent at RF and microwave frequencies.

Not only the x-direction contributes to the conductance of the wire. In practice, the y-direction will also contribute, increasing the effective skin depth and thereby lowering the high-frequency resistance. This is visualised for the differential mode of a differential transmission line with h$\gg$d in Fig. 2.15.

In [11] an empirical formula is given for the effect of conductor thickness, intended for wires with a w/t ratio close to unity. The skin depth contribution in the y-direction is related to the skin depth in the x-direction according to

$$\delta_y = \delta_x \frac{w}{t} \tag{2.54}$$

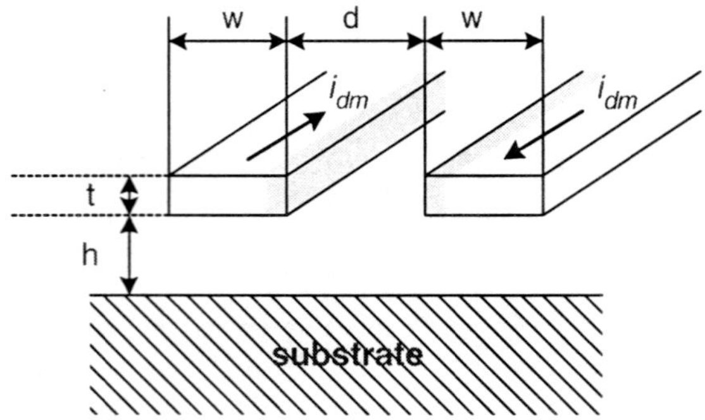

Fig. 2.15 Conductor thickness contributes to the conductance, increasing the effective skin depth

The total effective skin depth δ^* equals the sum of the skin depths in the x- and y-directions:

$$\delta^* = \delta_x + \delta_y = \delta_x\left(1 + \frac{w}{t}\right) \tag{2.55}$$

This empirical correction results in a frequency-independent correction factor. A linear correction term, proportional to the line width for the lateral skin effect, was also used in [2] in which an excellent fit for measured versus modelled series resistance was demonstrated using $\delta^* = \delta_x \cdot (1 + \mathrm{w}/20 \cdot 10^{-6})$. This correction was demonstrated to be accurate for line widths between 1 µm and 40 µm, so also for $\mathrm{w/t} \gg 1$. It can be argued that the correction factor for thickness effects should account for the actual, frequency-dependent, skin depth to avoid the part of the conductor within the skin depth being counted twice. The skin depth correction factor $\delta_y{}'$ in the y-direction, replacing equation (2.54), will then be

$$\delta_y' = \delta_x \frac{w - \delta_x}{t} \tag{2.56}$$

This leads to the corrected total effective skin depth δ'

$$\delta' = \delta_x + \delta_y' = \delta_x\left(1 + \frac{w - \delta_x}{t}\right) \tag{2.57}$$

The difference between δ^* and δ' is significant around the skin effect corner frequency f_δ. At frequencies $f \gg f_\delta$, the effective skin depth in the x-direction will be small with respect to the wire width, $\delta_x \ll w$, and then the correction factors for the skin effect in the y-direction, equations (2.54) and (2.56), will give almost identical results.

To summarise, the skin depth δ of a wire of infinite dimension in the skin effect direction depends on the frequency f and the resistivity ρ, see equation (2.48). In the case of a practical wire with a finite size in the skin depth direction (here width w in the x-direction) and a height t, the effective skin depth δ_x will depend on the skin depth δ and the width w of the wire, see equation (2.51) and Fig. 2.13. In the case of the same practical wire with a finite width w and a height t, thickness effects in the *y*-direction will also contribute to the conductance, resulting in the total effective skin depth δ^*, see equation (2.55). An alternative correction factor has been proposed to avoid the conductor area within the skin depth being counted twice, see equation (2.57).

It is possible to optimise the geometry of a line to minimise the series resistance. To minimise the high-frequency series resistance for a coplanar differential transmission line, the lines should be as thick as possible. The top metal layer will usually be the thickest available metal, and is therefore the preferred choice.

2.5 Resistivity-Frequency Mode Chart for a Microstrip Line

Depending on the transmission line configuration, the semiconductor resistivity can play a major role in the transmission line characteristics. The influence of the semiconductor on the transmission line properties is most pronounced in the case of microstrip lines implemented in a metal-insulator-semiconductor (MIS) configuration (see Fig. 2.16). The behaviour of such lines has been thoroughly analysed in [12] and [13], of which this section provides a summary.

There are three fundamental operating modes for such a configuration: dielectric quasi-TEM mode, skin effect mode and slow-wave mode. These modes are a function of the frequency and substrate resistivity, and can be visualised in a resistivity-frequency mode chart. The transitions between the three modes are a function of the skin depth δ, substrate resistivity ρ_{sub} and permittivity ε_{rsub}, insulator permittivity ε_{r1}, and insulator and semiconductor thickness b_1 and b_2.

Using equation (2.48), the characteristic frequency f_δ for the skin effect in the semiconductor layer, where the skin depth δ equals the substrate thickness b_2, can be derived:

$$f_\delta = \frac{\rho_{sub}}{\pi \mu b_2^2} \tag{2.58}$$

The frequency f_δ represents the limit at which the magnetic field fully penetrates the substrate layer; for $f < f_\delta$, the skin depth is larger than the substrate thickness b_2.

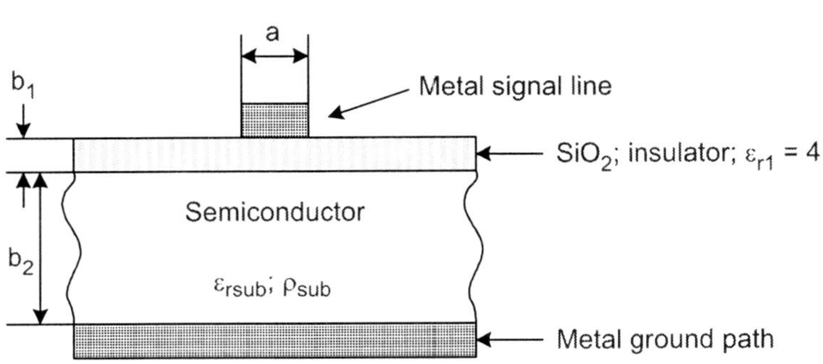

Fig. 2.16 Metal-insulator-semiconductor microstrip line configuration

The dielectric relaxation frequency f_ε of the substrate follows from equation (2.47)

$$f_\varepsilon = \frac{1}{2\pi\rho_{sub}\varepsilon_{rsub}} \tag{2.59}$$

In the case of $f > f_\varepsilon$ the substrate will act as a dielectric, and the resulting signal transport mode is referred to as the dielectric quasi-TEM mode. The speed v_{TEM} of the electrical signal in this mode is for $b_2 \gg b_1$ mainly determined by the permittivity of the substrate layer, according to

$$v_{TEM} = \frac{c}{\sqrt{\frac{b_1+b_2}{\frac{b_1}{\varepsilon_{r1}}+\frac{b_2}{\varepsilon_{rsub}}}}} \tag{2.60}$$

A typical value is $v_{TEM} \approx c/\sqrt{12}$.

The relaxation frequency f_s of the interfacial polarization is defined as

$$f_s = \frac{b_1}{b_2}\frac{1}{2\pi\rho_{sub}\varepsilon_{r1}} \tag{2.61}$$

In the case of $f < f_s$ the substrate will be mainly resistive, and the characteristic frequency for the skin effect will determine the signal transport mode. At $f < f_\delta$ the slow-wave mode will occur, at $f > f_\delta$ the skin effect mode. The three operating modes are visualised in the resistivity-frequency mode chart (see Fig. 2.17, where $b_1 = 1\,\mu m$, $b_2 = 200\,\mu m$, $\varepsilon_{r1} = 4$ and $\varepsilon_{rsub} = 12$).

The electric (E) and magnetic (H) fields concentrate in different areas depending on the operating mode. The basic field configurations are shown in Fig. 2.18.

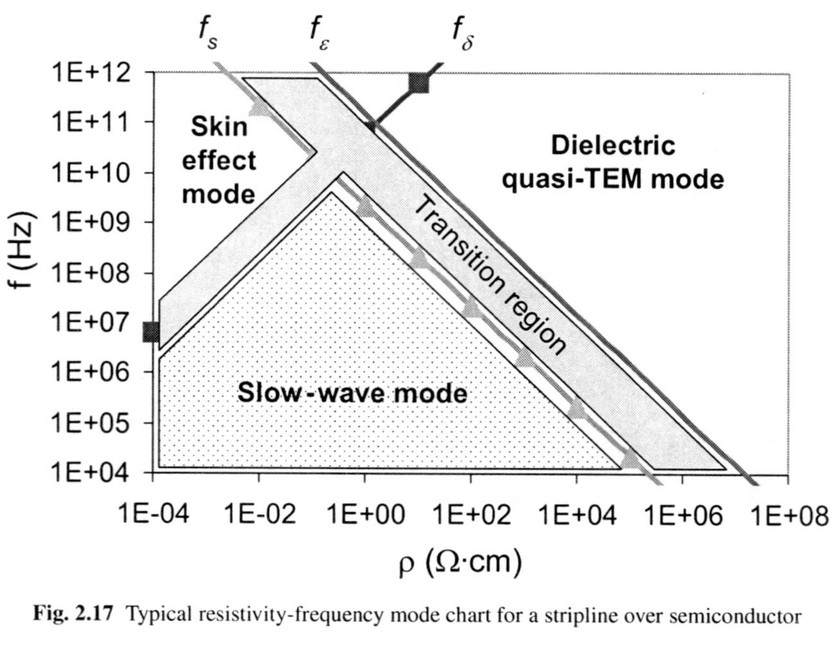

Fig. 2.17 Typical resistivity-frequency mode chart for a stripline over semiconductor

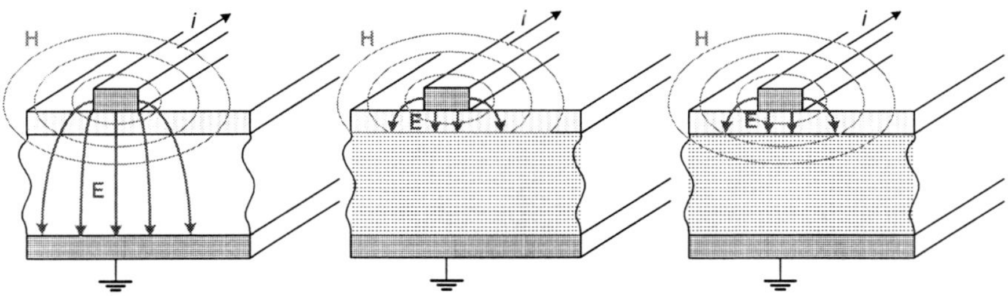

Fig. 2.18 Electric and magnetic field lines in the various fundamental operating modes

In the slow-wave mode the electric field lines do not penetrate the semiconductor whereas the magnetic field lines fully penetrate it. The separation between electric and magnetic fields leads to a combination of high line capacitance and high line inductance, increasing the line delay t_d. The slow-wave mode frequency range typically extends to at most a few GHz, depending on the substrate resistivity (see Fig. 2.17). In the case of thinner substrates, the characteristic frequency of the skin effect f_δ will shift to higher frequencies, extending the slow-wave range to higher frequencies.

In the slow-wave mode, the permittivity is increased to a value $\varepsilon_{s0} = \varepsilon_0 \cdot \varepsilon_{r1} \cdot (b_1 + b_2)/b_1$; the permeability equals $\mu = \mu_0$. This leads to a signal speed of

$$v_{slow-wave} = \frac{c}{\sqrt{\varepsilon_{r1} \frac{b_1+b_2}{b_1}}} \tag{2.62}$$

At a large b_2/b_1-ratio, the slowing factor becomes significant. Note that this ratio is typically in the range from 30 to 70 in modern RF IC processes.

In the skin effect mode, the low resistivity makes the substrate act as a lossy conductor with a relatively small skin depth. As in the slow-wave mode, the permittivity is increased to a value $\varepsilon_{s0} = \varepsilon_0 \cdot \varepsilon_{r1} \cdot (b_1 + b_2)/b_1$. The permeability now also increases, to a value $\mu = \mu_0 \cdot (b_1 + \delta/2)/(b_1 + b_2)$. The signal speed, assuming $b_1 \ll \delta/2$, now follows via

$$v_{skineffect} = \frac{c}{\sqrt{\varepsilon_r \mu_r}} = \frac{c}{\sqrt{\varepsilon_{r1} \frac{b_1+b_2}{b_1} \frac{b_1+\delta/2}{b_1+b_2}}} \approx \frac{c}{\sqrt{\varepsilon_{r1} \frac{\delta/2}{b_1}}} \tag{2.63}$$

The signal speed is a function of the skin depth δ and thus a function of the frequency f, as follows from equation (2.48):

$$v_{skineffect} \approx \frac{c}{\sqrt{\frac{\varepsilon_{r1}}{2 \cdot b_1} \left(\frac{\rho}{\pi f \mu} \right)^{1/4}}} \tag{2.64}$$

By normalising the frequency to the frequency characteristic of the skin effect in the substrate layer f_δ from equation (2.58), this result can be rearranged to

$$v_{skineffect} \approx \frac{c}{\sqrt{\varepsilon_{r1}\frac{b_2}{2\cdot b_1}}\left(\frac{f_\delta}{f}\right)^{1/4}} \tag{2.65}$$

In the skin effect mode, although $f > f_\delta$, the slowing factor may still be significant.

For each of the three fundamental operating modes an equivalent circuit can be derived capturing the behaviour of the line (see Fig. 2.19). In all the models, the metal signal and ground line series resistance have been ignored. All modes include the insulator capacitance C_1 and insulator/air inductance L_1. The substrate, modelled by the $C_s//G_s$-network, can be simplified for the different operating modes. In the skin effect mode, the substrate behaves in a resistive manner, as a lossy conductor that can be approximated as an ideal ground plane for the electric field. This simplifies the parallel network. In the slow-wave mode, the substrate behaves in a resistive manner and consequently C_s may be ignored. In the skin effect and slow-wave modes, the magnetic field concentrates in the substrate, resulting in losses (represented by the series resistors $R_s(f)$ and $R_\delta(f)$) and a frequency-dependent inductance term $L(f)$.

The element values per unit of length are summarised in Table 2.1.

Only in the dielectric quasi-TEM mode do the line characteristics behave almost independently of the frequency. Moreover, the losses per unit of length are relatively low. Losses are even lower in the slow-wave mode, often at the cost of a larger area. Besides, the delay per unit length is large in the slow-wave mode which can be either

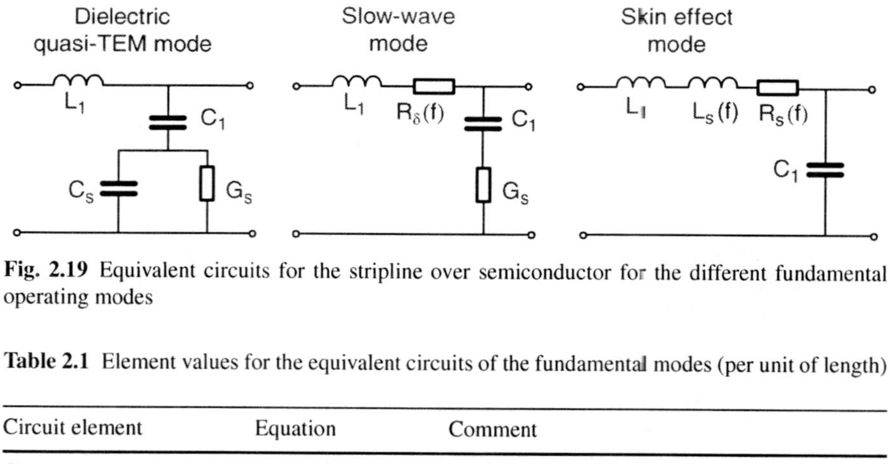

Fig. 2.19 Equivalent circuits for the stripline over semiconductor for the different fundamental operating modes

Table 2.1 Element values for the equivalent circuits of the fundamental modes (per unit of length)

Circuit element	Equation	Comment
C_1	$\varepsilon_0\varepsilon_{SiO_2}\frac{a}{b_1}$	Insulator capacitance. Frequency-independent
C_s	$\varepsilon_0\varepsilon_{Si}\frac{a}{b_2}$	Substrate capacitance; Frequency-independent
L_1	$\mu_0\frac{b_1}{a}$	Insulator/air inductance; Frequency-independent
$L_s(f)$	$\mu_0\frac{\delta}{2a}$	Inductance due to semiconductor; $\sim 1/\sqrt{(f)}$
G_s	$\frac{a}{\rho_{Si}b_2}$	Substrate conductance; Frequency-independent
$R_\delta(f)$	$\frac{4}{3}\pi\mu_0 f\frac{b}{a}\frac{f}{f_\delta}$	Resistance due to semiconductor; $\sim f^2$
$R_s(f)$	$2\pi f L_s = \pi\mu_0 f\frac{\delta}{a}$	Resistance due to semiconductor; $\sim\sqrt{(f)}$

an advantage (e.g., when used to implement shorted stub inductors) or a disadvantage (e.g., when the signal delay across the interconnect must be minimized).

To conclude, depending on the frequency and substrate resistivity, three fundamental modes exist for signal transport in a stripline according to the configuration shown in Fig. 2.16: dielectric quasi-TEM mode, slow-wave mode and skin effect mode. Using the resistivity-frequency mode chart shown in Fig. 2.17, the operating mode can be determined. The classification between the modes depends mainly on the substrate properties, via the dielectric relaxation frequency of the substrate layer f_ε, and on the skin effect, via the characteristic frequency for the skin effect in the substrate layer f_δ. There is a transitional region between the modes in which accurate line properties are difficult to predict.

The skin effect mode does not occur in a resistivity range typical of modern SiGe and RF-CMOS IC processes, $\rho_{sub} \approx 10\text{–}200\ \Omega \cdot \mathrm{cm}$. Nor does the skin effect mode occur in high-resistivity substrates as in GaAs and InP processes. The skin effect mode may occur in standard (digital) CMOS processes, in which low-resistivity substrates are used to avoid latch-up. In GaAs and InP processes, as a result of the high-resistivity, only the dielectric quasi-TEM mode occurs in the microstrip configuration studied. At the frequencies of interest for 10–40 Gb/s and RF circuits implemented in the SiGe technology applied in this book, signal transport via striplines occurs either in the slow-wave mode or in the transitional region between slow-wave mode and quasi-TEM mode. In both cases, the substrate properties are of major importance for the line characteristics. In practical circuits, not only the substrate, but the entire environment of the interconnect under study plays a role in the line properties. This makes stripline configurations often unattractive for complex high bit-rate and RF circuit design. A reduced effect of the substrate is expected in the case of other line configurations such as differential lines and coplanar configurations.

2.6 Preferred Transmission Line Configurations

The following considerations play a role in choosing a configuration for on-chip RF or microwave interconnect for high bit-rate applications (in arbitrary order).

- The transmission line should have a well-defined and controlled characteristic impedance and delay over the frequency range in which the signal has spectral content.
- The line should be shielded from the substrate using a low-ohmic grounded shield for three reasons: to minimise the coupling of signal to the substrate layer; to minimise the sensitivity to pick up signal from the substrate; and to minimise ground path losses.
- The shielding to other lines should be as good as possible.
- The signal attenuation should be low over a wide frequency range, at least up to 0.7 times the bit-rate. Low loss implies a resistive characteristic impedance, enabling simple resistive source and load impedance matching. Across an even larger frequency range the group delay should be constant. This is necessary to

minimise the line's jitter generation, since group delay variation over frequency will cause pattern-dependent zero-crossings and jitter.
- Slow-wave effects should be avoided. These effects are usually not interesting for broadband applications due to delay variations across frequency. In the case of differential signals, these considerations hold for both differential and common modes.
- The line should be implemented in an acceptable chip area, which means that the total width of the configuration should be small. Depending on the chip complexity, crossing of other (unrelated) lines should be acceptable with minimum impact to the line characteristics.
- The line characteristics should be predictable and reproducible at high yield.
- The line length should be as short as possible. In this section, it is however assumed that transmission line modelling is required, according to equation (2.45).

Low loss at high frequencies can be obtained when the electric field lines cover a large part of the signal line perimeter. This has different implications for single-ended lines and differential line configurations, depending on where the return path current flows.

Crossing of other lines with negligible impact on the electrical line characteristics can be implemented if at least three metal layers are available. The signal line(s) can then be implemented in the top metal, and a shield can be placed in the middle metal layer. Then, other signals can cross in the lowest metal layer.

Since the transmission lines are typically a few-100 μm or more, the yield may drop significantly when minimum design rule widths and spacings are applied. In order to prevent yield loss due to the transmission lines, some margin on top of the minimum layout design rules should be applied.

The presence of metal tiling fill patterns can cause asymmetry in differential configurations, reduce the characteristic impedance and increase the delay. They are usually unwanted effects. If allowed, it is therefore preferable to keep transmission lines free of tiling.

When compared with CPW, microstrip configurations have several drawbacks. A microstrip requires a wide ground plane and therefore does not satisfy the requirement of a small chip area. Nearby interconnect is not allowed since it will impact the characteristic impedance and delay, while crosstalk may also be significant. A coplanar configuration has favourable properties in terms of high-frequency loss, chip area and frequency dependence of Z_0 and slowing factor. This is because fewer electric field lines penetrate the substrate: there is a nearby low-ohmic ground return path. Even better signal transfer properties are obtained when such a coplanar configuration is shielded from the substrate using a low-ohmic shield such as an ac-grounded high-dope or metal layer. The shield can be connected to a supply line or ground line, whichever provides the best supply interference rejection. Note however that the supply and ground must be ac-shorted, requiring a low-ohmic supply network (e.g., on-chip supply decoupling capacitors).

Record-low loss, 0.3 dB/mm at 50 GHz, was demonstrated in [9] with the CPW over ground plane configuration by using a 40 μm wide signal line placed above a 16 μm thick oxide layer above a metal ground plane, large coplanar signal to ground

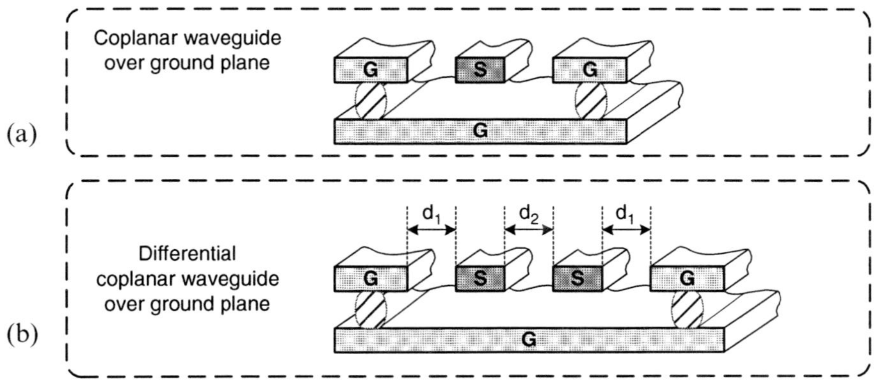

Fig. 2.20 Proposed transmission line configurations, single-ended (a) and differential (b)

spacing and wide ground lines. Using such a combination of CPW and microstrip technologies, practical values for the characteristic impedance in the range 40–90 Ω are feasible as demonstrated in [9]. Thus, the criteria for RF interconnect are best fulfilled with the configurations shown in Fig. 2.20.

The coplanar ground lines are shorted to the ground plane at regular intervals, thereby providing excellent shielding from other lines, circuits and the substrate. The low-ohmic metal ground ensures a frequency-independent parallel impedance between signal and ground. If coplanar and lateral spacings are equal, an optimum distribution of field lines across the surface of the signal line is obtained, resulting in minimum series resistance at high frequencies. In the differential configuration, the spacing between the signal lines d_2 may be chosen to differ from spacing d_1. This can be exploited to design both the differential and the common-mode characteristic impedance independently.

2.7 Applying the Skin Effect Formulas to a SiGe BiCMOS Process

In a cross-connect switch IC, a large number of transmission lines are needed for the distribution of all the broadband signals. In the switch matrix, long interconnects are needed in rows as well as columns, which makes it impossible to implement all the transmission lines in the thickest available metal layer. Therefore, it is of interest to analyse the skin effect in both the thickest available metal layer and in the ‘second-best’ metal layer, which will usually be the metal layer beneath the top metal layer. In this section, the Philips QUBiC4G SiGe BiCMOS technology [16] will be used as an example to analyse the frequency-dependent series resistance of differential transmission lines in the top two metal layers, Metal6 and Metal5. The differential-mode frequency-dependent series resistance of two typical differential transmission line configurations is calculated:

1. A differential transmission line in the 'best' metal layer, length $l = 1\,\text{mm}$, consisting of two signal wires, each $\text{w} = 5\,\mu\text{m}$ wide and $\text{t} = 3\,\mu\text{m}$ thick
2. A differential transmission line in the 'second-best' metal layer, length $l = 1\,\text{mm}$, consisting of two signal wires, each $\text{w} = 5\,\mu\text{m}$ wide and $\text{t} = 2\,\mu\text{m}$ thick

Although the metal back-end is often referred to as being aluminium ($\rho_{Al} = 2.7 \cdot 10^{-8}\Omega \cdot \text{m}$), the actual material is a composite of different metals. The exact value of the resistivity ρ of the metal layers can be found in the technology-dependent design manual if values for layer thickness t and sheet resistance R_{sq} are provided, via

$$\rho = R_{sq} \cdot t \tag{2.66}$$

With the technology used, this gives as resistivity for Metal5 $\rho_5 = 3.0 \cdot 10^{-8}\Omega \cdot \text{m}$ and for Metal6 $\rho_6 = 3.18 \cdot 10^{-8}\Omega \cdot \text{m}$. The skin depth δ for a theoretical transmission line with infinite width w follows from equation (2.48), where $\mu = \mu_0 = 1.26 \cdot 10^{-6}\,\text{H/m}$: for Metal5, $\delta_5 = 0.0872/\sqrt{(f)}\text{m}$; for Metal6, $\delta_6 = 0.0898/\sqrt{(f)}\text{m}$.

To account for the finite width $\text{w} = 5\,\mu\text{m}$ the effective skin depth in the x-direction, δ_x, is calculated using equation (2.51):

$$\delta_{x,5} = \delta(1 - e^{\frac{-w}{\delta}}) = \frac{0.0872}{\sqrt{f}}(1 - e^{\frac{-w \cdot \sqrt{f}}{0.0872}}) \tag{2.67}$$

$$\delta_{x,6} = \delta(1 - e^{\frac{-w}{\delta}}) = \frac{0.0898}{\sqrt{f}}(1 - e^{\frac{-w \cdot \sqrt{f}}{0.0898}}) \tag{2.68}$$

After correction for side effects in the y-direction, using equations (2.54) and (2.55), the total effective skin depth δ^* becomes for Metal5 and Metal6:

$$\delta_5^* = 3.5 \cdot \delta_{x,5} = \frac{0.305}{\sqrt{f}}(1 - e^{-5.73 \cdot 10^{-5}\sqrt{f}}) \tag{2.69}$$

$$\delta_6^* = 2.65 \cdot \delta_{x,5} = \frac{0.238}{\sqrt{f}}(1 - e^{-5.57 \cdot 10^{-5}\sqrt{f}}) \tag{2.70}$$

The skin effect corner frequency f_δ can be found from δ^*, since at f_δ, $\delta^* = \text{w}$. At $f \leq f_\delta$, the series resistance of the line is the dc resistance; above f_δ, the series resistance increases with a slope $\sqrt{(f)}$. In this example, this results in $f_{\delta,5} \approx 3.5\,\text{GHz}$ for Metal5 and $f_{\delta,6} \approx 1.9\,\text{GHz}$ for Metal6. The resulting frequency-dependent series resistances of the Metal5 and Metal6 transmission lines are shown in Fig. 2.21. The series resistance with correction for the contribution in the y-direction has been shown for both the approach described in [11] (solid lines $\text{R}_5(f)$ and $\text{R}_6(f)$) and the proposed alternative correction factor used in equation (2.57) (dashed lines $\text{R}'_5(f)$ and $\text{R}'_6(f)$). The series resistance shown is valid for the differential mode.

At $f = 10\,\text{GHz}$, the series resistance for Metal5 is a factor of 1.6 higher and that for Metal6 a factor of 2.1 higher relative to the dc resistance. This increase in resistance is relevant for 10 Gb/s signals. Note that the resistance of the two lines is almost identical at frequencies above 10 GHz. The series resistance should be

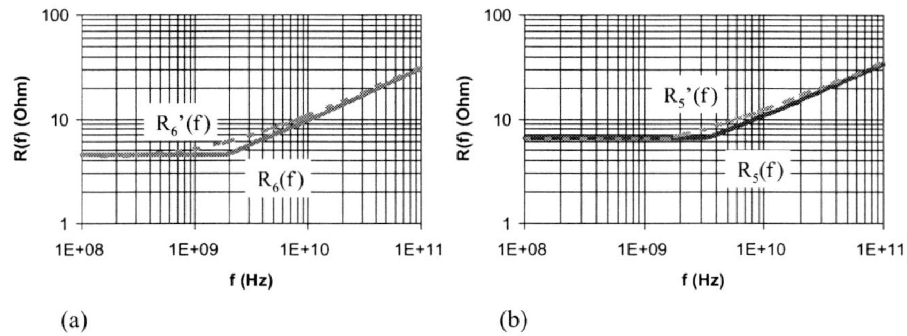

Fig. 2.21 Series resistance of example differential transmission lines implemented in Metal6 (a) and Metal5 (b). Each transmission line consists of two wires of 5 μm width and length 1 mm. Solid lines are based on equation (2.55), dashed lines are based on equation (2.57)

related to the differential characteristic impedance, typically $Z_{0dm} = 100\ \Omega$. When the 1 mm line is terminated to avoid reflections, the signal attenuation due to the line series resistance is approximately 0.8 dB at 10 GHz and approximately 1.6 dB at 40 GHz.

From these examples it is evident that the skin effect corner frequency f_δ typically lies at a few GHz, and the skin effect may consequently play an important role in the high-frequency loss of transmission lines in 10 and 40 Gb/s applications.

2.8 Models Including Skin Effect

As demonstrated in Section 2.7, equivalent circuit models including the skin effect are needed. This means that the component values of R and L shown in Fig. 2.3 become frequency-dependent, leading to the equivalent circuit of Fig. 2.22.

The parallel capacitance C is determined by the permittivity of the dielectric layers. The skin effect plays no role in the value of the parallel capacitance C. The shunt conductance G is related to the loss tangent of the dielectric layers, which is also independent of the skin effect.

Literature [3] presents graphs of the resistance $R(f)$ and inductance $L(f)$ measured for example geometries of transmission lines on a flex foil. When single elements $R(f)$ and $L(f)$ per section are used, defined as a function of frequency, only small-signal (ac-) simulations are supported, and this approach is consequently not adequate for the design of broadband circuits. A solution to this problem is to replace the series network of $R(f)$ and $L(f)$ by a more complex network of resistors and inductors, fitted to the (measured) frequency-dependent behaviour of the line, as shown in Fig. 2.23 [15].

The rationale behind this approach is that the impedance Z of the parallel network of a resistor R_p and inductor L_p can be written as

$$Z = R_s + j\omega L_s \tag{2.71}$$

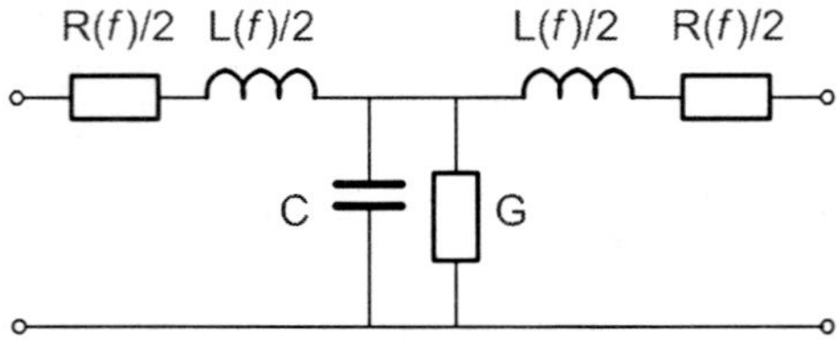

Fig. 2.22 Equivalent circuit model for one section of a transmission line including skin effect

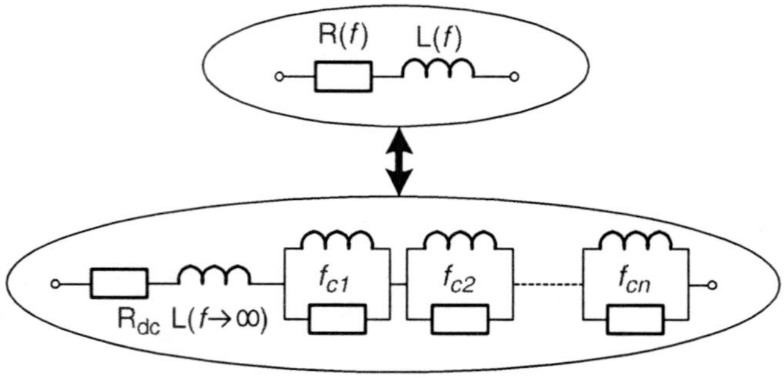

Fig. 2.23 Replacing the series network $R(f) + j\omega L(f)$ by an equivalent circuit with frequency-independent element values

with

$$R_s = \frac{R_p}{1 + R_p^2/\omega^2 L_p^2} \tag{2.72}$$

and

$$L_s = \frac{L_p}{1 + \omega^2 L_p^2/R_p^2} \tag{2.73}$$

The impedance Z represents a frequency-dependent R_s and L_s series network. The inductance L_s decreases above $\omega_c = R_p/L_p$ by a slope of $-40\,\text{dB/dec}$, the resistance R_s increases up to ω_c by 40dB/dec. By cascading sections with different cut-off frequencies ω_c it is possible to fit the network impedance to measurements. In practice, only a few $R_p//L_p$ sections will usually be needed to provide an accurate fit between model and measurements across the frequency range of interest. For example, in [2], only two sections are used to obtain an accurate fit for on-chip interconnect up to 20 GHz.

As equation (2.72) shows, the low-frequency series resistance of an $R_p//L_p$ network equals zero. Thus, a single series-resistor (e.g., R_{dc} in Fig. 2.23) is needed to represent the low-frequency series resistance. In a similar way, the inductance contribution of the $R_p//L_p$ networks approaches zero at $f \to \infty$. A single series-inductor (e.g., $L(f \to \infty)$ in Fig. 2.23) is needed to represent the high-frequency inductance limit.

2.9 Signal Transfer Across a Transmission Line

To analyse the importance of impedance matching a transmission line for broadband applications, the differential-mode voltage gain $A_{dm} = v_{o,dm}/v_{i,dm}$ and common-mode voltage gain $A_{cm} = v_{o,cm}/v_{i,cm}$ of a differential transmission line are analysed using the approach shown in Fig. 2.24. The source resistance R_s and load resistance R_l are varied simultaneously across a range of approximately $0.5 \cdot Z_0 \ldots 2 \cdot Z_0$, for differential and common modes.

The equivalent circuit for the transmission line can be generated using an electromagnetic (EM) simulator, such as Philips' Fasterix or Agilent's Momentum. In the case of a lossless transmission line, the characteristic impedance Z_0 will be real (e.g., $\mathrm{Im}(Z_0) = 0$). If then $R_l = R_s = Z_0$, the voltage gain becomes frequency-independent. An equivalent circuit for a differential transmission line above a metal ground plane, implemented in Philips' SiGe BiCMOS QUBiC4G IC process, has been developed using Fasterix. The transmission line configuration was implemented with signal lines in the top metal layer above a Metal1 ground plane, according to the preferred configuration described in Section 2.6. The simulation results obtained for the circuit of Fig. 2.24 using this line are shown in Fig. 2.25. Skin effect and radiation losses were included in this simulation; the Metal1 ground layer was assumed to be ideal.

The low-frequency transmission coefficient was in all cases approximately −6 dB, resulting from the resistive division of the signal across the source and load impedance. The differential-mode characteristic impedance was approximately 130 Ω, since at $R_s = R_l = 130\ \Omega$ the gain to the input is maximally flat, and the transfer to the output shows minimum loss. At $f = 20$ GHz, the gain is maximally sensitive to mismatch in source and load impedance while at $f = 40$ GHz the gain is almost independent of the source and load impedance. This can be explained via the wavelengths of these frequencies in relation to the line length. At a 40 GHz signal the wavelength in

(a)

(b)

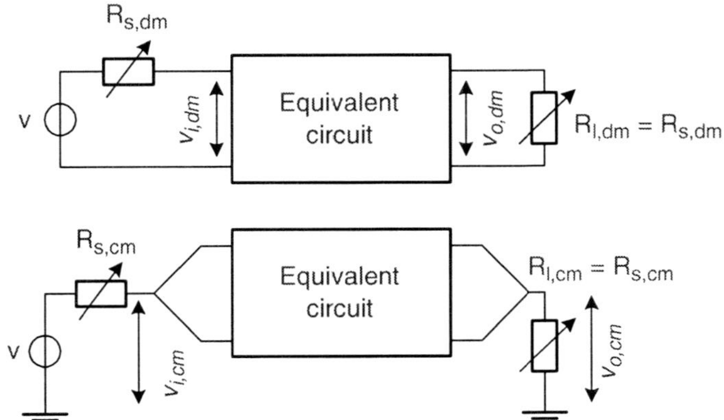

Fig. 2.24 Circuit for extracting the differential-mode (a) and common-mode (b) characteristics of a transmission line

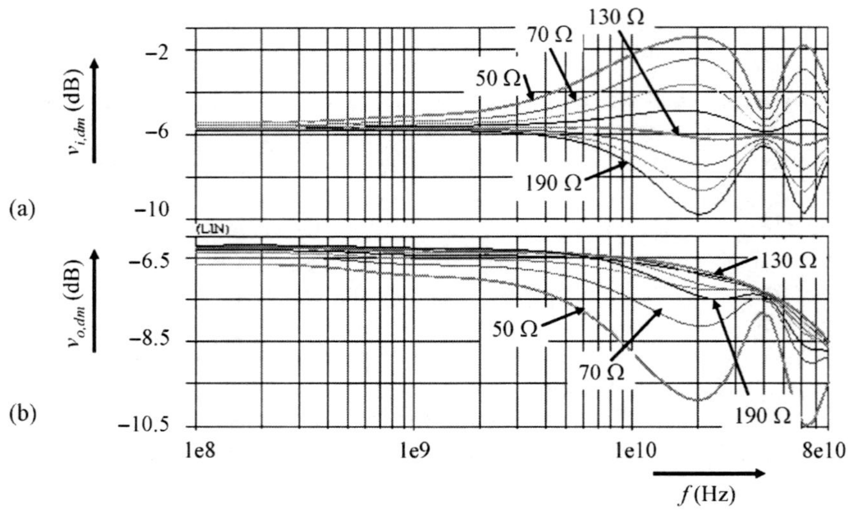

Fig. 2.25 Fasterix simulation result obtained for a 2-mm long Metal6 differential transmission line, differential mode, for source and load resistance values of 50 Ω to 190 Ω in increments of 20 Ω. Graph (a) shows the signal amplitude at the input of the line; graph (b) shows the signal amplitude at the output of the line

Metal6 equals $\lambda(40\text{GHz}) = 4\,\text{mm}$. A flat frequency-response is found because the 2 mm line length corresponds to $\lambda(40\text{GHz})/2$, see also equation (2.34). Similarly, a 2-mm line corresponds to $\lambda(20\text{GHz})/4$. Thus, 20 GHz signals that are (partially) reflected at one end of the line due to mis-termination arrive in anti-phase at the other end of the line.

Note that these results show that a 2-mm long transmission line is maximally sensitive to source and load mismatch at $f = 20\,\text{GHz}$, and it is therefore not recommended to apply such a length in 40 Gb/s applications.

The increase in the line series resistance at high frequencies results in an increase in the signal attenuation to the output. By considering the matched situation, $R_s = R_l = 130\ \Omega$, the series resistance $R(f)$ can be derived from the signal attenuation. For example, at 40 GHz there is a gain of -7.3 dB between signal source and output. Thus, the series resistance of both signal lines (each 2 mm in length) follows from $10^{-7.3/20} = R_l/(R(40\text{GHz}) + R_s + R_l)$, resulting in $R(40\text{GHz}) = 40.9\ \Omega$. This result is in agreement with the analytical result presented in Fig. 2.21a (in Fig. 2.21 the line is 1 mm long).

2.10 Interconnect Test Structures

In this section measurement results obtained for a single-ended and a differential on-chip transmission line will be presented. The lines were designed according to the preferred configurations described in Section 2.6.

2.10.1 Single-Ended Transmission Line

A single-ended transmission line as proposed in Fig. 2.20a was implemented in Philips' QUBiC4G technology, a SiGe BiCMOS process with five metal layers [16]. The coplanar lines were implemented in the top metal layer, the ground plane was implemented using a grounded highly doped (20 Ω/□) n-type buried layer. The line width and spacings were 5 μm, resulting in a 50 Ω characteristic impedance according to Fasterix simulations. A chip photo is shown in Fig. 2.26.

The line length was 2.2 mm, the total width of the ground-signal-ground (GSG) configuration was 25 μm. Tiling was avoided in the transmission line area. The line was analysed using a 13.5 GHz Agilent 2-port network analyser and Cascade GSG wafer probes. The measurement set-up up to the probe tips was calibrated using a Cascade general-purpose calibration substrate. Open and short de-embedding structures were implemented on the same chip [17]. The measured s-parameter data were used to define a 2-port in the Spectre circuit simulator. The approach presented in Fig. 2.24 was applied to find the characteristic impedance (derived from Fig. 2.27) and delay (derived from Fig. 2.29). In this analysis, the voltage gain from the source v (in front of the source resistance) to the input of the line was defined as $A_i = v_i/v$; the voltage gain to the output of the line was defined as $A_o = v_o/v$.

Fig. 2.26 Photomicrograph of a single-ended transmission line implemented in Philips' QUBiC4G process

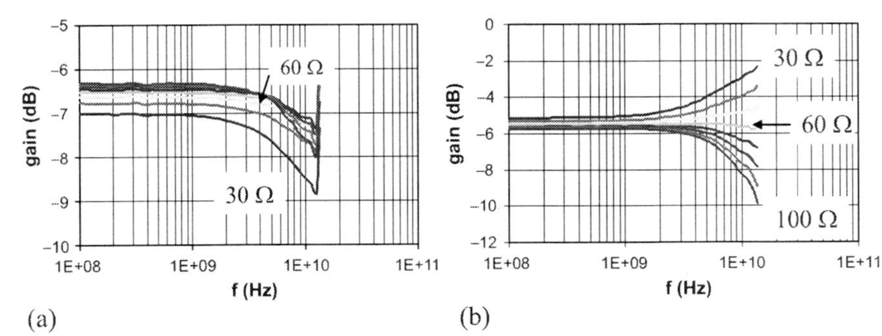

Fig. 2.27 Voltage gain A_o to the output (a) and A_i to the input (b) of the GSG line for $R_s = R_l$ between 30 Ω and 100 Ω in 10 Ω increments. Results are based on measured transmission line data

As can be seen in Fig. 2.27b, the gain A_i to the input is almost flat at $R_s = R_l \approx 60\ \Omega$. Thus, the characteristic impedance of the line is $Z_0 \approx 60\ \Omega$. In the matched condition, the low frequency gain to the output equals $A_{o,m}(f < 1\,\text{GHz}) = -6.5\,\text{dB}$, corresponding to a line series resistance of 6.8 Ω. The subscript 'm' in $A_{o,m}$ refers to the condition in which the source and load resistance are matched to the line impedance. At 13 GHz, the gain to the output of the line drops to $A_{o,m}(f = 13\,\text{GHz}) = -7.3\,\text{dB}$.

It is possible to extract the frequency-dependent series resistance of the line, $R(f)$, from the gain $A_{o,m}$ using the following equation for the matched condition:

$$R(f) = \frac{R_l}{10^{\frac{A_0}{20}}} - R_s - R_l \tag{2.74}$$

With $R_s = R_l = 60\ \Omega$, this gives the results shown as 'Eq. cct' in Fig. 2.28. An excellent fit for the skin effect corner frequency is obtained.

The fact that the characteristic impedance of the line was higher than expected (e.g., measured 60 Ω, expected 50 Ω) and the series resistance was 60% higher than expected (see Fig. 2.28) is due to a problem during IC fabrication (e.g., reduced metal thickness).

The line delay can be extracted from the group delay or from the voltage gain via the frequency at which maximum reflections occur. Since this frequency exceeds the highest measured frequency, the phase information is used to extract the line delay. The measured phase transfers to the line input and output, extracted via Spectre circuit simulations using the approach presented in Fig. 2.24, are shown in Fig. 2.29. At $R_s = R_l \approx 60\ \Omega$, the input impedance of the terminated transmission line behaves resistively. The characteristic impedance will consequently equal $Z_0 = 60\ \Omega$, as also found via the voltage gain analysis. The effective permittivity $\varepsilon_{r,eff}$ can be extracted from the line delay. The phase difference between the input and output of the line at $R_s = R_l = 60\ \Omega$ equals 58° at 12 GHz, corresponding to a delay of $t_d = 6.92\,\text{ps/mm}$, resulting in $\varepsilon_{r,eff} = 4.31$.

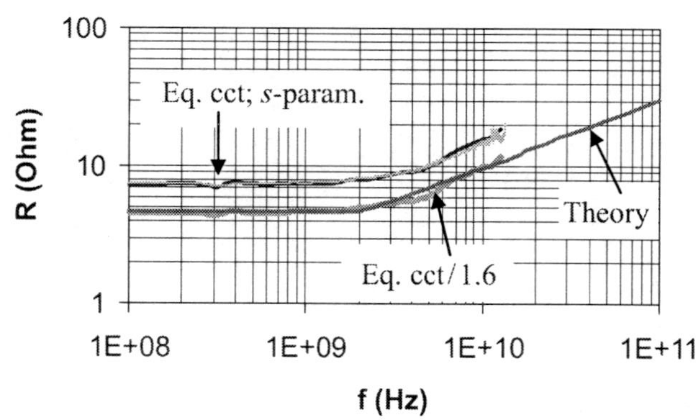

Fig. 2.28 Derived series resistance of the GSG line. For reference, the theoretical result presented in Fig. 2.21 (corrected for the line length) is also shown

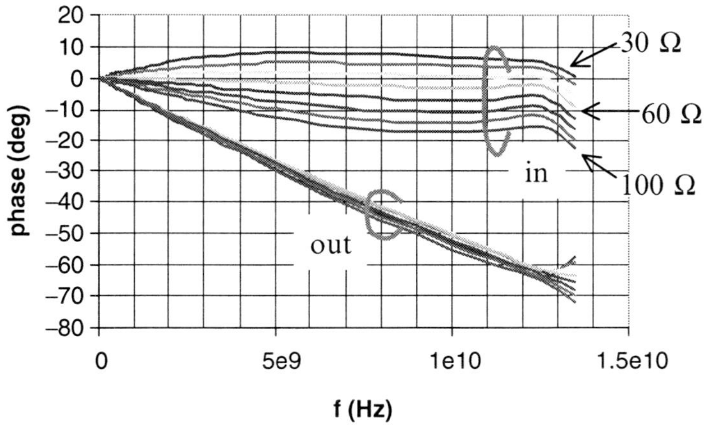

Fig. 2.29 Measured phase transfer to the input and output of the transmission line at $R_s = R_l$ between 30 Ω and 100 Ω in 10 Ω increments

So far, the characteristics of the transmission line have been analysed assuming resistive terminations. The characteristic impedance may however include an imaginary part due to the series resistance R and parallel loss G, as follows from equation (2.20). The complex characteristic impedance can be extracted directly from the measured *s*-parameters using the approach described in [10]:

$$Z_0 = \sqrt{Z_s^2 \frac{(1+S_{11})^2 - S_{21}^2}{(1-S_{11})^2 - S_{21}^2}} \tag{2.75}$$

Here, Z_s is the source (and load) resistance of the measurement set-up at which the *s*-parameters are obtained (usually 50 Ω in the case of single-ended configurations) and Z_0 represents the complex characteristic impedance. This equation has been implemented in a Mathematica program, which post-processes the measured (de-embedded) *s*-parameters. The complex characteristic impedance resulting for the QUBiC4G GSG line is shown in Fig. 2.30. The results are more accurate than the approximation to Z_0 found via Spectre circuit simulations.

The equivalent transmission line model resulting for this example GSG line is shown in Fig. 2.31. The model provides a delay of $t_d = 1$ ps per section, a characteristic impedance of $Z_0 = 60\ \Omega$, and a series resistance of 6.8 Ω/mm. Since the delay equals 6.92 ps/mm, one section represents a line length of 145 μm. At $f = 16$ GHz, there will be ten sections per wavelength.

2.10.2 Differential Transmission Line

A differential ground-signal-signal-ground (GSSG) transmission was implemented in the Philips' QUBiC4G process with five metal layers [16]. The line was designed

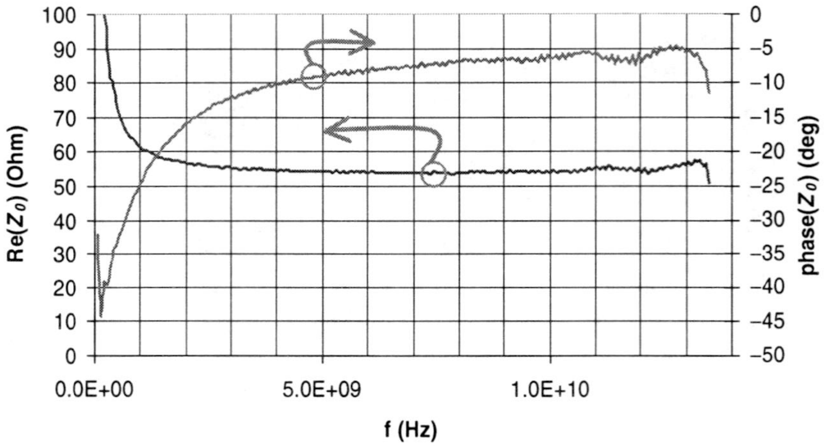

Fig. 2.30 Re(Z_0) and phase(Z_0) for the QUBiC4G GSG line, extracted from measured *s*-parameter data

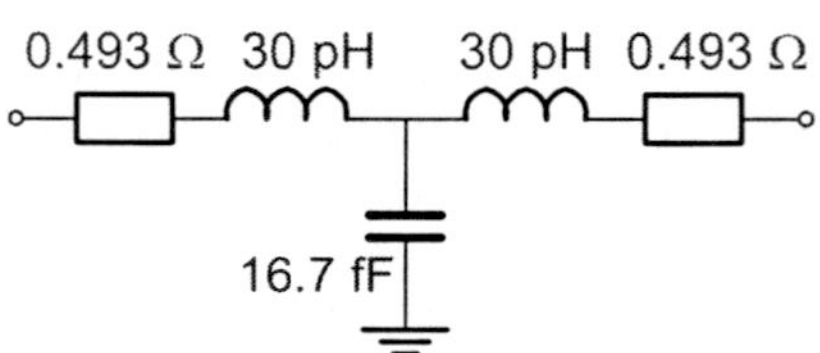

Fig. 2.31 Equivalent transmission line model for the example GSG line. One section is shown, representing a line length of 145 μm

Fig. 2.32 Photomicrograph of a differential transmission line in Philips' QUBiC4G process

as proposed in Fig. 2.20b, with the coplanar lines implemented in the 3-μm thick top metal layer, a line length of 2.2 mm, line widths and spacing 5 μm, above a Metal1 ground plane. The total transmission line width is 35 μm.

The chip photomicrograph is shown in Fig. 2.32. Cascade GSGSG wafer probes were used for evaluation in combination with an Agilent 4-port network analyser, allowing characterisation up to 20 GHz.

Open and short de-embedding structures were implemented on the same chip. Tiling is avoided in the transmission line area.

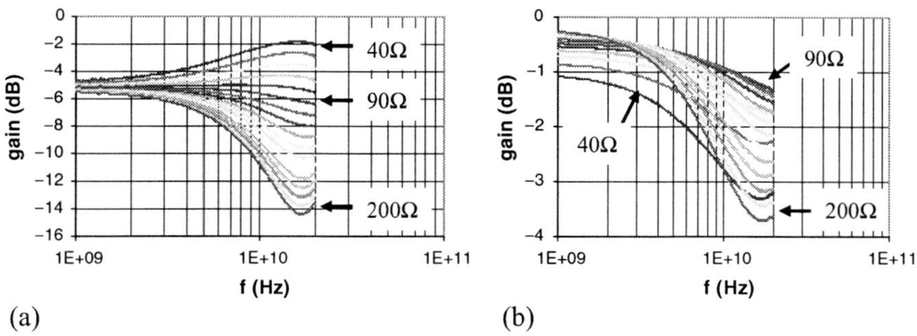

Fig. 2.33 Differential voltage gain to the input (a) and output (b) of the GSSG transmission line at $R_s = R_l$ between 40 Ω and 200 Ω in 10 Ω increments. The results are based on transmission line data obtained after calibration and de-embedding

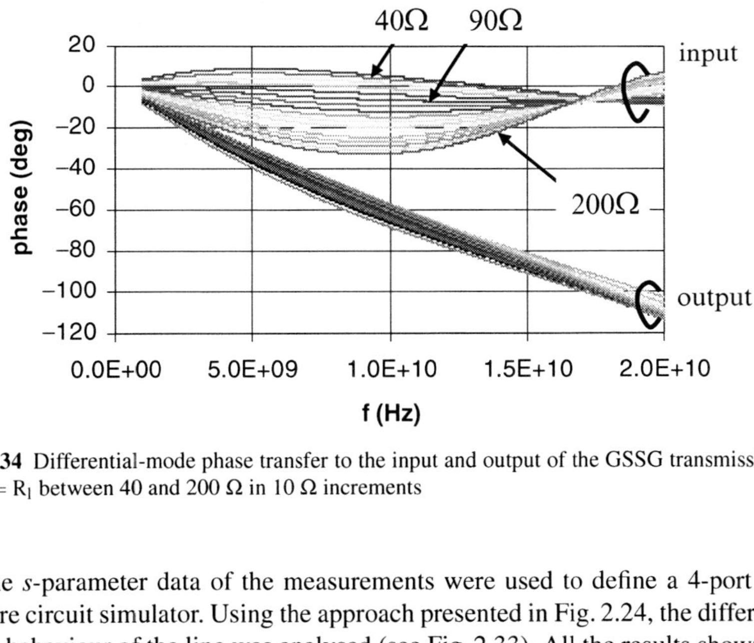

Fig. 2.34 Differential-mode phase transfer to the input and output of the GSSG transmission line at $R_s = R_l$ between 40 and 200 Ω in 10 Ω increments

The *s*-parameter data of the measurements were used to define a 4-port in the Spectre circuit simulator. Using the approach presented in Fig. 2.24, the differential-mode behaviour of the line was analysed (see Fig. 2.33). All the results shown were obtained after calibration and open-short de-embedding. The voltage gain to the input and output was maximally flat at $R_s = R_l$ near 90 Ω, indicating the differential-mode characteristic impedance of the line.

The line delay was obtained from the phase difference between the input and output (see Fig. 2.34).

The phase transfer to the input of the line remained close to zero at $R_s = R_l =$ 90 Ω, a phase of zero corresponds to a purely resistive input impedance. The linear phase relation between the input and output signals results in a constant group delay of $t_{dm} = -(d\varphi/d\omega) = 14.43$ ps across 2.2 mm or 6.56 ps/mm, corresponding to $\varepsilon_{r,eff,dm} = 3.87$.

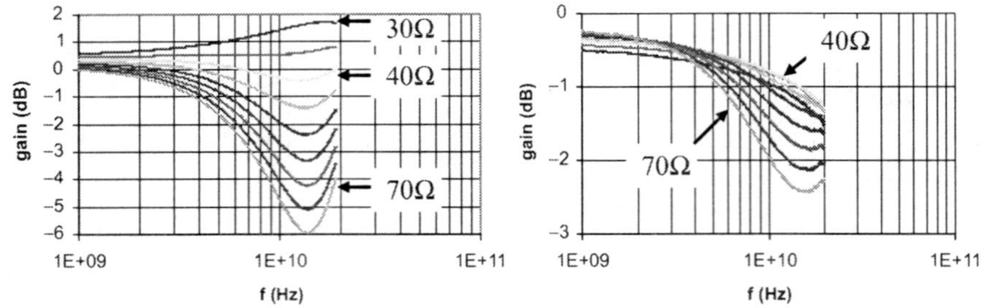

(a) (b)

Fig. 2.35 Common-mode voltage gain to the input (a) and output (b) of the GSSG transmission line at $R_s = R_l$ between 30 Ω and 70 Ω in 5 Ω increments. The results are based on transmission line data after calibration and de-embedding

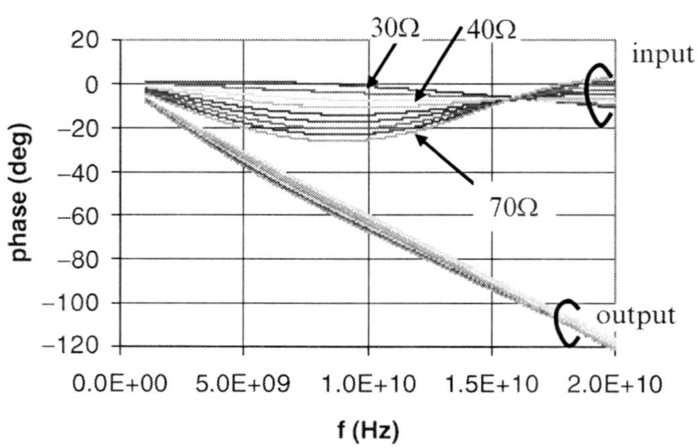

Fig. 2.36 Common-mode phase transfer to the input and output of the GSSG transmission line at $R_s = R_l$ between 30 Ω and 70 Ω in 5 Ω increments

The common-mode line characteristics were derived in a similar way, according to the approach shown in Fig. 2.24. The results are shown in Fig. 2.35 (gain) and Fig. 2.36 (phase). The measured common-mode characteristic impedance was $Z_{0cm} = 40\ \Omega$; the common-mode delay, derived from the phase difference between the input and output signals, was $t_{cm} = 7.09\,\text{ps/mm}$, corresponding to $\varepsilon_{r,\text{eff,cm}} = 4.53$. Note that the common-mode signals travel at a slower speed compared to the differential-mode signals, due to the different configuration of the field lines.

With GSSG lines, as with GSG lines, it is possible to derive the characteristic impedance directly from the measured *s*-parameters. The data from all four ports needs to be rearranged to a suitable format in which the differential-mode parameters are separated from the common-mode parameters. This transform is described in [14], and is included in the control software of the Agilent 4-port network analyser:

$$\left[\begin{array}{cc|cc} S_{dd11} & S_{dd12} & S_{dc11} & S_{dc12} \\ S_{dd21} & S_{dd22} & S_{dc21} & S_{dc22} \\ \hline S_{cd11} & S_{cd12} & S_{cc11} & S_{cc12} \\ S_{cd21} & S_{cd22} & S_{cc21} & S_{cc22} \end{array}\right] = [\mathbf{T}] \cdot \left[\begin{array}{cccc} S_{11} & S_{12} & S_{13} & S_{14} \\ S_{21} & S_{22} & S_{23} & S_{24} \\ S_{31} & S_{32} & S_{33} & S_{34} \\ S_{41} & S_{42} & S_{43} & S_{44} \end{array}\right] \tag{2.76}$$

The 2×2 matrices $[S_{dd}]$, $[S_{cc}]$, $[S_{cd}]$ and $[S_{dc}]$ represent the differential-mode parameters, the common-mode parameters, the common-mode to differential-mode conversion, and the differential-mode to common-mode conversion, respectively.

The transform matrix [**T**] is found by splitting the single-ended input signals applied during the analysis of S_{11} to S_{44} into differential and common-mode terms. In the case of a symmetrical transmission line, all the coefficients of matrices $[S_{cd}]$ and $[S_{dc}]$ are equal to 0.

Equation (2.75) can be applied to the matrix $[S_{dd}]$ to find the complex differential-mode characteristic impedance. The results are shown in Fig. 2.37.

The results presented in Fig. 2.37 are in line with our expectations. The characteristic impedance is approximately $Z_{0dm} \approx 85\ \Omega$, while the phase increases from $-45°$ at low frequencies at which $\omega L \ll R(f)$ to about $-10°$ at $f = 10\,\text{GHz}$. The nearly resistive behaviour of the characterisitc impedance Z_{0dm} indicates low loss of the line.

The differential-mode lumped element values for the line are extracted from the differential-mode de-embedded s-parameters using the following equations [11]:

$$K = \left\{ \frac{\left(S_{11}^2 - S_{21}^2 + 1\right)^2 - \left(2S_{11}\right)^2}{(2S_{21})^2} \right\}^{1/2} \tag{2.77}$$

$$\gamma = \frac{1}{l} \ln \left\{ \frac{1 - S_{11}^2 + S_{21}^2}{2S_{21}} \pm K \right\} \tag{2.78}$$

When the characteristic impedance Z_0 (equation (2.75)) and propagation constant γ (equations (2.77) and (2.78)) have been determined, the model parameters R, L, G and C follow from the definitions for Z_0 (equation (2.20)) and γ (equation (2.17)):

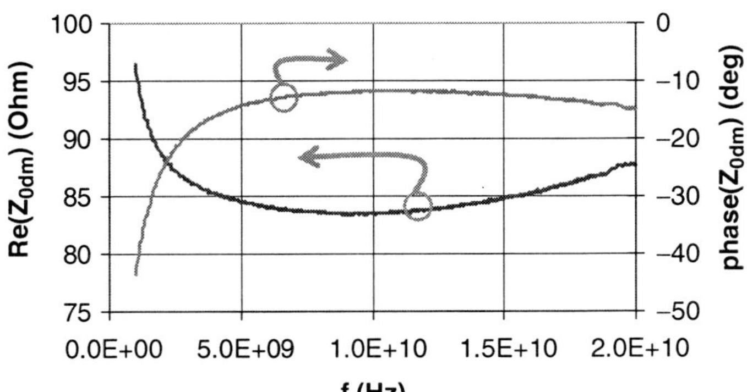

Fig. 2.37 Differential-mode $\text{Re}(Z_{0dm})$ and $\text{phase}(Z_{0dm})$ derived from measured 4-port s-parameters

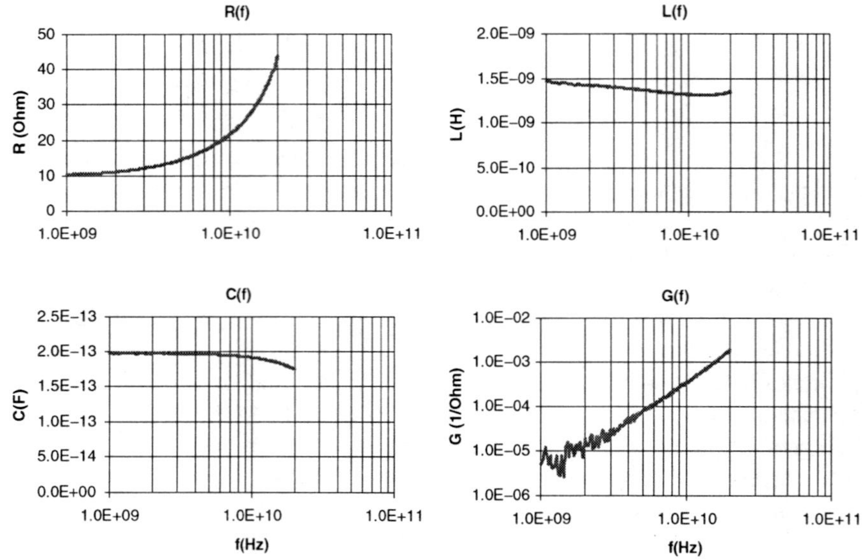

Fig. 2.38 Differential-mode R(f), L(f), C(f) and G(f), all derived from measured 4-port s-parameters

$$R = \mathrm{Re}\left(\gamma Z_0\right) \tag{2.79}$$

$$L = \mathrm{Im}\left(\gamma Z_0\right)/\omega \tag{2.80}$$

$$C = \mathrm{Im}\left(\gamma / Z_0\right)/\omega \tag{2.81}$$

$$G = \mathrm{Re}\left(\gamma / Z_0\right) \tag{2.82}$$

These equations yield the line parameters per metre. The results for the differential mode of the GSSG line are shown in Fig. 2.38, which gives the parameters for the total line length. All results are in line with our expectations.

The parallel loss G(f) that has so far been ignored is indeed of minor significance up to 20 GHz. The parallel conductance increases by 40 dB/dec at $f > 1$ GHz. This frequency dependence can be explained via a series resistance R_s associated with the capacitance of the RLC model, which can be translated into a parallel equivalent resistance R_p via

$$R_p = (Q^2 + 1)R_s \tag{2.83}$$

with

$$Q = \frac{1}{2\pi f R_s C_s} \tag{2.84}$$

A frequency-independent series resistance R_s thus translates into a parallel equivalent resistance R_p that drops by 40 dB/decade at frequencies at which $Q > 1$.

Equation (2.75) can also be applied to the matrix $[S_{cc}]$ to find the complex common-mode characteristic impedance. The results are shown in Fig. 2.39.

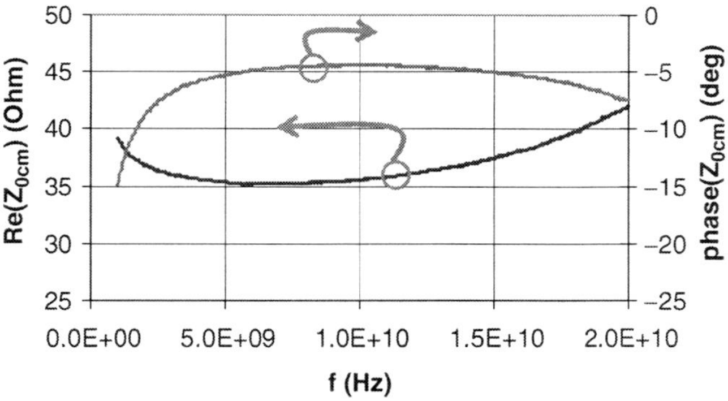

Fig. 2.39 Common-mode $\mathrm{Re}(Z_{0cm})$ and $\mathrm{phase}(Z_{0cm})$ derived from measured 4-port s-parameters

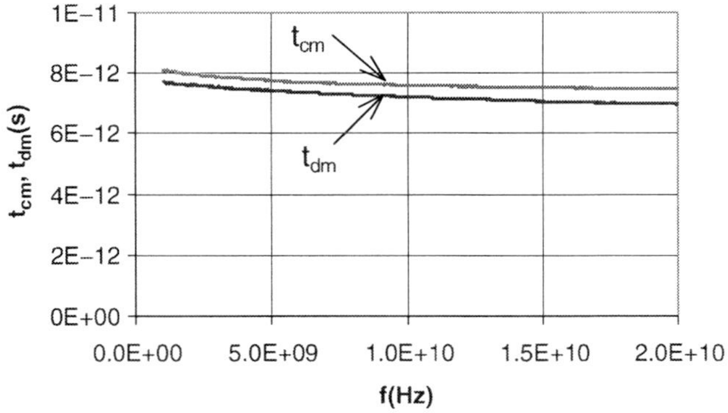

Fig. 2.40 Differential and common-mode delay per mm, derived from measured s-parameters

The common-mode characteristic impedance is equal to $Z_{0cm} = Z_{0dm}/4$ in the case of two uncoupled signal lines. In this example, the ratio is $Z_{0dm}/Z_{0cm} \approx 85/37 = 2.3$, demonstrating that there is significant coupling between the signal lines.

The frequency-dependent delay for differential and common-mode is derived from the s-parameters via $t_d = \partial(\mathrm{Im}(\gamma))/\partial\omega$. The common-mode delay is slightly different from the differential-mode delay, as illustrated in Fig. 2.40, due to a difference in $\varepsilon_{r,\mathrm{eff}}$ between the differential and the common mode, which is mainly due to the passivation layer. The medium-resistivity p-type substrate plays no role in the line characteristics due to the Metal1 ground layer underneath the coplanar line.

The equivalent circuit model for the GSSG line can now be defined using equations (2.39) to (2.42). First, these equations need to be inverted, resulting in:

$$C_g = \frac{t_{cm}}{2Z_{0cm}} \tag{2.85}$$

$$C_c = \frac{t_{dm}}{Z_{0dm}} - \frac{t_{cm}}{4Z_{0cm}} \tag{2.86}$$

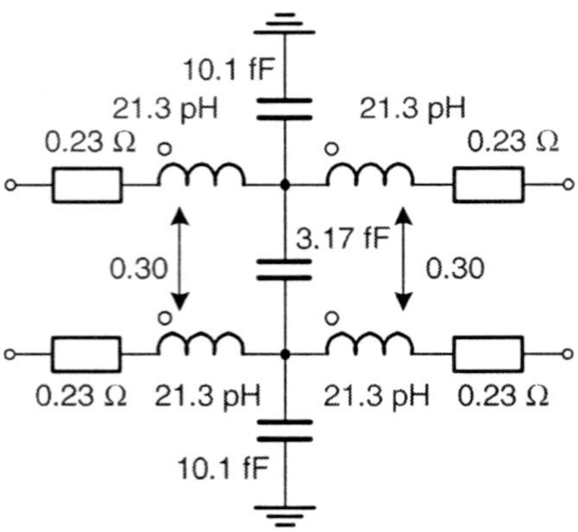

Fig. 2.41 Equivalent circuit for one section of the GSSG transmission line, representing a length of 100 μm

$$L = Z_{0cm} t_{cm} + \frac{Z_{0dm} t_{dm}}{4} \tag{2.87}$$

$$k = \frac{1}{L}\left(Z_{0cm} t_{cm} - \frac{Z_{0dm} t_{dm}}{4}\right) \tag{2.88}$$

The resulting model for a section of 100 μm length is shown in Fig. 2.41. The following data were used: (obtained from Fig. 2.37) $Z_{0dm} = 85\ \Omega$ (from Fig. 2.39) $Z_{0cm} = 37\ \Omega$ (from Fig. 2.40) $t_{dm} = 7.0\,\text{ps/mm}$, $t_{cm} = 7.5\,\text{ps/mm}$, and (from Fig. 2.38 at $f = 10\,\text{GHz}$) a series resistance of 20 Ω for the total line length. The model is valid near $f = 10\,\text{GHz}$.

2.11 Modelling and Considerations of Digital Interconnect

Typical operating frequencies for microwave and many digital application areas are above 1 GHz and on-chip line lengths may be substantial with respect to the wavelength of the signals on the lines, and we would therefore expect similar interconnect models to be used. Wiring density is however of utmost importance for digital ICs, and interconnect widths and spacing are consequently typically close to minimum design rules for the majority of the lines in digital ICs. The output impedance Z_{drv} of the digital circuits driving the lines is also typically high (e.g., $Z_{drv} \gg Z_0$) while the lines are mainly capacitively loaded and terminated by the inputs of CMOS gates.

Effects that are a major concern in digital applications are signal delay and crosstalk. Crosstalk to neighbouring parallel lines can limit the allowable line length. The signal delay is dominated by RC delays and is a function of potential signal transitions on neighbouring lines. Worst-case delay is obtained when the neighbouring lines are simultaneously driven with a signal transition of the opposite polarity. An overview of line modelling and design for digital applications is given in [18].

The complexity of the line models for digital interconnect depends on the line length. In its simplest form, each line is represented by a single capacitance C_l. Such models can easily be generated via parasitic extraction software routines. A more accurate line model will be required when the line resistance R_l becomes significant with respect to the output impedance of the line driver Z_{drv}. The line resistance can be included as a single lumped resistance. This approach, however, overestimates the line delay, because the line resistance and capacitance are distributed. A more accurate delay prediction is obtained via $t_d = Z_{drv} * C_t + R_l * C_l/2$, where C_t represents the total load capacitance. The factor 1/2 accounts for the distributed effect. A further refinement is obtained using a distributed RC model. However, the line delay from a distributed RC model is proportional to the square of the length, because both the line resistance R_l and the capacitance C_l are proportional to the length. For a correct delay representation, a multi-section RLC (transmission line) model is required.

2.12 Circuit and Interconnect Design Flow

Single-ended and differential circuits are both widely used in high bit-rate functions. There is therefore a need for single-ended and differential interconnects. For the broadband applications considered in this book, the line delay needs to be minimised, so slow-wave effects are unwanted. Several arguments (see Section 2.6) lead to the proposed on-chip transmission line configurations shown in Fig. 2.20. The proposed configurations can be realised in any IC technology, provided that at least two metal layers are available. For example, the proposed transmission line configurations and models can also be used in CMOS technologies.

For differential transmission lines, it is essential that the circuit models capture both common-mode and differential line characteristics. Knowledge of line impedance and delay is essential for circuit design. A simple model that offers sufficient degrees of freedom to correctly model line impedance and delay is given in Fig. 2.6. Equations (2.39) to (2.42) relate the model to the line impedance and delay. Using the model shown in Fig. 2.6, circuit and interconnect design can be done in parallel while already capturing the impact of the interconnect on the circuit behaviour at an early stage in the design. This leads to a design flow that is significantly different from the traditional design flow shown in Fig. 2.2. The design flow used for circuit design in this book puts the interconnect design and modelling at a central place, as shown in Fig. 2.42. Starting from the fact that the transmission lines inevitably exhibit delay and can be designed across a relatively narrow range of characteristic impedance levels (e.g., 30–150 Ω, with 50 Ω single-ended or 100 Ω differential widely used), the circuit and interconnect designs are made in parallel. Even in initial circuit simulations, transmission line interconnect models are included for lines that are *anticipated* to require such models, as discussed in Section 2.3.

When interconnect test structures are available, the interconnect models can be updated according to the measured *s*-parameters. Equations (2.79)–(2.82) can then

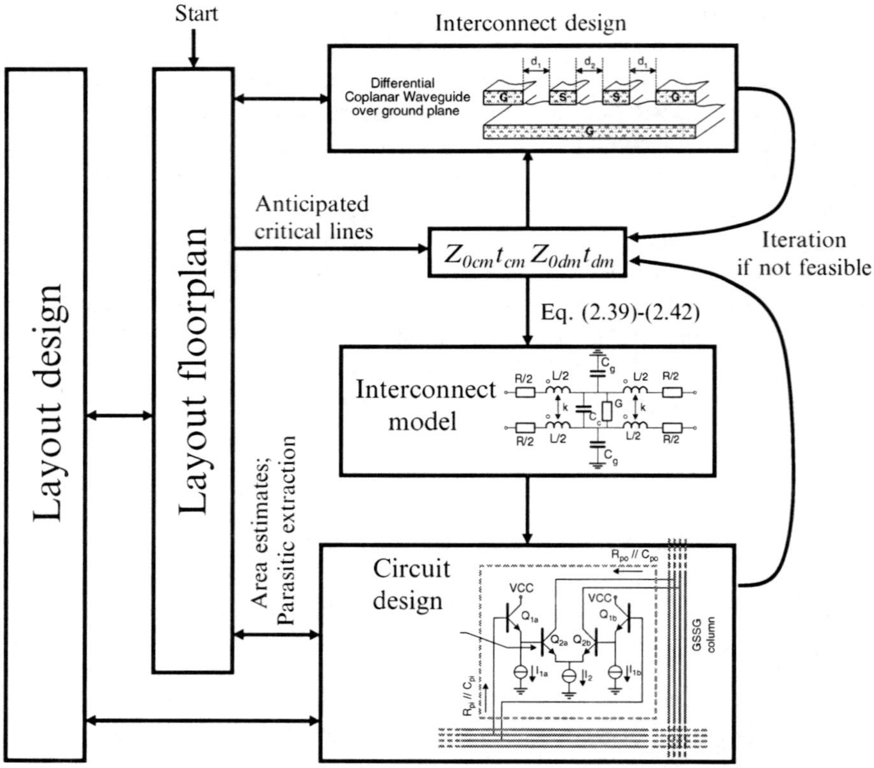

Fig. 2.42 Circuit and interconnect design flow used in this book

be used to extract the element values for the equivalent model for differential and common modes.

The design flow shown in Fig. 2.42 will be applied to the design of the cross-connect switch in Chapter 4 and the design of the InP PRBS generator presented in Chapter 6.

2.13 Conclusions and Outlook

The preferred transmission line configuration for RF, microwave and broadband applications is the CPW implemented in the thickest available metal layer, above a metal ground layer. Such an implementation can be used in single-ended and differential applications involving one or two signal lines, respectively. The ground layer provides shielding from the substrate. The substrate properties consequently play no role in the characteristic impedance and delay of the line. Slow-wave effects do not occur in these lines.

Transmission line modelling is required when the total line delay t_d exceeds $t_r/10$, with t_r being the shortest rise-time of the signals on the line.

When losses are low, the characteristic impedance is real and can be accurately approximated by $Z_0 = \sqrt{(L/C)}$ while the delay is accurately approximated as $t_d = \sqrt{(LC)}$, with L and C being the lumped values for the line. Losses are included via a series resistance R. Parallel losses are usually of minor significance and may be ignored, so that the equivalent circuit model for the single-ended configuration is the RLC model.

Due to the skin effect, the line resistance R and inductance L are frequency-dependent. In example on-chip geometries, the skin effect corner frequency is in the range of 1–10 GHz, and consequently the skin effect needs to be included in line models for ICs intended for 10–40 Gb/s applications. The RLC model can be extended to include the skin effect by replacing the RL network by multiple R//L sections with different corner frequencies geared to the measured or calculated line parameters.

In the case of differential configurations, the differential-mode line parameters Z_{0dm} and t_{dm} will usually differ from the common-mode line parameters Z_{0cm} and t_{cm}. The ratio (Z_{0dm}/Z_{0cm}) for uncoupled lines such as two single-ended CPW lines is 4. In most GSSG configurations this ratio will be between 1 and 4. The equivalent circuit model for the differential configuration is the RLMCG model. When losses are ignored (e.g., $R = G = 0$), this model provides 4 degrees of freedom (e.g., L, M, C_c between the signal lines and C_g from each signal line to ground), and can thus always be fitted to a set of line parameters Z_{0dm}, t_{dm}, Z_{0cm} and t_{cm}. The mutual inductance term M represents the coupling between the signal lines.

Differential transmission lines can be analysed using a 4-port network analyser. The resulting (4×4) s-parameter matrix can be transformed to four matrices, each of size (2×2). These (2×2) matrices represent the differential-mode line parameters, common-mode line parameters, differential- to common-mode conversion and common- to differential-mode conversion. The model parameters $R(f)$, $L(f)$, $G(f)$ and $C(f)$ can be derived from the s-parameter data, as can other line characteristics such as the (complex) characteristic impedance Z_0, delay t_d and attenuation.

In interconnect configurations shielded from the substrate by a metal ground shield the signal speed v depends on the slowing factor $\sqrt{(\varepsilon_{r,eff})}$, which is determined mainly by the dielectric layers surrounding the metal layers. The use of SiO_2 as a dielectric results in a typical signal speed $v \approx c/2$, with c being the speed of light. With this approach, the delay across a transmission line can be approximated on the basis of the signal speed and line length.

The phase of Z_0 of a low-loss transmission line starts at $-45°$ at low frequencies and rises to $0°$ at high frequencies. The line loss can be estimated on the basis of the phase of Z_0. In broadband applications, little delay variation over frequency is important for minimising jitter generation.

At a line length corresponding to $\lambda/4$, the sensitivity to source and load impedance mismatch reaches its maximum. The reflections are more significant at the input (source side) of the line than at the output.

In the near future, the transfer from aluminium to copper or gold interconnect will reduce the line attenuation, but will play no significant role with respect to the line characteristic impedance and delay. The use of low-k dielectrics finds its origin in digital ICs, in which it is used to minimise crosstalk or to allow a narrower line pitch at a given maximum allowable crosstalk to neighbouring wires. These low-k dielectrics are also interesting for microwave applications. Transmission lines with a higher characteristic impedance and thus higher inductance per unit length become practical. For a transmission line intended as inductance, e.g., shorted stub, a high characteristic impedance reduces the required line length. In the case of inductors, the self-resonant frequency will be higher. However, the effect of the passivation layer (with high ε_r) on transmission lines using top metal layer will be significant.

In CMOS and BiCMOS processes there is a trend towards an increased number of interconnect layers for denser routing. The top metal layers are designed to handle the higher currents needed for increased power dissipation at reduced supply voltages. So, while the lowest metal layers are designed for reduced pitch and thickness, the top metal layers are thicker. The low-loss on-chip interconnect configurations discussed in this chapter will therefore remain suitable for use in the foreseeable future.

References

1. A. Deutsch, G.V. Kopcsay, et al., "When are transmission-line effects important for on-chip interconnections," in *Proc. Electronic Components and Technology Conference*, May 1997, pp. 704–712.
2. B. Kleveland, X. Qi, et al., "High-frequency characterisation of on-chip digital interconnects," *IEEE J. Solid-State Circuits*, vol. 37, No. 6, June 2002, pp. 716–725.
3. A. Balakrishnan, C.M. Carpenter, "Analyses and design of head-preamplifier connections in read-write channels for magnetic rigid-disk drives," *IEEE Trans. Magn.*, vol. 34, No. 1, January 1998, pp. 24–29.
4. Tektronix application note, differential impedance measurements with the tektronix 8,000B series instruments, [Online]. Available: http://www.tektronix.com/oscilloscopes
5. K.S. Lowe, "Bufferless broadcasting: a low power distributed circuit technique for broadcasting 10-Gb/s chip input signals," *IEEE J. Solid State Circuits*, vol. 32, No. 10, October 1997, pp. 1551–1555.
6. W. Dürr, U. Erben, A. Schüppen, H. Dietrich, H. Schumacher, "Investigation of microstrip and coplanar transmission lines on lossy silicon substrates without backside metallization", *IEEE Trans. Microwave Theory Tech.*, vol. 46, No. 5, May 1998, pp. 712–715.
7. M. Pfost, H.-M. Rein, T. Holzwarth, "Modeling substrate effects in the design of high-speed Si-bipolar IC's," *IEEE J. Solid-State Circuits*, vol. 31, No. 10, October 1996, pp. 1493–1501.
8. T.S.D. Cheung, J.R. Long, K. Vaed, et al., "On-chip interconnect for mm-wave applications using an all-copper technology and wavelength reduction," *ISSCC Dig. Tech. Papers*, 2003, pp. 396–397.
9. B. Kleveland, C.H. Diaz, et al., "Exploiting CMOS reverse interconnect scaling in multigigahertz amplifier and oscillator design," *IEEE J. Solid-State Circuits*, vol. 36, No. 10, October 2001, pp. 1480–1488.
10. W.R. Eisenstadt, Y. Eo, "S-parameter-based IC interconnect transmission line characterization," *IEEE Trans. Comp., Hybrids, Manufact. Technol.*, vol. 15, No. 4, August 1992, pp. 483–490.

11. Y. Eo, W.R. Eisenstadt, "High-speed VLSI interconnect modeling based on S-parameter measurements," *IEEE Trans. Comp., Hybrids, Manufact. Technol.*, vol. 16, No. 5, August 1993, pp. 555–562.
12. H. Hasegawa, M. Furukawa, H. Yanai, "Properties of microstrip line on Si-SiO_2 system," *IEEE Trans. Microwave Theory Tech.*, vol. MTT-19, No. 11, November 1971, pp. 869–881.
13. H. Hasegawa, S. Seki, "Analysis of interconnection delay on very high-speed LSI/VLSI chips using an MIS microstrip line model," *IEEE Trans. Microwave Theory Tech.*, vol. MTT-32, No. 12, December 1984, pp. 1721–1727.
14. D.E. Bockelman, W.R. Eisenstadt, "Combined differential and common-mode scattering parameters: theory and simulation," *IEEE Trans. Microwave Theory Tech.*, vol. 43, No. 7, July 1995, pp. 1530–1539.
15. H.A. Wheeler, Formulas for the skin effect, *Proceedings of the I.R.E.*, Sept. 1942, pp. 412–424.
16. P. Deixler, R. Colclaser, et al., "QUBiC4G: a $f_T/f_{max} = 70/100$ GHz 0.25 μm low power SiGe-BiCMOS production technology with high quality passives for 12.5Gb/s optical networking and emerging wireless applications up to 20GHz," in *Proc. IEEE BCTM*, 2002, pp. 201–204.
17. P.J. van Wijnen, H.R. Claessen, et al., "A new straightforward calibration and correction procedure for on waferhigh frequency S-parameter measurements (45MHz–18GHz)," in *Proc. IEEE BCTM*, 1987, pp. 70–73.
18. A. Deutsch, P.W. Coteu, G. Kopcsay, et al., "On-chip wiring design challenges for gigahertz operation," *Proc. IEEE*, vol. 89, No. 4, April 2001, pp. 529–555.
19. O. Kromat, U. Langmann G. Hanke, W.J. Hillery, "A 10-Gb/s Silicon bipolar IC for PRBS testing," *IEEE J. Solid State Circuits*, vol. 33, No. 1, January 1998, pp. 76–85.

Chapter 3
Device Metrics

3.1 Introduction

The performance constraints of transistors define fundamental circuit limitations. For example, by relating circuit performance to widely accepted technology parameters one can predict the impact of a new technology on applications. This chapter reviews device metrics that will be used for circuit design in the rest of this book. The metric that is most widely used for the evaluation of an IC process is f_T, representing the transition frequency or unity current gain bandwidth of the transistors. The f_T can be used to estimate the gain-bandwidth product of a basic amplifier circuit as shown in Fig. 3.1a.

In Fig. 3.1b, C_{in} represents the base-emitter capacitance and gm the transconductance. It is assumed that the two transistors of the amplifier are identical and biased at the same currents. The small-signal voltage gain $A = v_2/v_1$ derived from the equivalent circuit is $A = -gm \cdot R_L$; the bandwidth B is $B = 1/(2\pi R_L C_{in})$. The gain-bandwidth product is $|A \cdot B| = gm/(2\pi C_{in})$, which corresponds to the f_T of the transistors. So, the f_T can be used to estimate the gain-bandwidth product or the first-order low-pass response of the amplifier of Fig. 3.1a. However, the transistor model used to calculate the small-signal gain of the amplifier of Fig. 3.1a ignores many aspects that are important for high-frequency circuit design, such as the base series resistance, the base-collector capacitance, etc. Therefore, it is important to analyse the relevance of the widely used metrics for high-frequency circuit design, and to develop new metrics that provide similar insights while using a more appropriate transistor model.

In Section 3.3 the definitions of widely accepted small-signal device metrics such as f_T and f_{max} will be reviewed. The available bandwidth f_A, representing the -3 dB bandwidth of a differential amplifier, will also be reviewed. Although metric f_A is not frequently used in the literature, differential amplifiers are widely used in broadband ICs, and therefore f_A is important for high bit-rate circuit design. For example, f_A is a useful parameter for relating the bandwidth of the RF path of the cross-connect switch described in Chapter 4 to technology parameters.

H. Veenstra and J.R. Long, *Circuit and Interconnect Design for RF and High Bit-Rate Applications,*
doi: 10.1007/978-1-4020-6884-3_3,

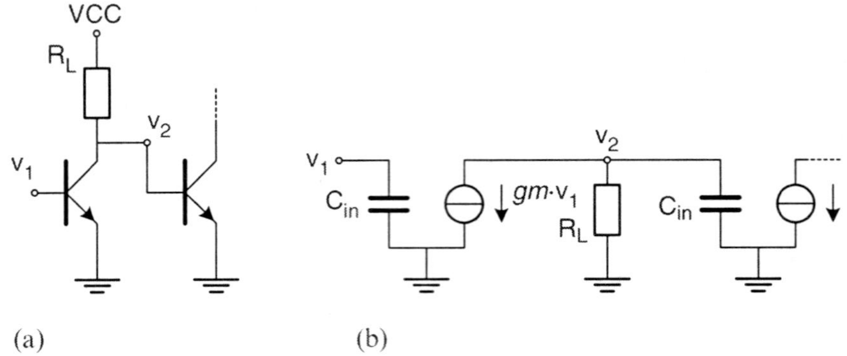

Fig. 3.1 Amplifier (a) and simplified small-signal equivalent circuit (b)

The available bandwidth f_A can be subdivided into two contributions: the input bandwidth f_V and the output bandwidth f_{out}. The input bandwidth f_V represents the bandwidth from the input voltage source to the collector current conversion for a grounded collector terminal. The output bandwidth f_{out} represents the bandwidth of the collector current to output voltage conversion in the grounded base connection using a (bias-dependent) load resistance for a specified low-frequency gain. Analysis of the input bandwidth f_V and the output bandwidth f_{out} yields valuable insight into their relative contributions to f_A. Such information provides guidelines for the optimisation of (next-generation) IC processes targeting high bit-rate applications. So, while f_A is a useful parameter for circuit design, f_V and f_{out} are most relevant for process optimisation.

In Section 3.3 a new technology parameter, f_{cross}, will be introduced that will be useful for relating the maximum attainable oscillation frequency of an LC-VCO to technology parameters in Chapter 7. At frequencies below f_{cross}, a cross-coupled differential pair provides a negative parallel equivalent input resistance. Consequently, a cross-coupled differential pair will ensure undamping at frequencies below, but not above, f_{cross}.

All device metrics are expressed in terms of common-emitter y-parameters, which can be derived from s-parameters. Since s-parameters are routinely measured during process development, they are widely available. Deriving device metrics from measured s-parameters has the advantage that no (time-consuming) parameter derivation is necessary to obtain the metrics. Besides, inaccuracies resulting from parameter fitting and phenomena not included in the model are avoided.

In Section 3.4 the device metrics will be evaluated for a simplified transistor model, to obtain practical formulas that can be used in circuit design.

Several trade-offs have to be made during the definition of an IC process. For exmple, in Chapter 1 it was already mentioned that BV_{CEO} can be traded off against f_T. A high f_A is required to obtain the widest bandwidth circuits for high bit-rate applications. What this implies for the IC process will be analysed in Section 3.5. Since IC processes are usually optimised using f_T and f_{max} as figures of merit (FOM), the relationship between f_T, f_{max} and f_A will be analysed in Section 3.6.

Several recently published IC technologies will be compared in Section 3.7. Trends in the fields of transistor device parameters and passives and the process back-end will be highlighted. The reduction in feature size in combination with increased current densities results in a steady increase in transistor self-heating with successive technology generations. Therefore a discussion of self-heating will be provided in Section 3.7.2.

In this chapter, definitions and comparisons focus on bipolar npn transistors. The results are applicable to widely used bipolar IC processes such as Si, SiGe, SiGe:C, GaAs HBT and InP HBT. Since the focus of this book is on high bit-rate and (mainly) large-signal circuits, noise and distortion of transistors are not analysed.

3.2 Miller Effects

Since the term 'Miller effect' will be widely used in relation to device metrics and circuit design in this book, the employed terminology should be explained. The input impedance of an amplifier depends on the feedback impedance Z applied between the output and input terminals and the open-loop gain A (see Fig. 3.2).

Assuming an infinite input impedance and zero output impedance for the open-loop amplifier circuit, the input impedance Z_{in} in closed loop equals

$$Z_{in} = \frac{Z}{1+A} \tag{3.1}$$

If the feedback impedance Z is a capacitor, seen from the input the capacitance will look $(A+1)$ times larger. This effect, widely known as the Miller effect, was first reported by John Miller in 1920 [1]. A Miller effect can also occur seen from the output, which is important for the derivation of the output bandwidth. The current through the feedback impedance Z can be related to the output voltage v_o via:

$$i = v_o \frac{1+1/A}{Z} \tag{3.2}$$

These two Miller effects will be widely used below. When considering the Miller effect on the input impedance, the commonly used term *Miller effect* will be used. The Miller effect on the output impedance will be referred to as the *output Miller effect*.

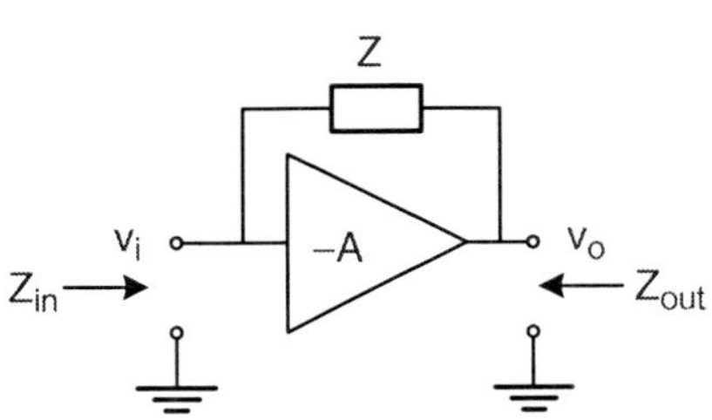

Fig. 3.2 Amplifier with open-loop gain – A and feedback impedance Z

3.3 Definitions Based on *y*-Parameters

In this section, *y*-parameters will be used to define several small-signal transistor parameters. All the described analyses were performed for an npn transistor with four terminals: collector (c), base (b), emitter (e) and substrate (s) (see Fig. 3.3). Since pnp transistors are often not needed nor available for high-speed design, they will not be considered.

In general, in a 4-terminal device the voltage and current relationships will form a 4×4 matrix:

$$\begin{pmatrix} i_e \\ i_b \\ i_c \\ i_s \end{pmatrix} = \begin{pmatrix} y_{ee} & y_{eb} & y_{ec} & y_{es} \\ y_{be} & y_{bb} & y_{bc} & y_{bs} \\ y_{ce} & y_{cb} & y_{cc} & y_{cs} \\ y_{se} & y_{sb} & y_{sc} & y_{ss} \end{pmatrix} \cdot \begin{pmatrix} v_e \\ v_b \\ v_c \\ v_s \end{pmatrix} \tag{3.3}$$

To simplify calculations, the substrate network (between the substrate and the ground) will often be ignored in the analyses. In the case of GaAs and InP IC processes the high-resistivity of the substrate allows one to treat the substrate network as being an open circuit. This effectively reduces the number of terminals of the transistor to 3 (c, b and e) since $y_{es} = y_{bs} = y_{cs} = y_{ss} = 0$. In the case of SiGe IC processes, if the substrate network is to be ignored, the substrate must be connected to the ground, forcing $v_s = 0$. The assumption that the substrate is grounded can in practice be approximated by placing a sufficient number of substrate contacts close to the collector. In the case of differential circuits in a symmetrical layout, placing the transistors close together creates a low-impedance differential network between the substrate terminals of the transistors. The substrate contacts will then be important mainly for the common-mode impedance of the substrate network.

If a SiGe transistor is treated as a 3-terminal device (with terminals c, b and e), the collector to substrate network can be removed from the transistor model, but must still be included in the calculations. Assuming $v_s = 0$, the collector-substrate network for each transistor can be moved from the transistor model to (in parallel to) the collector load impedance.

Since the substrate is shielded from the base by the collector, it is reasonable to assume that $y_{bs} = y_{sb} = 0$. For the same reason, $y_{es} = y_{se} = 0$. The collector-substrate impedance (often represented by a capacitance C_{cs}) is usually ignored or moved from the transistor model to (i.e., in parallel to) the collector load impedance. In both cases this results in transistor *y*-parameters $y_{sc} = y_{cs} = 0$. In the common-emitter

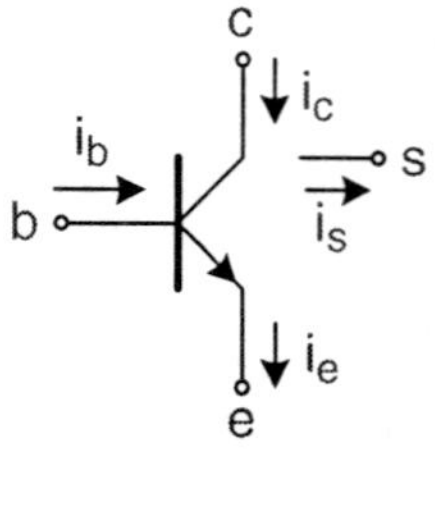

Fig. 3.3 4-Terminal npn device

configuration the emitter is also grounded, so $v_e = 0$. As a result, the y-matrix for the common-emitter configuration reduces to the following 2×2 matrix:

$$\begin{pmatrix} i_b \\ i_c \end{pmatrix} = \begin{pmatrix} y_{bb} & y_{bc} \\ y_{cb} & y_{cc} \end{pmatrix} \cdot \begin{pmatrix} v_b \\ v_c \end{pmatrix} = \begin{pmatrix} y_{11} & y_{12} \\ y_{21} & y_{22} \end{pmatrix} \cdot \begin{pmatrix} v_b \\ v_c \end{pmatrix} \tag{3.4}$$

For any equivalent transistor model the common-emitter y-parameters are derived using the following four relationships:

$$y_{11} = \left.\frac{i_b}{v_b}\right|_{v_c=0} ; \; y_{12} = \left.\frac{i_b}{v_c}\right|_{v_b=0} ; \; y_{21} = \left.\frac{i_c}{v_b}\right|_{v_c=0} ; \; y_{22} = \left.\frac{i_c}{v_c}\right|_{v_b=0} \tag{3.5}$$

The condition $v_c = 0$ implies that the collector is grounded for the calculation of y_{11} and y_{21}. Similarly, the base needs to be grounded for the calculation of y_{12} and y_{22}. The term y_{11} represents the input admittance, y_{21} represents the forward transadmittance, y_{12} is the feedback transadmittance and y_{22} represents the output admittance.

3.3.1 Unity Current Gain Bandwidth f_T

The unity current gain bandwidth or f_T of an npn transistor can be expressed in terms of common-emitter y-parameters. For derivation of the quantity f_T the base is driven by an ac current source and the collector is ac-grounded so that $v_c = 0$ (see Fig. 3.4).

The dc sources I_e and V_{cb} set the bias condition while the ac current source i_b provides the ac excitation. The inductor is used to create a high ac impedance in parallel with the base-collector junction. $|\omega L| \gg |1/\omega C|$ must be chosen in the frequency range in which the current gain is analysed to obtain the proper ac settings (Fig. 3.4b). The capacitor is used to sink the ac collector current i_c. The quantity f_T is derived from the current gain $|i_c/i_b|$. Using the common-emitter y-parameter equations (3.4) with $v_c = 0$, the current gain can be expressed as:

$$\left|\frac{i_c}{i_b}\right| = \left|\frac{y_{21}}{y_{11}}\right| \tag{3.6}$$

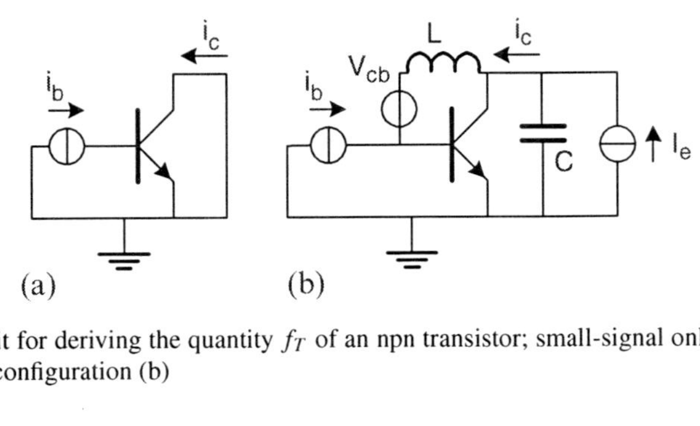

Fig. 3.4 Circuit for deriving the quantity f_T of an npn transistor; small-signal only (a) and small-signal plus dc configuration (b)

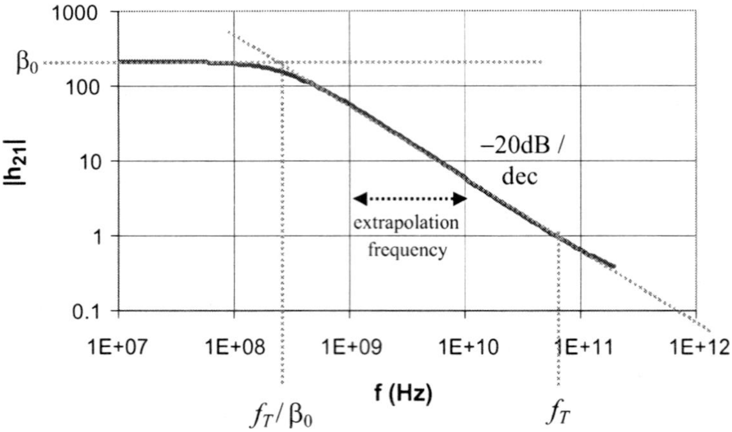

Fig. 3.5 Definition of f_T. The absolute value of the current gain shown is valid for a single bias condition

The f_T of the transistor is defined as the extrapolated frequency, with the magnitude of the current gain, $|i_c/i_b| = |y_{21}/y_{11}| = |h_{21}|$, equalling unity (=0dB). Extrapolation by -20dB/decade corresponds to ignoring the feedforward via the base-collector capacitance C_{bc} in y_{21}. An example of a current gain curve is shown in Fig. 3.5. The low-frequency current gain is represented by β_0. The curve shows a typical $|h_{21}|$ as a function of frequency together with the asymptotic response.

The zero in the current gain due to C_{bc} causes a reduced slope for $|y_{21}/y_{11}|$ near f_T. By definition, the f_T value is derived from the asymptotic response, with the extrapolation frequency chosen in the frequency range between f_T/β_0 and f_T, at a frequency at which the slope of the current gain is -20dB/decade.

Figure 3.5 shows a frequency sweep at a given bias condition, resulting in a single point of the f_T-curve at the corresponding operating point. The f_T-curve, showing f_T as a function of the bias condition, may be generated from a bias sweep at a single (extrapolation) frequency f_x. Provided that at $f = f_x$ the current gain shows a 20 dB/decade roll-off, the f_T is obtained via extrapolation from the current gain β at the extrapolation frequency: $f_T(\mathrm{I}) = f_x \cdot \beta(\mathrm{I})$. An example of an f_T-curve is shown in Fig. 3.6.

The quantity f_T depends on the collector-base voltage V_{cb} due to various effects such as the voltage dependency of the collector-base junction capacitance C_{bc} and quasi-saturation. It must therefore always be specified at what V_{cb} an f_T-value is obtained. A typical value for which f_T is reported in the literature is $V_{cb} = 1$ V. Sometimes peak-f_T values are reported that can only be obtained at high V_{cb} (near BV_{CEO}). In low power, and hence low supply voltage circuit design, the f_T at $V_{cb} = 0$ V is of more importance. In the definition of f_T, the collector-substrate impedance C_{cs} is shorted and consequently plays no role. The (extrinsic) base series resistance R_b has only a minor effect on f_T since only the current gain is considered.

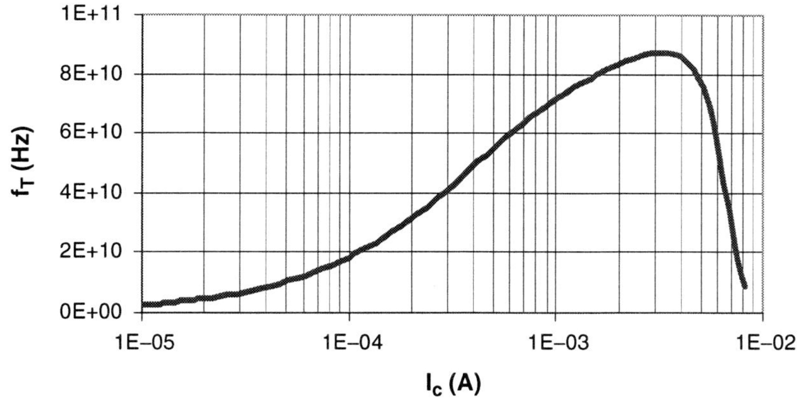

Fig. 3.6 Example of an f_T-curve. The curve is valid for a 0.5×4.7 npn biased at $V_{cb} = 0\,V$ and is generated using an extrapolation frequency $f_x = 4\,GHz$

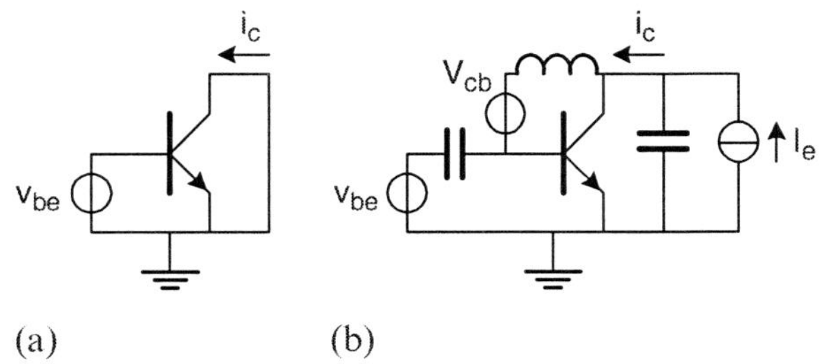

Fig. 3.7 Schematic for deriving the quantity f_V of an npn transistor; small-signal only (a) and small-signal plus dc configuration (b)

3.3.2 Input Bandwidth f_V

For the derivation of the input bandwidth f_V, the collector is shorted and the source impedance is ignored. Since the collector is shorted, no Miller effect of the base-collector capacitance occurs (assuming that the collector series resistance R_c is zero). Therefore, f_V may seem fairly irrelevant for circuit design. However, the quantity f_V is an important parameter for technology optimisation because it represents the dominant contribution to the available bandwidth f_A at high current densities, as will be shown below.

The input bandwidth or f_V of an npn can be expressed in terms of common-emitter y-parameters. For the derivation of the quantity f_V, the base is driven by an ac voltage source and the collector is ac-grounded so that $v_c = 0$ (see Fig. 3.7).

The dc sources V_{cb} and I_e set the bias condition of the npn while the ac source v_{be} provides the ac excitation. The inductor acts as a choke, the capacitor as a decoupling to absorb the collector current i_c while forcing v_c to equal zero. The quantity f_V represents the $-3\,dB$ bandwidth of the collector current $|i_c|$ and is a function of the collector-base voltage V_{cb} due to various effects such as the voltage dependency

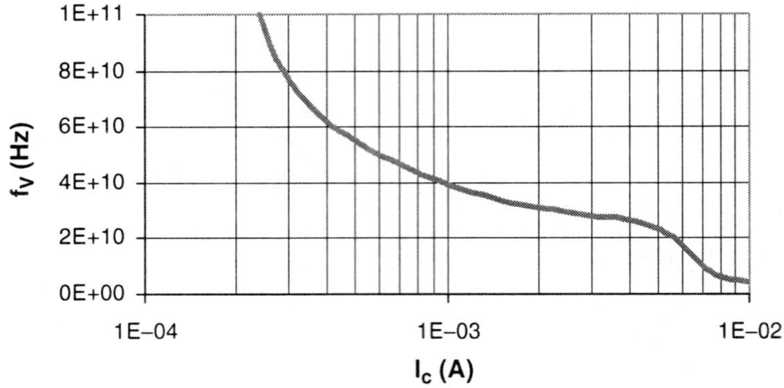

Fig. 3.8 Example of an f_V-curve. The 0.5×4.7 npn transistor is biased at $V_{cb} = 0\,V$ and reaches its peak-f_T at $I_c = 3.2\,mA$

of the collector-base junction capacitance C_{bc} and quasi-saturation. The quantity f_V is in this book defined at $V_{cb} = 0$ unless otherwise indicated. In terms of common-emitter y-parameters f_V equals the $-3\,dB$ bandwidth of $|y_{21}|$, as follows from equation (3.4), where $v_c = 0$. An example of an f_V-curve is shown in Fig. 3.8.

In practice, the base is always driven via a non-zero source impedance. An additional transfer function will then exist from the source (in front of the source impedance) to v_{be}; the quantity f_V always represents the $-3\,dB$ bandwidth of the transconductance $|i_c/v_{be}|$. Note that while the f_T-curve can be derived using a bias sweep at a single extrapolation frequency, derivation of the f_V-curve requires a frequency sweep per bias point.

The input bandwidth f_V decreases with an increasing collector current I_c. This is due to the increase in the diffusion capacitance component of C_{be} at increasing V_{be} while the base series resistance is almost independent of V_{be}. A low base resistance R_b is therefore essential for obtaining a high f_V. For example, transistors with a double base contact have a considerably higher f_V than those with a single base contact.

3.3.3 Output Bandwidth f_{out} and Available Bandwidth f_A

The available bandwidth f_A represents the bandwidth of a differential amplifier and is an important parameter for broadband circuit design. For example, it is not always optimum to bias the transistors in a differential pair amplifier at peak-f_T. The f_A-curve provides an important guideline for determining the optimum bias point for the transistors in a differential pair and consequently provides important information for the design of broadband circuits. Moreover, the peak-f_A value is a strong indicator of the highest bit-rate that can be supported in broadband circuits. The f_A can be subdivided into two contributions, namely the input bandwidth f_V and the output bandwidth f_{out}. Analysis of the relative contributions of f_V and f_{out}

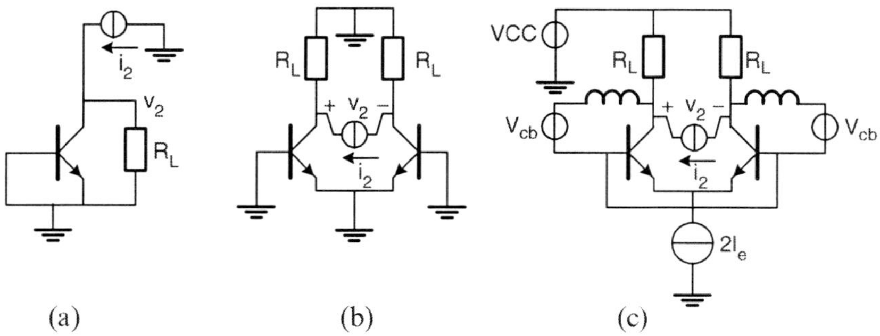

Fig. 3.9 Definition of the output bandwidth f_{out}; single-ended (a) and differential configurations (b), (c). Configuration (c) includes dc biasing

to f_A provides valuable feedback for technology improvement. So, like f_V, f_{out} is a useful parameter for process development.

For the derivation of the output bandwidth f_{out} and the available bandwidth f_A, a load resistance $\mathrm{R_L} = 1/\mathrm{Y_L}$ needs to be connected to the collector port. The output bandwidth f_{out} is defined as the bandwidth of the voltage at the output port on the condition that a current is driven into the collector while the base terminal is ac-grounded. This configuration is shown in Fig. 3.9.

The output bandwidth f_{out} equals the $-3\,\mathrm{dB}$ bandwidth of the output voltage $|v_2|$. So, the output bandwidth f_{out} is defined by the parallel impedance of $\mathrm{R_L}$ with $1/y_{22}$. In practice, the way in which the substrate is contacted may play a role in f_{out} because the substrate impedance is part of y_{22}. The value of f_{out} further depends on the supply voltage VCC (via the collector-substrate junction capacitance which contributes to y_{22}) and the collector-base voltage $\mathrm{V_{cb}}$ (via the collector-base junction capacitance).

For the derivation of the available bandwidth f_A, the input is usually driven by a voltage source, as in Fig. 3.10, but this is not strictly necessary. The available bandwidth f_A is defined by the $-3\,\mathrm{dB}$ bandwidth of the voltage gain, $|\mathrm{A}| = |v_{out}/v_{in}| = |v_2/v_1|$, with $\mathrm{R_L}$ being chosen so that the low-frequency gain equals a predefined value. Usually a value of 10 (=20 dB) is chosen, although f_A may also be specified for a different low-frequency gain. Note that the load resistance $\mathrm{R_L}$ depends on the bias condition; at a higher bias current a lower load resistance will be required to keep the low-frequency gain constant. In some CMOS processes the output conductance $\mathrm{g_{ds}}$ may severely limit the gain, to such an extent even that a low-frequency gain of 10 may not be feasible. This will not usually be a problem for bipolar transistors.

The small-signal voltage gain A can be derived from the common-emitter y-parameters (3.4) using $i_2 = -v_2 \cdot \mathrm{Y_L}$. This gives

$$A = \frac{-y_{21}}{y_{22} + Y_L} \tag{3.7}$$

Thus, f_A follows from the $-3\,\mathrm{dB}$ bandwidth of $|\mathrm{A}|$, with A being given by equation (3.7). If the input is driven by a voltage source, as in Fig. 3.10, the bandwidth

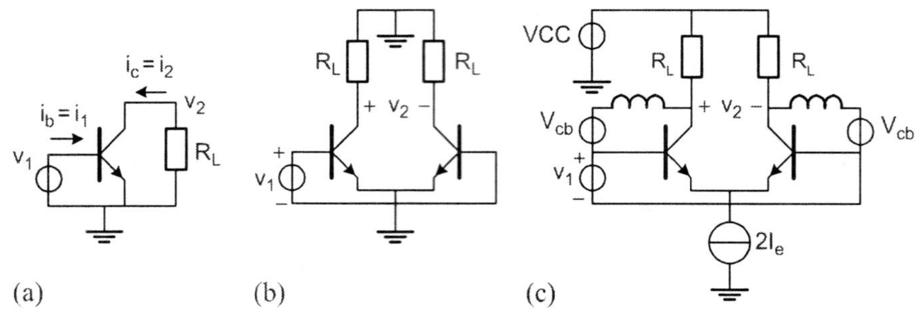

Fig. 3.10 Definition of the available bandwidth f_A; single-ended (a) and differential configurations (b), (c). Configuration (c) includes dc biasing

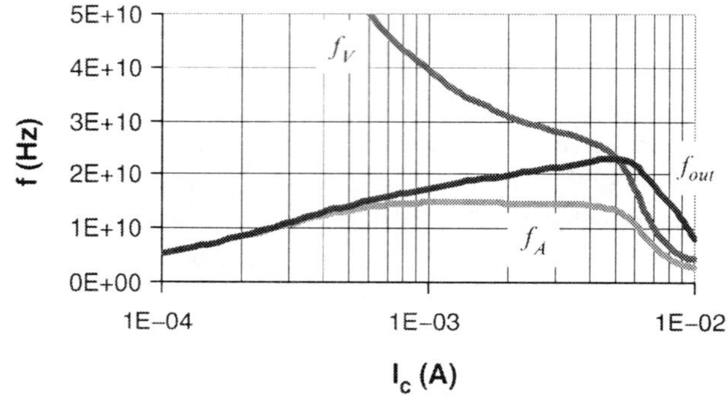

Fig. 3.11 Example f_{out}, f_V and f_A-curves for a 0.5×4.7 npn transistor biased at $V_{cb} = 0\,V$; f_A and f_{out} are for 20 dB low-frequency gain. The transistor reaches its peak-f_T at $I_c = 3.2\,mA$

of the output voltage $|v_{out}| = |v_2|$ represents the f_A. If the input is driven by a source with an arbitrary source impedance, f_A is found via the -3 dB bandwidth of the gain $|A| = |v_{out}/v_{in}| = |v_2/v_1|$.

The numerator of equation (3.7) represents the input bandwidth, since the f_V equals the -3 dB bandwidth of $|y_{21}|$, while the denominator represents the output bandwidth f_{out} of the resistively loaded common-emitter configuration at ac-grounded input (base) terminal. So, both the input and the output bandwidths play an equally important role in the f_A of a transistor, although one of the two will typically dominate at a given bias condition. This is shown in Fig. 3.11, which shows an example f_A-curve for a 20 dB low-frequency gain.

At low bias currents, a high load resistance R_L is needed to achieve 20 dB low-frequency gain, resulting in a low output bandwidth. At the same time, the input bandwidth will be relatively high due to the low base-emitter capacitance C_{be}. So, at low bias currents the output bandwidth f_{out} will usually dominate f_A. With an increasing bias current, the input bandwidth will decrease due to an increase in C_{be}, while the output bandwidth will increase due to a decrease in the load resistance R_L. These effects are clearly visible in Fig. 3.11 at bias currents below 4 mA.

In SiGe IC processes the peak-f_A will usually occur at a lower collector current density than the peak-f_T. In the f_A-curve of Fig. 3.11, the ratio of the current densities of peak-f_T and peak-f_A is $J_{c,fAp}/J_{c,fTp} \approx 2$. So, biasing a differential pair at peak-f_T may not result in the maximum bandwidth. The ratio of the current densities of peak-f_T and peak-f_A depends heavily on the base resistance and collector-base capacitance, as will be shown below.

In practical broadband circuit configurations, the amplifier output will be loaded by an impedance Z_{load} that may usually be represented by a parallel network of load capacitance C_{load} and load resistance R_{load}. The load resistance R_{load} will reduce the low-frequency gain. The amplifier load resistance R_L may be compensated to achieve the desired low-frequency gain under loaded condition. The load capacitance C_{load} is seen in parallel with the amplifier output capacitance C_p and may reduce the bandwidth of the loaded amplifier significantly with respect to f_A. So, in circuit configurations, the output bandwidth may become the dominant factor in the amplifier bandwidth. However, a low-frequency gain of 20 dB is not often required. For example, CML logic circuits typically use a small-signal low-frequency gain of 2–4. To emphasize the importance of the output bandwidth, analysis of f_A at 20 dB low-frequency gain will yield a meaningful indicator for the design of broadband circuits.

The quantity f_A is applicable to both bandwidths of the single-ended and differential amplifier configurations of Fig. 3.10. Due to the virtual ground at the common-emitter node of the differential pair, the common-emitter y-parameter analysis remains valid for the differential pair configuration, in which each collector is loaded by a load resistor with value R_L. The gain A then refers to the differential-mode voltage gain.

3.3.4 *Negative Resistance of a Cross-Coupled Differential Pair f_{cross}*

A commonly used circuit topology is a cross-coupled differential pair as shown in Fig. 3.12. Such a circuit provides a negative input resistance (for a certain frequency range) and may consequently be used in, for example, oscillator circuits and latches.

Due to the virtual ground at the common-emitter node, the common-emitter y-parameters may be used for calculations. Use is made of a differential input signal using two identical voltage sources ($v_1/2$) in series, so that a virtual ground will also exist between the two voltage sources (see Fig. 3.12). The differential input admittance $Y_i = i_1/v_1$ is now analysed as follows, using the common-emitter relations from equation (3.4):

$$i_1 = i_4 + i_6 = (y_{22} - y_{21})\frac{v_1}{2} + (y_{11} - y_{12})\frac{v_1}{2} \tag{3.8}$$

For the input admittance this gives:

$$Y_i = \frac{i_1}{v_1} = \frac{1}{2}(y_{11} - y_{12} - y_{21} + y_{22}) \tag{3.9}$$

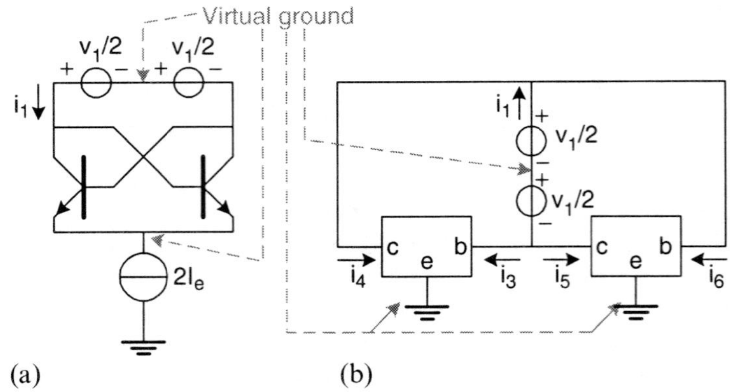

Fig. 3.12 Analysis of the input impedance of a cross-coupled differential pair; schematic (a) and definitions for calculations using common-emitter *y*-parameters (b)

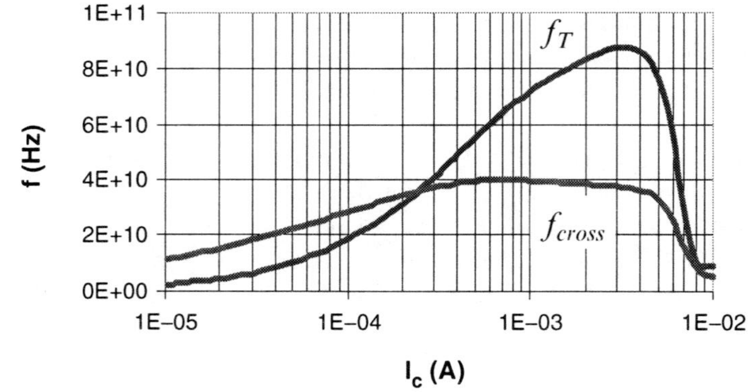

Fig. 3.13 Example of an f_{cross}-curve for a 0.5 × 4.7 npn transistor biased at $V_{cb} = 0\,V$. For comparison, the f_T-curve of the same transistor is also shown. The transistor reaches its peak-f_T at $I_c = 3.2\,mA$

Note that the collector to substrate network has been ignored in the 2-port *y*-parameter equations. The collector to substrate network with impedance $Z_{cs} = 1/Y_{cs}$ is important for a cross-coupled differential pair because a $2 \cdot Z_{cs}$ series network is connected parallel to the differential input impedance, or

$$Y_i = \frac{i_1}{v_1} = \frac{1}{2}(y_{11} - y_{12} - y_{21} + y_{22} + Y_{cs}) \tag{3.10}$$

The device metric f_{cross} refers to the highest frequency at which the parallel equivalent input impedance of the cross-coupled differential pair still provides undamping. The undamping follows from the real part of the input admittance $\mathrm{Re}(Y_i)$. At $f < f_{cross}$ the input admittance has a negative real part (e.g., $\mathrm{Re}(Y_i) < 0$); f_{cross} occurs where $\mathrm{Re}(Y_i) = 0$. A typical f_{cross}-curve is shown in Fig. 3.13. For the derivation of the f_{cross}-curve, a frequency sweep is required per bias point. Note that the metric f_{cross} is almost independent of the bias current for an order of magnitude vari-

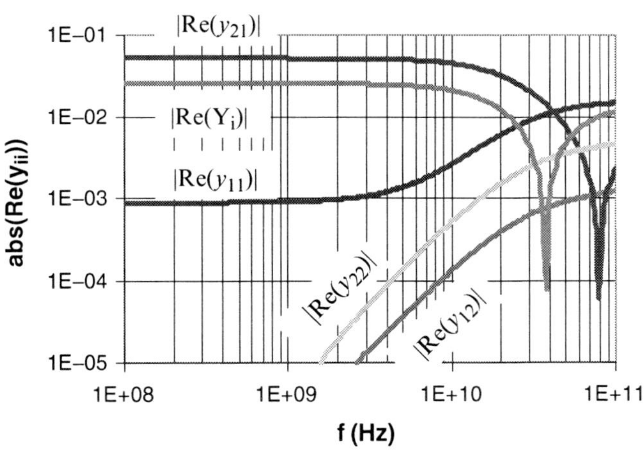

Fig. 3.14 Contributions $|\mathrm{Re}(y_{ii})|$ and $|\mathrm{Re}(\mathrm{Y_i})|$. The 0.5×4.7 npn transistor is biased at peak-f_T ($\mathrm{I_c} = 3.2\,\mathrm{mA}$); f_{cross} occurs where $\mathrm{Re}(\mathrm{Y_i}) = 0$ ($f_{cross} = 37.7\,\mathrm{GHz}$)

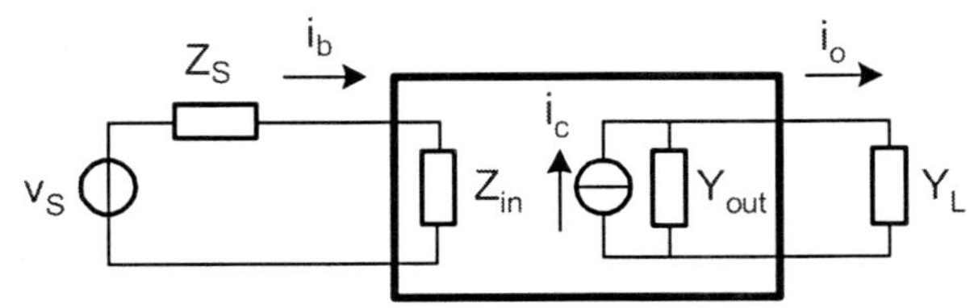

Fig. 3.15 Definitions for the calculation of the power gain. The model is valid at a single frequency

ation in collector current. This (typical) behaviour of f_{cross} will be explained below on the basis of approximate formulas derived for a simplified transistor model.

As follows from equation (3.9), all 4 y-parameters play a role in f_{cross}. Figure 3.14 shows an example of the real parts of all 4 y-parameters across a frequency sweep biased at peak-f_T. As can be seen, $\mathrm{Re}(y_{21})$ and $\mathrm{Re}(y_{11})$ are the most significant contributions at $f = f_{cross}$. Thus, intuitively, both f_V and f_T define f_{cross}.

3.3.5 Maximum Oscillation Frequency f_{max}

The available power gain G_A of the transistor in common-emitter configuration driven by a source impedance at the base terminal $\mathrm{Z_s} = 1/\mathrm{Y_s}$ and terminated by a load impedance at the collector terminal $\mathrm{Z_L} = 1/\mathrm{Y_L}$ is the ratio (Power available at the output)/(Power available from the source). Using Fig. 3.15, the available power values can be derived.

The model can be used for calculating the maximum oscillation frequency f_{max} on the basis of either the maximum available gain $f_{max}(G_{max})$ or the unilateral gain $f_{max}(U)$. To calculate the unilateral gain, the collector-base capacitance C_{bc} is ignored. Ignoring C_{bc} simplifies tuning of the source and load impedances for maximum power gain since the ports can be tuned without influencing each other. However, $f_{max}(U)$ gives an optimistic value that is less relevant to circuit performance than $f_{max}(G_{max})$. Since f_{max}-values in publications often refer to $f_{max}(U)$, the $f_{max}(U)$ value is often used for benchmarking. In the following analyses f_{max} refers to $f_{max}(G_{max})$.

While the output admittance of the transistor Y_{out} may be ignored in most small-signal calculations, it is essential to include it in the calculation of the available power at the output. The available power from the source P_{avs} is delivered at an input power match, so at $Z_{in} = Z_s{}^*$, where $Z_s{}^*$ is the complex conjugate of Z_s. So,

$$P_{avs} = \frac{|v_s|^2}{4\,\mathrm{Re}(Z_s)} = \frac{i_b \cdot i_b^*}{4\,\mathrm{Re}(Y_s)} \tag{3.11}$$

The available power at the output P_{avo} is delivered to a matched load, where $Z_L = Z_{out}{}^*$, and equals

$$P_{avo} = \frac{i_c \cdot i_c^*}{4\,\mathrm{Re}(Y_L)} \tag{3.12}$$

The notations i_c^* and i_b^* refer to the complex conjugate of i_c and i_b, respectively. The available power gain can in general be expressed as [3]

$$G_A = \frac{P_{avo}}{P_{avs}} = \frac{i_c \cdot i_c^* \cdot \mathrm{Re}(Y_S)}{i_b \cdot i_b^* \cdot \mathrm{Re}(Y_L)} \tag{3.13}$$

The power gain has a maximum G_{max} under simultaneous input and output match, so at $Z_s = Z_{in}{}^*$ and $Z_L = Z_{out}{}^*$, with Z_{in} and Z_{out} being the input and output impedances of the two-port terminated by Z_s and Z_L. The input impedance Z_{in} for a two-port terminated with a load impedance Z_L can be derived from the y-parameter equations (3.4) in combination with the condition imposed by the output impedance $i_c = -v_c \cdot Y_L$:

$$\begin{aligned} i_b &= y_{11} v_b + y_{12} v_c \\ i_c &= y_{21} v_b + y_{22} v_c = -Y_L v_c \\ Z_{in} &= v_b / i_b \end{aligned} \tag{3.14}$$

Solving this gives for the input impedance:

$$Z_{in} = \frac{1}{y_{11} - \dfrac{y_{12} y_{21}}{y_{22} + Y_L}} \tag{3.15}$$

In a similar way, the output impedance Z_{out} for a two-port driven by a source impedance Z_s equals [3]:

$$Z_{out} = \frac{1}{y_{22} - \dfrac{y_{12} y_{21}}{y_{11} + Y_s}} \tag{3.16}$$

The conditions for a simultaneous input and output match, $Z_s = Z_{in}{}^*$ and $Z_L = Z_{out}{}^*$, can be combined with the equations for input and output impedance to find the maximum power gain G_{max}. This yields [4]:

$$G_{\max} = \left|\frac{y_{21}}{y_{12}}\right| \cdot (k - \sqrt{k^2 - 1}) \tag{3.17}$$

with

$$k = \frac{2\,\mathrm{Re}(y_{11})\,\mathrm{Re}(y_{22}) - \mathrm{Re}(y_{21}y_{12})}{|y_{12}y_{21}|} \tag{3.18}$$

Factor k is referred to as the stability factor and should fulfil the condition $k > 1$ for stability. Note that G_{max} is undefined for $k < 1$, but when $G_{max} \approx 1$, k will always be larger than unity for practical devices. The unity power gain frequency, at which the power gain G_{max} is 0 dB, defines $f_{max}(G_{max})$.

An oscillator requires positive feedback. To ensure that oscillation is sustained at f_{max}, no power may be lost in the feedback loop. So, f_{max} represents the highest frequency at which oscillation is possible. Since the simultaneous input and output port matching conditions that apply to G_{max} cannot be assumed for most oscillator circuits, the practical maximum oscillation frequency remains well below f_{max}.

3.4 Approximate Formulas for the Device Metrics

In this section the device metrics for the widely used simplified transistor model shown in Fig. 3.16 will be analysed. The main goal is to gain insight into the relevance of the most important device parameters for the device metrics.

The base series resistance has been included in the model as a single lumped (extrinsic) term R_b. The impedance between the base and the emitter has been modelled by a capacitance C_{be}. This model is consequently intended for the frequency range in which the current gain $\beta = i_c/i_b$ shows a roll-off of 20 dB per decade, or $f > f_T/\beta_0$. At lower frequencies $f << f_T/\beta_0$, a resistance $R_{be} = \beta_0/gm$ needs to be added parallel to C_{be}. The equivalent model includes only the most relevant transistor parameters to provide simple relationships between the device metrics and the most important transistor parameters. The emitter series resistance, although important for some of the metrics, has been omitted as it significantly

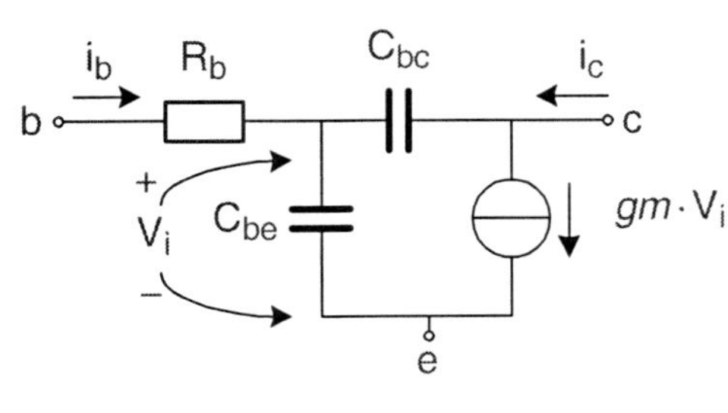

Fig. 3.16 Simplified small-signal transistor model for a transistor in a common-emitter configuration used to derive approximate formulas for the device metrics

complicates the y-parameter equations [2], thereby complicating the link between transistor parameters and device metrics. Appendix A of the book describes how the common-emitter y-parameters are to be derived for a transistor model including an emitter series resistance R_e, a base series resistance R_b and a collector series resistance R_c.

In the simplified transistor model shown in Fig. 3.16, gm represents the transconductance obtained via differentiation $gm = dI_c/dV_{be}$. In the case of an exponential relationship between I_c and V_{be} as in equation (3.19) the gm equals:

$$I_c = I_{c0} \cdot e^{\frac{qV_{be}}{kT}} \tag{3.19}$$

$$gm = \frac{dI_c}{dV_{be}} = \frac{qI_c}{kT} = \frac{I_c}{V_T} \tag{3.20}$$

Here, V_T equals the thermal voltage; $V_T = kT/q$. At room temperature ($T = 300\,K$) the transconductance can be approximated by $gm = qI_c/kT \approx 38.6 \cdot I_c$.

In the following, the common-emitter y-parameters of the simplified small-signal transistor model will be derived using equation (3.5). To determine y_{11}, the collector is ac-grounded, the base is driven by a voltage v_b and the base current i_b is calculated. This gives:

$$y_{11} = \frac{j\omega(C_{be} + C_{bc})}{j\omega R_b(C_{be} + C_{bc}) + 1} \tag{3.21}$$

To determine y_{21}, the collector is ac-grounded, the base is driven by a voltage v_b and the collector current i_c is calculated. This gives:

$$y_{21} = \frac{gm - j\omega C_{bc}}{j\omega R_b(C_{be} + C_{bc}) + 1} \tag{3.22}$$

To determine y_{12}, the base is ac-grounded, the collector is driven by a voltage v_c and the base current i_b is calculated. This gives:

$$y_{12} = \frac{-j\omega C_{bc}}{j\omega R_b(C_{be} + C_{bc}) + 1} \tag{3.23}$$

To determine y_{22}, the base is ac-grounded, the collector is driven by a voltage v_c and the collector current i_c is calculated. This gives:

$$y_{22} = \frac{j\omega C_{bc}(1 + gm \cdot R_b) + (j\omega)^2 R_b C_{be} C_{bc}}{j\omega R_b(C_{be} + C_{bc}) + 1} \tag{3.24}$$

Note that capacitance C_{cs} has been omitted from the model shown in Fig. 3.16. When C_{cs} is included, all the common-emitter y-parameters will remain unchanged except y_{22}, for which a term must be added to include the susceptance of C_{cs}. When C_{cs} is included, y_{22} will become:

$$y_{22} = \frac{j\omega C_{bc}(1 + gm \cdot R_b) + (j\omega)^2 R_b C_{be} C_{bc}}{j\omega R_b(C_{be} + C_{bc}) + 1} + j\omega C_{cs} \tag{3.25}$$

With these results for the y-parameters, the device metrics for the simplified transistor model shown in Fig. 3.16 can be evaluated. This will be described in the next section.

3.4.1 Approximation for f_T

The unity current gain bandwidth f_T is obtained when the asymptotic response of the magnitude $|y_{21}/y_{11}|$ equals unity. Using equations (3.21) and (3.22) gives

$$\left|\frac{y_{21}}{y_{11}}\right| = \left|\frac{gm - j\omega C_{bc}}{j\omega(C_{be}+C_{bc})}\right| \tag{3.26}$$

The asymptotic response is found from $|gm/j\omega(C_{be}+C_{bc})| = 1$, and in the simplified models yields

$$f_T = \frac{gm}{2\pi(C_{be}+C_{bc})} \tag{3.27}$$

Equation (3.27) may in practice be very useful for estimating the base-emitter capacitance C_{be} at a given bias condition. The f_T and C_{bc} for a given transistor geometry at a given set of operating conditions (bias current and collector-base voltage) can usually be found in the technology design manual or it can be found from the operating point information in a circuit simulation. The transconductance gm can be found with the aid of equation (3.20). By combining the f_T, C_{bc} and gm, the corresponding C_{be} is obtained using equation (3.27).

The quantity f_T is an important FOM for the bandwidth of the widely used cascode stage, whose basic topology is shown in Fig. 3.17. A common-emitter input transistor Q_1 is loaded by a cascode transistor Q_2. In the case of a cascode transistor, also referred to as common-base stage due to the ac-grounded base, the emitter current i_e is the input and the collector current i_c is the output. Since $i_e = i_b + i_c$ and $i_c = \beta(\omega) \cdot i_b$, with (for $\omega > \omega_T/\beta_0$)$\beta(\omega) = -j\omega_T/\omega$, the following relationship for the current gain i_c/i_e can be derived:

$$\frac{i_c}{i_e} = \frac{\beta(\omega)}{\beta(\omega)+1} \approx \frac{-j\frac{\omega_T}{\omega}}{1-j\frac{\omega_T}{\omega}} = \frac{1}{1+\frac{j\omega}{\omega_T}} \tag{3.28}$$

So, f_T represents the $-3\,\mathrm{dB}$ bandwidth of the current transfer ratio of the cascode stage. In addition, the small-signal delay from input current to output current equals the transit time $\tau = 1/\omega_T$.

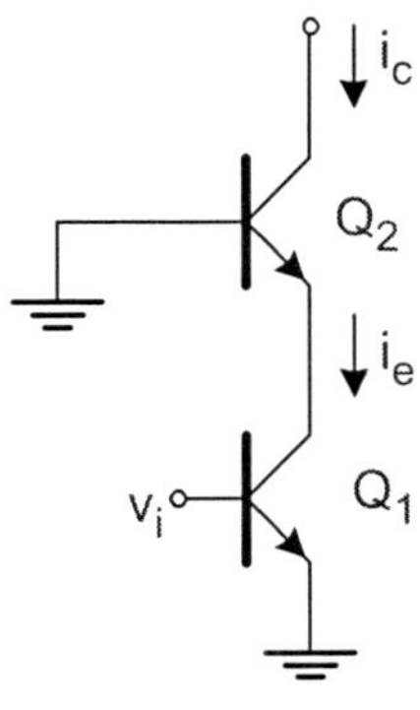

Fig. 3.17 Common-emitter input transistor Q_1 loaded by a cascode or common-base transistor Q_2

3.4.2 Approximation for f_V

The input bandwidth f_V follows from the -3dB bandwidth of $|y_{21}|$. An approximation can be obtained using equation (3.22). This equation shows a pole at $\omega_p = 1/(\mathrm{R_b}(\mathrm{C_{be}} + \mathrm{C_{bc}}))$ and a zero at $\omega_z = gm/\mathrm{C_{bc}}$. Since at typical operating currents $(\mathrm{C_{be}} + \mathrm{C_{bc}}) \gg \mathrm{C_{bc}}$ and $\mathrm{R_b} > 1/gm$, the zero is at a frequency $\omega_z \gg \omega_p$ and may be ignored. So, the input bandwidth f_V is approximately

$$f_V \approx \frac{1}{2\pi R_b(C_{be} + C_{bc})} \tag{3.29}$$

Using the result obtained for f_T with equation (3.27), f_V may also be written as

$$f_V \approx \frac{f_T}{gm \cdot R_b} \tag{3.30}$$

In a practical circuit configuration, the base terminal will be driven by a source resistance $\mathrm{R_s} > 0$. The input bandwidth for a transistor in a circuit application follows from equation (3.30), in which $\mathrm{R_b}$ is replaced by $(\mathrm{R_s} + \mathrm{R_b})$.

3.4.3 Approximation for f_{out}

The output bandwidth f_{out} follows from the output impedance $1/y_{22}$ in combination with the load resistance $\mathrm{R_L}$. From equation (3.25) it follows that, up to the input bandwidth f_V, the output impedance may be approximated by a capacitance $\mathrm{C_{22}} = \mathrm{C_{bc}}(1 + gm \cdot \mathrm{R_b}) + \mathrm{C_{cs}}$. So if the output bandwidth is lower than f_V, the output bandwidth can be approximated by

$$f_{out} = \frac{1}{2\pi R_L C_{22}} \tag{3.31}$$

In general, the output impedance $1/y_{22}$ can be mapped onto a parallel network $\mathrm{R_p}//\mathrm{C_p}$, in which both $\mathrm{R_p} = 1/\mathrm{Re}(y_{22})$ and $\mathrm{C_p} = \mathrm{Im}(y_{22})/\omega$ are frequency-dependent. The frequency dependence of the output capacitance $\mathrm{C_p}$ at $f > f_V$ affects the output bandwidth at low values of $\mathrm{R_L}$, where the output bandwidth may become equal to or larger than f_V. Using equation (3.25), the following results for $\mathrm{R_p}$ and $\mathrm{C_p}$ are derived:

$$\frac{1}{R_p} = Y_p = \mathrm{Re}(y_{22}) = \frac{-\omega^2 R_b C_{be} C_{bc} + \frac{\omega^2}{\omega_v} C_{bc}(1 + gm \cdot R_b)}{1 + \left(\frac{\omega}{\omega_v}\right)^2} \tag{3.32}$$

$$C_p = \frac{\mathrm{Im}(y_{22})}{\omega} = \frac{C_{bc}(1 + gm \cdot R_b) + \frac{\omega^2}{\omega_v} R_b C_{be} C_{bc}}{1 + \left(\frac{\omega}{\omega_v}\right)^2} + C_{cs} \tag{3.33}$$

Equation (3.32) shows the frequency dependency of R_p. At $f \ll f_V$, the denominator equals 1 so that the resistance value of R_p decreases by $-40\,\mathrm{dB/dec}$. At $f = f_V$, equation (3.32) simplifies to

$$\mathrm{Re}(y_{22}) = \frac{1}{2R_b}\frac{C_{bc}}{C_{be}+C_{bc}}\left(1+gm\cdot R_b - \frac{C_{be}}{C_{be}+C_{bc}}\right) \quad (3.34)$$

At $f \gg f_V$, equation (3.32) simplifies to a frequency-independent result given by:

$$\mathrm{Re}(y_{22}) = \frac{1}{R_b}\frac{C_{bc}}{C_{be}+C_{bc}}\left(1+gm\cdot R_b - \frac{C_{be}}{C_{be}+C_{bc}}\right) \quad (3.35)$$

Like the resistance, the capacitance may also be considered in different frequency regions around f_V. For $f \ll f_V$ this leads to

$$C_p = C_{bc}(1+gm\cdot R_b) + \frac{\omega^2}{\omega_v}R_bC_{be}C_{bc} + C_{cs} \overset{\omega\ll\omega_v}{\rightarrow} C_{bc}(1+gm\cdot R_b) + C_{cs} \quad (3.36)$$

The contribution of the base-collector capacitance C_{bc} is multiplied by a factor $(1+gm\cdot R_b)$. This apparent gain is here referred to as the output Miller effect (see Section 3.2) and can be minimised by minimising the base series resistance R_b. Note that the output Miller effect becomes increasingly important with an increasing collector current I_c (due to an increase in *gm*). Although it may not seem obvious at first glance, a low base resistance is therefore important for a high output bandwidth.

At $f = f_V$, the resulting output capacitance C_p equals

$$C_p = \frac{C_{bc}}{2}\left(1+gm\cdot R_b + \frac{C_{be}}{C_{be}+C_{bc}}\right) + C_{cs} \quad (3.37)$$

At $f \gg f_V$ the resulting output capacitance C_p equals

$$C_p = \left(\frac{\omega_v}{\omega}\right)^2 C_{bc}(1+gm\cdot R_b) + \frac{C_{be}C_{bc}}{C_{be}+C_{bc}} + C_{cs} \overset{\omega\gg\omega_v}{\rightarrow} \frac{C_{be}C_{bc}}{C_{be}+C_{bc}} + C_{cs} \quad (3.38)$$

The frequency dependence of R_p and C_p are visualised in Fig. 3.18. Note that the shape of the results holds for typical bias conditions, but the absolute values of gm, f_V and C_{be} are closely related to the bias condition.

In the presence of a non-zero source resistance R_s, the above analysis remains valid but R_b needs to be replaced by $R_b + R_s$. The main effects of the source resistance are a reduction in the input bandwidth and an increase in the output capacitance. With a non-zero source resistance R_s, the input bandwidth will decrease from f_V (valid for $R_s = 0$) to f_{VR}:

$$f_{VR} = \frac{1}{2\pi(R_b+R_s)(C_{be}+C_{bc})} \quad (3.39)$$

The parallel equivalent output capacitance for low frequencies ($f < f_{VR}$) will increase to C_{22R}:

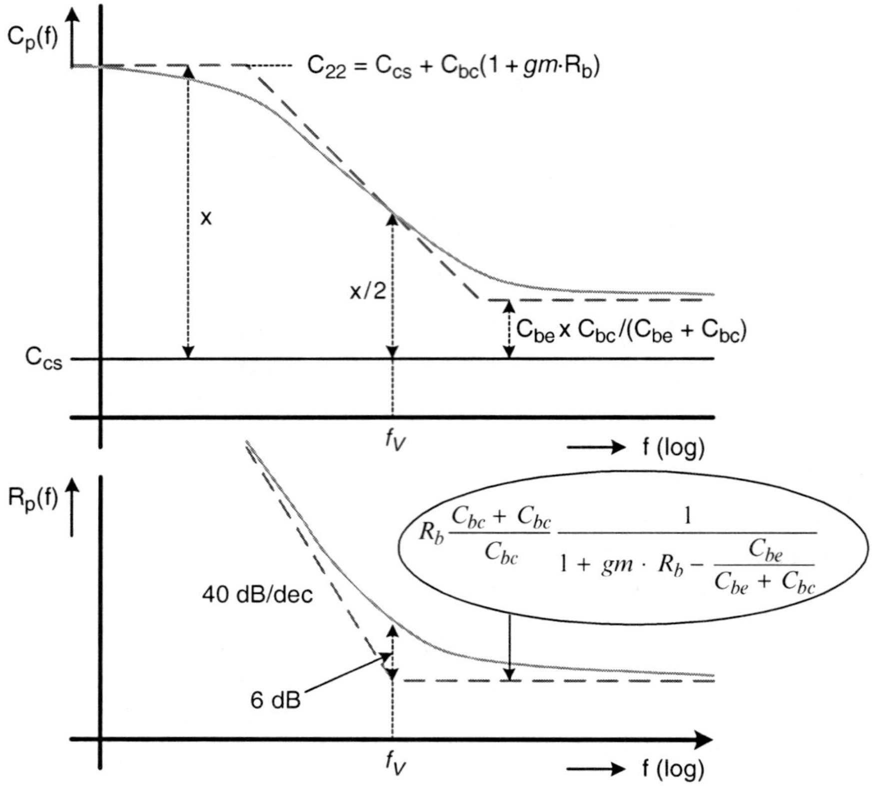

Fig. 3.18 Transistor output impedance $1/y_{22}$ mapped onto an equivalent network $R_p//C_p$

$$C_{22R} = C_{cs} + C_{bc}(1 + gm(R_b + R_s)) \tag{3.40}$$

To obtain a low output capacitance and a high input bandwidth in circuit configurations such as differential amplifiers, it is important to drive the differential amplifier with a low source resistance R_s. Comparison of R_s with the base resistance R_b will provide a good benchmark for the maximum allowable source resistance.

3.4.4 Approximation for f_A

Next, the available bandwidth $f_A = \omega_A/2\pi$ will be analysed.

The available bandwidth f_A follows from the voltage gain using equation (3.7). The voltage gain may be written as

$$A = -y_{21} \cdot \frac{1}{y_{22} + Y_L} \tag{3.41}$$

The first term represents the transconductance with input bandwidth f_V as given by equation (3.29). With increasing bias current this term becomes more dominant in the amplifier bandwidth. The second term represents a parallel load impedance of $1/y_{22}$ with $1/\mathrm{Y_L}$. With y_{22} from equation (3.24) this gives

$$A = \frac{-y_{21}}{y_{22}+Y_L} = -\frac{gm - j\omega C_{bc}}{j\omega C_{bc}(1+gm\cdot R_b)+(j\omega)^2 C_{be}C_{bc}R_b + Y_L\left(1+\frac{j\omega}{\omega_v}\right)} \tag{3.42}$$

The zero in equation (3.42) is at a frequency $\omega_z = gm/\mathrm{C_{bc}}$ which is typically higher than ω_A and may therefore be ignored. The metric f_A is then completely determined by the denominator. Using the definition of $\mathrm{C_{22}}$ as shown in Fig. 3.18, $\mathrm{C_{22}} = \mathrm{C_{bc}}(1+gm\cdot\mathrm{R_b})$, the amplifier bandwidth follows from

$$\omega_A\left(C_{22}+\frac{Y_L}{\omega_v}\right) = Y_L - \omega_A^2 R_b C_{be} C_{bc} \tag{3.43}$$

Equation (3.43) is first analysed by ignoring the quadratic term, which is true if $\omega_A \ll \sqrt{(\omega_v \mathrm{Y_L}(\mathrm{C_{be}}+\mathrm{C_{bc}})/\mathrm{C_{be}}\mathrm{C_{bc}})} = \sqrt{(\omega_v/\mathrm{R_L}\mathrm{C_{eq}})} = \sqrt{(\omega_v\cdot\omega_{eq})}$ with $\mathrm{C_{eq}}$ being the equivalent capacitance of the series connection of $\mathrm{C_{be}}$ and $\mathrm{C_{bc}}$, so $\mathrm{C_{eq}} \approx \mathrm{C_{bc}}$. This condition is satisfied if f_A has been analysed for sufficiently small low-frequency gain values, since they require a low value of $\mathrm{R_L}$. Then, the $-3\,\mathrm{dB}$ cut-off frequency is determined by

$$\omega_A = \frac{1}{R_L C_{22} + \frac{1}{\omega_v}} \tag{3.44}$$

This result shows that the amplifier bandwidth is approximately determined by a parallel network of two first-order time constants, $\tau_A = \tau_{out} + \tau_V$. Time constant $\tau_{out} = \mathrm{R_L C_{22}}$ is the time constant of the output capacitance $\mathrm{C_p}$ evaluated at low-frequency (see Fig. 3.18) times $\mathrm{R_L}$, and $\tau_V = 1/\omega_v$ is the time constant of the input bandwidth.

The contribution of the base-collector capacitance $\mathrm{C_{bc}}$ to $\mathrm{C_{22}}$ is multiplied by a factor $(1+gm\cdot\mathrm{R_b})$. This gain is here referred to as the output Miller effect; it can be minimised by minimising the base series resistance $\mathrm{R_b}$. Note that the output Miller effect becomes increasingly important with an increasing collector current $\mathrm{I_c}$ (due to an increase in *gm*). Therefore, the output bandwidth increase for f_A that would be expected with an increasing bias current (via a reduced $\mathrm{R_L}$) may be partly lost due to the output Miller effect. A low base resistance, important for the output bandwidth, is thus also important for f_A.

The output bandwidth $\omega_{out} = 1/\tau_{out} = 1/\mathrm{R_L C_{22}} = 1/(\mathrm{R_L}\cdot\mathrm{C_{bc}}(1+gm\cdot\mathrm{R_b}))$ can be plotted as a function of the bias current (see Fig. 3.19). Due to the output Miller effect, the desired increase in output bandwidth with an increasing bias current $\mathrm{I_c}$ will be lost at bias currents above $gm\cdot\mathrm{R_b} = 1$, or at $\mathrm{I_c} > 1/(40\cdot\mathrm{R_b})$. With $\mathrm{R_b} = 60\,\Omega$, this corresponds to $\mathrm{I_c} > 0.42\,\mathrm{mA}$.

It is interesting to observe the agreement of the f_{out}-curve of Fig. 3.19 with the output bandwidth curve shown in Fig. 3.11 at collector currents up to $\mathrm{I_{c,fTp}} = 3.2\,\mathrm{mA}$. The IC process represented in Fig. 3.11 is hampered by base resistance $\mathrm{R_b}$ that is too large, resulting in an output Miller effect in $\mathrm{C_{22}}$ that dominates at bias currents well below the bias current for peak-f_T, which in turn results in a relatively flat f_A-curve over the bias current.

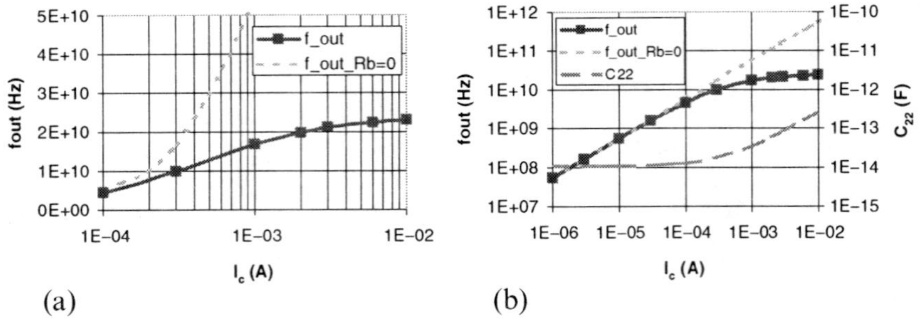

Fig. 3.19 Example of the influence of the output Miller effect on the output bandwidth at $R_b = 60\ \Omega$, $C_{bc} = 11$ fF. The desired increase in the output bandwidth with an increasing I_c ($R_b = 0$-line) is lost at high bias currents due to the output Miller effect. In Fig. (a) the curves are shown on a linear y ordinate to enable comparison with the curve of 3.11

To obtain a high peak-f_A, C_{22} should not start to rise at bias currents below the current for peak-f_T, $I_{c,fTp}$. This requires $gm \cdot R_b < 1$ at the collector current for peak-f_T, or a base resistance $R_b < 1/(40 \cdot I_{c,fTp})$. The base resistance plays an important role in defining the ratio of the current densities at peak-f_T and peak-f_A, as will be shown in Section 3.5.

In the case of f_A evaluated at increased low-frequency gain values, the quadratic term in equation (3.43) may not be ignored. The analysis of f_A is then as follows. Using the above equivalent circuit for the output admittance, y_{22} may be written as

$$y_{22} = \mathrm{Re}(y_{22}) + j \cdot \mathrm{Im}(y_{22}) = Y_p + j\omega C_p \tag{3.45}$$

Here, Y_p and C_p are as described by equations (3.32) and (3.33) respectively, and as graphically shown in Fig. 3.18. So, the amplifier voltage gain A can be written as

$$A = \frac{-y_{21}}{y_{22} + Y_L} = -\frac{gm - j\omega C_{bc}}{\left(1 + \frac{j\omega}{\omega_v}\right)(Y_p + Y_L + j\omega C_p)} = -\frac{gm - j\omega C_{bc}}{(Y_P + Y_L)\left(1 + \frac{j\omega}{\omega_v}\right)\left(1 + \frac{j\omega}{\omega_{out}}\right)} \tag{3.46}$$

In this equation, the term $C_p/(Y_p + Y_L) = \tau_{out} = 1/\omega_{out}$ is the time constant of the parallel network at the collector port, $C_p//R_p//R_L$ (see Fig. 3.20).

From equation (3.46) it follows that ω_v and ω_{out} are equally important role to the amplifier bandwidth. The zero in equation (3.46) is at $\omega_z = gm/C_{bc}$ which is typically higher than ω_A and may therefore be ignored. Depending on the ratio ω_v/ω_{out}, three situations can be distinguished. When $\omega_v >> \omega_{out}$, the amplifier bandwidth will be dominated by ω_{out} and hence $\omega_A \approx \omega_{out}$. When $\omega_v << \omega_{out}$, the amplifier bandwidth will be dominated by ω_v and hence $\omega_A \approx \omega_v$. When $\omega_v = \omega_{out}$, the amplifier bandwidth will be $\omega_A = \omega_{out} \cdot \tan(\pi/8)$, and will hence lie at $\omega_A = (\sqrt{2} - 1)\omega_{out}$.

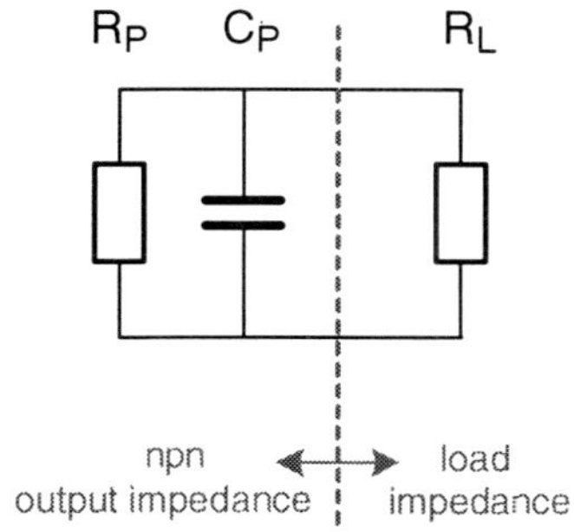

Fig. 3.20 Equivalent output circuit for the determination of f_A. The time constant of the output circuit equals τ_{out}

3.4.5 Approximation for f_{cross}

To analyse $\omega_{cross} = 2\pi f_{cross}$, the real part of the input admittance of a cross-coupled differential pair needs to be considered. From the evaluation of equation (3.9), the input admittance Y_i for the cross-coupled differential pair using the simplified transistor model of Fig. 3.16 follows:

$$\frac{y_{11} - y_{12} - y_{21} + y_{22}}{2} = \frac{j\omega C_{be} + j\omega C_{bc}(4 + gm \cdot R_b) - gm - \omega^2 R_b C_{be} C_{bc}}{2\left(1 + \frac{j\omega}{\omega_v}\right)} \tag{3.47}$$

At low frequencies, the differential input impedance of the cross-coupled differential pair simplifies to the well-known $Z_i = -2/gm$. At f_{cross}, the real part of Y_i is zero. This occurs at

$$\omega_{cross} = \sqrt{\frac{gm \cdot \omega_v}{C_{be} + C_{bc}\left(4 + gm \cdot R_b - \frac{C_{be}}{C_{be}+C_{bc}}\right)}} \tag{3.48}$$

Using $\omega_T = gm/(C_{be} + C_{bc})$, this result can be written as

$$\omega_{cross} = \sqrt{\omega_v \omega_T} \sqrt{\frac{C_{be} + C_{bc}}{C_{be} + C_{bc}\left(4 + gm \cdot R_b - \frac{C_{be}}{C_{be}+C_{bc}}\right)}} \overset{C_{be} \gg C_{bc}}{\approx} \sqrt{\omega_v \omega_T} \tag{3.49}$$

So, the frequency f_{cross} will be somewhat lower than $\sqrt{(f_V \cdot f_T)}$. At low bias currents, at which the condition $C_{be} \gg C_{bc}$ is not fulfilled, the approximate relation $\sqrt{(f_V \cdot f_T)}$ will over-estimate f_{cross}. Figure 3.21 shows both f_{cross} obtained in Spectre circuit simulation according to Fig. 3.12a and its approximation $\sqrt{(f_V \cdot f_T)}$. As can be seen, the approximation is valuable for indicating the trend in f_{cross} over an order of magnitude variation in the bias current, but in this example it has an error of about 30%. Since in practice a cross-coupled differential pair is typically operated at a bias current near $I_{c,fTp}$, the relation $f_{cross} \approx \sqrt{(f_V \cdot f_T)}$ is valuable for analysing the dominant contributions at typical bias current densities.

To obtain a high f_{cross}, a high f_V and f_T are simultaneously required. Using the equations for f_T (3.27) and f_V (3.29), the relation between f_{cross} and technology parameters can be expressed as

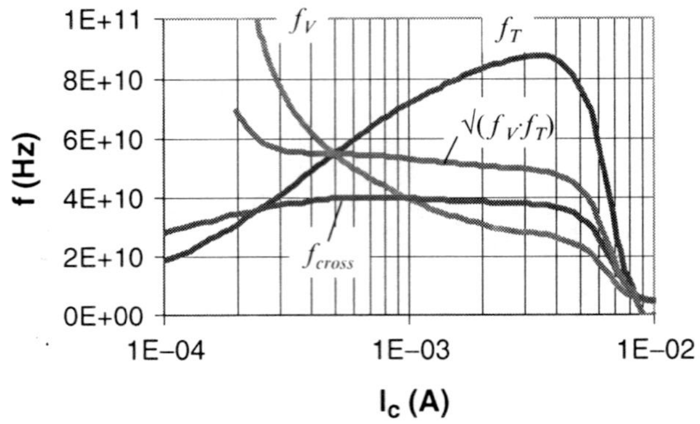

Fig. 3.21 Comparing f_{cross} with $\sqrt{(f_V \cdot \cdot f_T)}$ for a 0.5 × 4.7 npn biased at $V_{cb} = 0\,V$. The transistor reaches its peak-f_T at $I_c = 3.2\,mA$

$$f_{cross} \approx \sqrt{f_V \cdot f_T} = \frac{1}{2\pi(C_{be}+C_{bc})}\sqrt{\frac{gm}{R_b}} \tag{3.50}$$

The base series resistance R_b plays an important role in f_{cross}. Increasing the bias current does lead to an increase in gm but it simultaneously increases C_{be}, and therefore has little impact on f_{cross}. This explains why f_{cross} is relatively insensitive to bias current variations. Note that in the analysis of f_{cross}, the emitter series resistance R_e was ignored. In Chapter 7, the analysis of f_{cross} will be repeated for an arbitrary R_e.

3.4.6 Approximation for f_{max}

To obtain an approximate relation for f_{max}, the procedure proposed in [3] will be followed. Since the base-collector capacitance is included in the calculations, the analysis provides an approximation for f_{max} based on the maximum available gain G_{max} (and not the unilateral gain). First, the y-parameter relations are further simplified in the frequency range $f_V < f < f_T$. From these simplified y-parameters, approximate formulas for the input and output impedance of the transistor, Z_{in} and Z_{out}, are derived. Now that the conditions for maximum power gain are known, the power gain under matched conditions can be evaluated. Since f_{max} is, like f_T, an *extrapolated* figure of merit, it does not in any way restrict this analysis based on y-parameters that are only valid in a limited $(f_V < f < f_T)$ frequency range, provided that extrapolation is performed starting from a frequency in the range $f_V < f < f_T$.

Using $\omega > \omega_V$, parameter y_{11} from equation (3.21) may be simplified to

$$y_{11} = \frac{j\omega(C_{be}+C_{bc})}{j\omega R_b(C_{be}+C_{bc})+1} \approx \frac{1}{R_b} \tag{3.51}$$

Using $\omega > \omega_V$, parameter y_{12} from equation (3.23) may be simplified to

$$y_{12} = \frac{-j\omega C_{bc}}{j\omega R_b(C_{be}+C_{bc})+1} \approx -C_{bc}\cdot\omega_V \tag{3.52}$$

Using $\omega_V < \omega < gm/\mathrm{C_{bc}}$, parameter y_{21} from equation (3.22) may be simplified to

$$y_{21} = \frac{gm - j\omega C_{bc}}{j\omega R_b(C_{be}+C_{bc})+1} \approx \frac{\omega_T}{j\omega R_b} \tag{3.53}$$

Using $\omega_V < \omega < gm/\mathrm{C_{be}}$ and $gm\cdot\mathrm{R_b} > 1$, parameter y_{22} from equation (3.24) may be simplified to

$$y_{22} = \frac{j\omega C_{be}(1+gm\cdot R_b)+(j\omega)^2 R_b C_{be} C_{bc}}{j\omega R_b(C_{be}+C_{bc})+1} \approx \frac{C_{bc}(gm+\frac{1}{R_b})}{C_{be}+C_{bc}} \approx C_{bc}\cdot\omega_T \tag{3.54}$$

Using these approximations for the y-parameters, the relations for the input and output impedance of the transistor, equations (3.15) and (3.16), may be approximated in the frequency range $\omega_V < \omega < \omega_T$ by

$$Z_{in} = \frac{1}{\frac{1}{R_b} + \frac{C_{bc}\omega_V\omega_T}{j\omega R_b}\frac{1}{2\omega_T C_{bc}}} = \frac{1}{\frac{1}{R_b}\left(1-0.5j\frac{\omega_V}{\omega}\right)} \overset{\omega>\omega_V}{\approx} R_b \tag{3.55}$$

and

$$Z_{out} = \frac{1}{\omega_T C_{bc} + \frac{C_{bc}\omega_V\omega_T}{j\omega R_b}\cdot 0.5R_b} = \frac{1}{\omega_T C_{bc}\left(1-0.5j\frac{\omega_V}{\omega}\right)} \overset{\omega>\omega_V}{\approx} \frac{1}{\omega_T C_{bc}} \tag{3.56}$$

For maximum available power gain, the required source and load impedances are hence $\mathrm{Z_s} = \mathrm{R_b}$ and $\mathrm{Z_L} = 1/\omega_T\cdot\mathrm{C_{bc}}$. The maximum available power gain can now be evaluated using equation (3.13):

$$G_{\max} = \frac{i_c\cdot i_c^*\cdot \mathrm{Re}(Z_L)}{i_b\cdot i_b^*\cdot \mathrm{Re}(Z_s)} = \frac{|i_c|^2}{|i_b|^2}\cdot\frac{1}{\omega_T C_{bc} R_b} \tag{3.57}$$

To find i_c and i_b, the y-parameter relations in equations (3.51) to (3.54) are used in combination with the relations given by the source and load impedance $i_b = -v_b/\mathrm{Z_s}$ and $i_c = -v_c/\mathrm{Z_L}$. For the input port this yields:

$$i_b = \frac{1}{R_b}v_b - \omega_V C_{bc} v_c \equiv -\frac{1}{R_b}v_b \tag{3.58}$$

$$i_c = \frac{-j\omega_T}{R_b\omega}v_b + \omega_T C_{bc} v_c \equiv -\omega_T C_{bc} v_c \tag{3.59}$$

From relation (3.59) it follows that $v_c = \mathrm{j}v_b/2\mathrm{R_b}\omega\mathrm{C_{bc}}$ and v_c may be eliminated:

$$i_c = \frac{-j\omega_T}{2R_b\omega}v_b \tag{3.60}$$

So, the maximum available power gain of equation (3.57) under matched conditions becomes

$$G_{\max} = \frac{|i_c|^2}{|i_b|^2} \cdot \frac{1}{\omega_T C_{bc} R_b} = \frac{\omega_T^2}{4\omega^2} \cdot \frac{1}{\omega_T C_{bc} R_b} = \frac{\omega_T}{4\omega^2 C_{bc} R_b} \tag{3.61}$$

From this result it is clear that the maximum available power gain decreases quadratically with frequency, or -20dB per decade (for $\omega > \omega_V$). An example of a power gain curve for a single bias condition is shown in Fig. 3.22. The power gain shown represents the maximum stable gain. Figure 3.22 shows a frequency sweep under a given bias condition, resulting in a single point along the f_{max}-curve. The f_{max}-curve, showing f_{max} as a function of the bias condition, may be generated from a bias sweep at a single (extrapolation) frequency f_x. Provided that at $f = f_x$ the power gain shows a -20 dB/decade roll-off, the quantity f_{max} is obtained via extrapolation from the power gain G_{max} at the extrapolation frequency: $f_{max}(I) = f_x \cdot G_{max}(I)$. An example of an f_{max}-curve obtained in a Spectre circuit simulation is shown in Fig. 3.23.

The asymptotic frequency f_{max} at which $G_{max} = 1$ as derived from equation (3.61) is

$$f_{\max} = \sqrt{\frac{f_T}{8\pi R_b C_{bc}}} \tag{3.62}$$

So, for high f_{max} a low R_b and low C_{bc} are important. The quantity f_{max} depends on the collector-base voltage V_{cb} due to various effects such as the voltage dependency of the collector-base junction capacitance C_{bc} and quasi-saturation. It should always be specified at what V_{cb} the f_{max}-value is obtained. A typical value is V_{cb} is 1 V.

While the approximate equation (3.62) for f_{max} provides useful information on the IC process requirements for obtaining a high f_{max}, it should be realised that

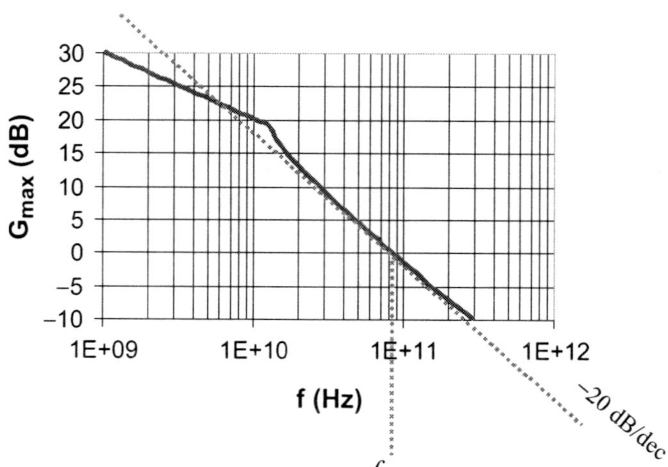

Fig. 3.22 Definition of f_{max}. The power gain G_{max} is shown for a 0.5×4.7 npn biased at peak-f_T and $V_{cb} = 0$V

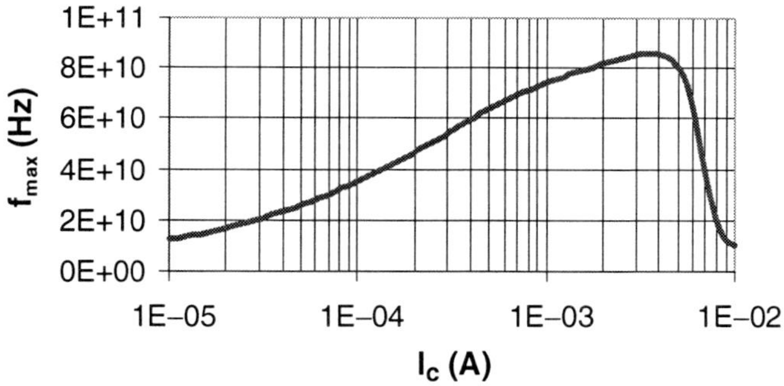

Fig. 3.23 Example of an f_{max}-curve for a 0.5×4.7 npn biased at $V_{cb} = 0\,V$. The transistor reaches its peak-f_T at $I_c = 3.2\,mA$

Table 3.1 Comparing published f_{max}-data with calculated data obtained using equation (3.62)

Literature	Emitter area $(\mu m)^2$	f_T (GHz)	f_{max} (GHz)	R_b (Ω)	C_{jc} (fF)	f_{max} using (3.62) (GHz)
Washio 2002 [6]	$0.2 \cdot 1.0$	76	180	120	1.9	115
Hashimoto 2002 [7]	$0.2 \cdot 1.0$	122	178	82	2.2	164
QUBiC4I	$0.5 \cdot 4.7$	87 ($V_{cb} = 0$)	85 ($V_{cb} = 0$)	60	11	73

the resulting f_{max} value may deviate significantly from published f_{max} data, since publications may refer to f_{max} values based on the unilateral gain. Unfortunately, only few publications specify which definition was used to obtain the published f_{max} values. For example, in [6] and [7], IC processes are described with sufficient detail to evaluate equation (3.62) (see Table 3.1). The last row, indicated as QUBiC4I, shows the numbers for the experimental IC process used to generate the example curves shown in this chapter. As can be seen, equation (3.62) provides a reasonably accurate fit for the QUBiC4I process and the process in [7], but not for the results in [6]. The ambiguity in the f_{max} definition may (partly) explain the discrepancy between the calculated and published f_{max} values in Table 3.1.

Metric f_{max} is relevant for the design of single-transistor low-noise amplifier (LNA) circuits [5]. If a given power gain P for the single-transistor LNA is required at a given frequency f_{LNA}, then from the quadratic roll-off of the power gain G with frequency (see Fig. 3.22) it follows that the minimum required f_{max} for the IC process is $f_{max} > \sqrt{(P)} \cdot f_{LNA}$.

3.5 Optimising a Technology for f_A

Since f_A is a good FOM for broadband applications, it is important to optimise f_A of a new technology intended for broadband applications. Usually, when a new technology is introduced, f_T and f_{max} are used to benchmark the performance improvement with current IC process generations. In this section, the technology requirements for achieving a high peak-f_A will be analysed. The analysis will be performed for a low-frequency gain of 10. A gain of 10 may seem high, but this puts extra emphasis on the output bandwidth. In current-mode logic for example, typical gain values of 2–4 are used. Since in the f_A-definition the output is only loaded with a load resistance R_L while in practical circuits the output will always be loaded by a subsequent stage, evaluating f_A for a low-frequency gain of 10 provides a good balance between the input and output contributions to f_A.

The development budget of a new process for mass-production applications is often limited and the new process must ensure low production costs per mm^2. It is therefore attractive to increase the performance of the npn transistor without scaling the lithography. The migration from the Philips IC process QUBiC4 to QUBiC4G and later QUBiC4X is based on this approach. A significant increase in f_T plus a small increase in f_{max} (without scaling the lithography) was realised by introducing first SiGe and later SiGe:C to the npn base. However, an increase in f_T and f_{max} does not necessarily lead to an increase in peak-f_A. In addition, the ratio of the current densities for peak-f_T and peak-f_A, $J_{c,fTp}/J_{c,fAp}$, may increase considerably, causing the f_A at the current density for peak-f_T to decrease.

In the following table, the effect of introducing SiGe and SiGe:C on the npn performance (of the Philips QUBiC4 IC process family) is summarised. All processes are based on the same lithography. In the table, Q4 refers to the Si production process described in [10]; Q4G refers to the SiGe production process described in [8] and Q4X refers to a SiGe:C predevelopment process.

All figures relate to a $0.5\,\mu m \times 4.7\,\mu m$ drawn emitter size except those in the last row, which relate to a drawn emitter scaled to $0.4\,\mu m$. The improvement in production tolerances over time enabled the use of a smaller minimum feature size in the latest IC process generation. Due to inside spacers, the effective emitter area equals the drawn emitter area reduced by the inside spacer width and length of $0.11\,\mu m$ per side. So, a $0.5\,\mu m \times 4.7\,\mu m$ drawn emitter corresponds to an effective emitter area of $0.28\,\mu m \times 4.48\,\mu m$.

Despite the improvements in f_T and f_{max}, the peak-f_A has barely improved over the three generations of IC processes. This can be explained as follows. In the first place, the increase in peak-f_T was accompanied by an increase in current density for peak-f_T. From QUBiC4 to QUBiC4X, the current density for peak-f_T, and hence also the gm at peak-f_T, increased by more than a factor of three. The base resistance R_b however remained more or less constant. Although the intrinsic transistor has improved significantly across the 3 IC process generations, the extrinsic part of the transistor has remained the same. So the extrinsic base resistance R_{bc} has not changed. Only the reduction in the emitter width from $0.5\,\mu m$ to $0.4\,\mu m$ has somewhat reduced the intrinsic part of the base resistance R_{bv} in the Q4X process. In the second place, the base-collector capacitance C_{bc} has increased considerably due to the increased collector doping. These two effects have an impact on the output capacitance C_{22}:

$$C_{22} = C_{cs} + C_{bc}(1 + gm \cdot R_b) \tag{3.63}$$

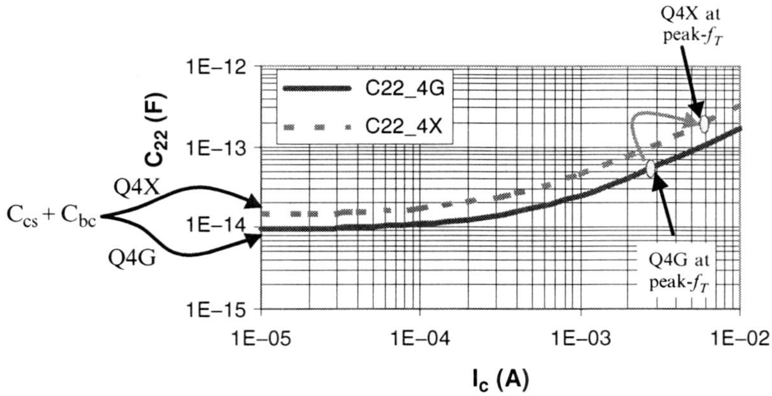

Fig. 3.24 Comparing the output capacitance C_{22} of the SiGe (Q4G) and SiGe:C (Q4X) process variants for a 0.5×4.7 npn transistor biased at $V_{cb} = 0$ V. The value of C_{22} at the current density for peak-f_T has increased by about a factor of 3, as indicated by the arrows

Table 3.2 Extracting f_A and its contributions. All FOMs are at $V_{cb} = 0$ V

Process	f_T (GHz)	f_{max} (GHz)	$J_{c,fTp}$ (mA/μm²)	R_b (Ω)	C_{bc} (fF)	C_{cs} (fF)	f_V (GHz) at pkf$_T$	f_{out} (GHz) at pkf$_T$	f_A (GHz) at pkf$_T$	$gm \cdot R_b$ at pkf$_T$	Peak-f_A (GHz)
Q4	33	60	1.4	94	4	2.4	19	33	13	1.7	14.6
Q4G	61	73	2.0	58	7	2.4	26	24	13	2.3	15.2
Q4X 0.5	117	84	4.4	69	12	2.5	17	17	9	7.2	13.0
Q4X 0.4	109	90	4.4	61	10	2.4	22	20	10	5.0	15.9

In Fig. 3.24, the increase in output capacitance realised from the SiGe process QUBiC4G to the SiGe:C process QUBiC4X has been visualised using the figures given in Table 3.2 and $gm = 38.6 \cdot I_c$. The output capacitance C_{22} has increased somewhat at low currents due to the increase in C_{bc}. The output Miller effect causes C_{22} to increase at bias currents at which $gm \cdot R_b > 1$. The increase in output capacitance C_{22} in turn causes the output bandwidth to flatten-off at currents beyond the point at which $gm \cdot R_b = 1$. The increased level of C_{22} for the QUBiC4X process causes a *reduction* in f_{out} when compared at peak-f_T.

At currents at which $gm \cdot R_b > 1$, a further increase in current leads to a reduction in f_V (due to the increase of the diffusion capacitance contribution to C_{be}) while the output bandwidth f_{out} no longer increases (due to the increase in C_{22}). So, the current density for peak-f_A is defined mainly by the base resistance and occurs at the point at which $gm \cdot R_b \approx 1$. In a first approximation, the current density for peak-f_A does consequently not shift because R_b remains constant. Since the introduction of SiGe and SiGe:C to the IC process has increased the current density for peak-f_T, the ratio $(J_{c,fTp}/J_{c,fAp})$ has also increased.

To conclude, not reducing R_b and increasing C_{bc} while scaling the current density for peak-f_T and hence an increased ratio $(J_{c,fTp}/J_{c,fAp})$ has important consequences for circuit design. The transistors of a differential pair need to be biased across the different generations of IC processes at *similar* current densities to achieve the best broadband performance. This also means that the circuits implemented in the IC processes of the newer generation do not profit much from the increase in peak-f_T, since not many transistors will be biased at peak-f_T. In some cases, the increase in f_T will lead to an increase in performance (e.g., in the case of common-base stages, see Section 3.4.1), but this will usually not lead to a significant improvement in overall performance. If all current densities in the circuits are scaled in the same ratio as the increase in current density for peak-f_T in the newer IC process, the same circuit may perform worse than before. This is because the f_A at peak-f_T may decrease due to the reduced output bandwidth, as demonstrated in Table 3.2 for some of the Philips QUBiC IC processes.

To benefit from the improved FOMs of a new process generation for broadband circuit design, an increase in f_A is desired. This may be obtained by a reduction in base resistance, a reduction in base-collector capacitance or both. The ultimate goal for the base resistance is to reduce it to such an extent that at peak-f_T, $gm \cdot R_b \leq 1$.

Figure 3.25 compares the FOMs of the QUBiC4G and QUBiC4X processes.

As can be seen, the current density at peak-f_A does not change significantly across the three process generations. The output bandwidth is the dominant factor in f_A in all the process variants, especially in the QUBiC4X process because f_{out} no longer increases at currents exceeding $I_c \approx 1\,\text{mA}$. In a first-order approximation, f_A is derived from f_V (equation (3.30)) and f_{out} (equations (3.31) and (3.63)) as follows:

$$\frac{1}{f_A} = \frac{gm \cdot R_b}{f_T} + \frac{2\pi |A|}{gm}(1 + gm \cdot R_b)C_{bc} + \frac{2\pi |A|}{gm}C_{cs} \tag{3.64}$$

Equation (3.64) provides valuable information for the optimisation of the f_A at peak-f_T. When operating at peak-f_T, gm may be assumed to be independent of R_b and C_{bc}. Also, when changing R_b and/or C_{bc}, the peak-f_T does not change significantly. Figure 3.26 shows an example plot based on equation (3.64) showing how f_A depends on R_b and C_{bc} in the QUBiC4X technology. In this technology, the base resistance is dominated by the extrinsic part. The arrows indicate how f_A would change in the case of a reduction by a factor of 2 in the extrinsic R_b and C_{bc}. Note that the calculated f_A at peak-f_T (e.g., equation (3.64) with the values given in the bottom row of Table 3.2 yields $f_A = 10.5\,\text{GHz}$) is in close agreement to the 10 GHz obtained in a Spectre circuit simulation using the MEXTRAM 504 model.

A reduction in R_b is advantageous for both f_V and f_{out}; a reduction in C_{bc} is mainly important for f_{out}. So, reducing R_b requires the most attention with respect to the IC process under study. Reducing the base resistance also helps to reduce the minimum noise figure of the transistor. Note that several companies have recently begun to implement techniques for reducing the extrinsic base resistance. These techniques are referred to as 'raised extrinsic base' [17] or 'elevated extrinsic base' [24]. It is interesting to observe that, based on the relation for f_{max} (3.62), the sensitivities of f_{max} to R_b and C_{bc} variation are not equal. Because a reduction in C_{bc} results in an increase in f_T, reducing C_{bc} is a slightly more effective measure

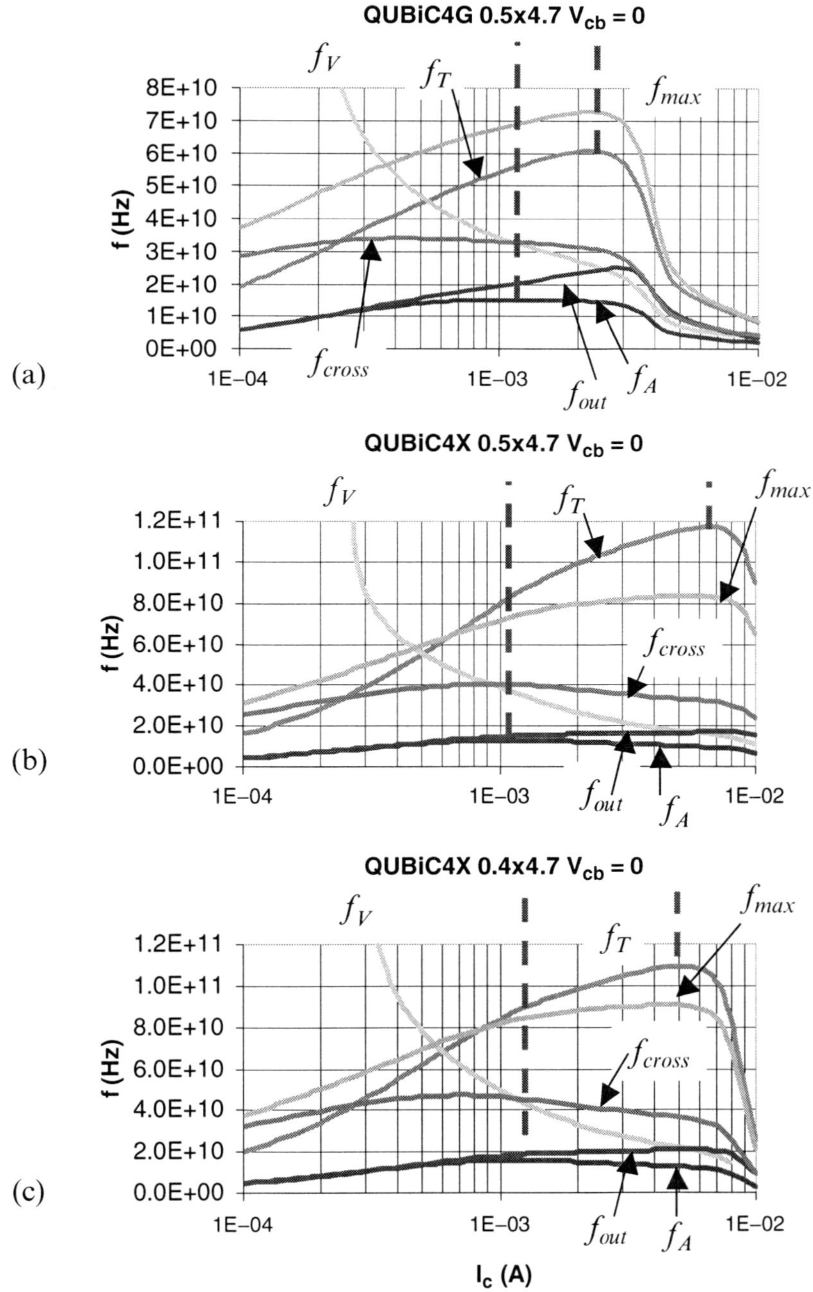

Fig. 3.25 Comparing the FOMs of three process generations. The vertical dotted lines indicate the currents at peak-f_A and peak-f_T

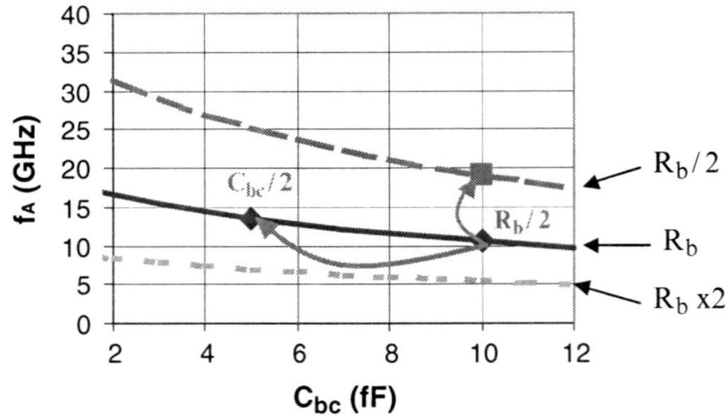

Fig. 3.26 Effect of R_b and C_{bc} on the f_A at peak-f_T in the example SiGe:C process

for increasing f_{max} than reducing R_b. While a reduction of C_{bc} is most effective for f_{max}, a reduction of R_b is more important for f_A.

3.6 Relationship between f_A, f_T and f_{max}

Although f_A is a good FOM for broadband applications, IC processes are usually optimised and benchmarked on the basis of f_T and f_{max}. So, it is important to understand the relationship between the various FOMs.

On the basis of the approximate formula for f_{max} (3.62), the following relationship exists:

$$R_b C_{bc} = \frac{f_T}{8\pi f_{\mathrm{max}}^2} \tag{3.65}$$

The equation for ω_A (3.44), at a given low-frequency gain, shows that ω_A is the result of a parallel configuration of the input bandwidth ω_V with the output bandwidth $1/(R_L \cdot C_{22}) = 1/(R_L \cdot C_{bc}(1 + gm \cdot R_b))$. Since $|A| = gm \cdot R_L$, the following relationship exists:

$$\frac{1}{\omega_A} = \frac{1}{\omega_V} + R_L C_{bc}(1 + gm \cdot R_b) = \frac{1}{\omega_V} + \frac{|A|}{gm} C_{bc} + \frac{|A| \cdot f_T}{8\pi f_{\mathrm{max}}^2} \tag{3.66}$$

The term $1/\omega_V$ represents the input bandwidth, the term $|A|/gm \cdot C_{bc}$ the output bandwidth in the case $R_b = 0$ and the last term the output Miller effect due to the base resistance. It hence follows from equation (3.66) that in an IC process with a high f_T and a low f_{max} (e.g., $f_{max} < f_T$) the output Miller effect will have a dominant impact on f_A, in particular when considering high gain values. This was for example observed in Fig. 3.11.

In Section 3.5 it was mentioned that if the aim is to minimise the output Miller effect, the ultimate goal must be to realise $gm_p \cdot R_b \leq 1$ with gm_p being the effective

gm when biased at peak-f_T. If the emitter series resistance R_e is known, gm_p (at room temperature) may be approximated using

$$gm_p \approx \frac{38.6 I_{c,fTp}}{1 + 38.6 I_{c,fTp} \cdot R_e} \tag{3.67}$$

Here, $I_{c,fTp}$ is the collector current at peak-f_T. The base resistance can be estimated if the collector-base capacitance C_{bc}, f_T and f_{max} values have been determined using equation (3.65), assuming that f_T and f_{max} reach their peak values at the same current densities. So, the condition $gm_p \cdot R_b \leq 1$ corresponds to

$$gm_p \cdot R_b = \frac{1}{8\pi} \frac{f_T}{f_{\max}^2} \frac{gm_p}{C_{bc}} \leq 1 \tag{3.68}$$

The condition given by equation (3.68) will usually not be fulfilled. The greater the value of $gm_p \cdot R_b$, however, the more the current density for peak-f_T will deviate from the current density for peak-f_A. In addition, when equation (3.68) yields a higher value, f_{out} will be more dominant in f_A. So, the results of equation (3.68) together with the f_A at peak-f_T provide a good benchmark for comparing the fit of IC technologies for broadband applications.

Table 3.3 gives the $gm_p \cdot R_b$ result of equation (3.68) for a number of technologies. f_V, f_{out} and f_A at peak-f_T were calculated using data provided in the literature. If f_{max} is based on the unilateral gain, the base resistance derived using equation (3.65) is optimistic. This can be seen in the two bottom rows in the table, in which the derived R_b-value is approximately a factor of 2 lower than the published data on R_b. If f_{max} is based on the maximum available gain, as in the top two rows, equation (3.65) provides an accurate estimation for R_b.

Since the emitter sizes for the different processes in Table 3.3 are not identical, the absolute values of R_b and C_{bc} must first be normalised before they can be compared. In the table, f_A is not the peak-f_A but the value at the current density for peak-f_T. It is interesting to observe that the processes described in [25] and [6] have an almost identical f_A at peak-f_T, despite the large difference in f_T and f_{max}. Since the process described in [6] achieves $gm_p \cdot R_b < 1$, the output Miller effect plays an

Table 3.3 Extracting f_A at peak-f_T and its contributions

Process	f_T (GHz)	f_{max} (GHz)	$I_{c,fTp}$ (mA)	C_{bc} (fF)	R_e (Ω)	R_b (Ω)	R_b using (3.65) (Ω)	gm_p using (3.67) (A/V)	f_V at pkf_T (3.30) (GHz)	f_{out} at pkf_T (3.31) (3.63) (GHz)	f_A at pkf_T (GHz)	$gm \cdot R_b$ at pkf_T
Q4G	61	73	2.24	7	10.9	58	65	0.045	23.6	28.3	12.9	2.90
Q4X .4	109	90	5.01	10	5.07	61	54	0.098	18.3	22.3	10.1	5.23
[25]	200	230	3	5.5	3.5	50	27	0.082	48.5	46.6	23.8	2.25
[6]	76	180	0.7	1.9	27	120	49	0.016	40.5	45.5	21.4	0.77

insignificant role, and the output bandwidth does not saturate at bias currents up to peak-f_T.

The fact that the f_V at peak-f_T is approximately equal to f_{out} at peak-f_T in all the processes in Table 3.3 does not mean that the peak-f_A occurs at the same current as the peak-f_T, since the output bandwidth already saturates at a current before the peak-f_T if $gm_p \cdot R_b > 1$.

When introducing a new IC technology, it is not sufficient to increase only the f_T. In fact, an increase in f_T is often obtained by an increase in collector doping (increasing C_{bc}), shifting the peak-f_T to higher current densities (increasing gm_p). As follows from equation (3.66), the increase in C_{bc} reduces the output bandwidth at peak-f_T. The increase in f_T should be accompanied by a reduction in R_b to increase the input and output bandwidths f_V and f_{out} at peak-f_T, too.

3.7 Trends in Device Metrics: A Comparison of Recent Technologies

Data quoted in different publications are usually difficult to compare because different formats, different conditions and/or different definitions for the device metrics may be used. In this section, a comparison is made using data relating to Philips production and pre-production IC processes of the QUBiC family. These processes are intended for RF applications. All the metrics of these processes are based on the same definitions and conditions. The comparison covers the time frame 1998–2004.

3.7.1 Trends Relating to Device Metrics

Table 3.4 summarises the main trends relating to the Philips' QUBiC family [11].

The table highlights several trends that are valid not only for Philips, but also for many other companies.

A high-speed transistor requires a narrow base. In SiGe IC processes the base layer is epitaxially grown, whereas in homojunction Si IC processes the base is typically implanted. Inclusion of Ge in the base also improves the high-speed performance, thanks to the narrow bandgap in the neutral base region. A high concentration of Ge is beneficial for high-speed performance. The percentage of Ge is however limited due to the strain induced by the relatively large Ge atoms.

After the base layer has been formed, base dopant diffusion should be kept to a minimum to maintain the narrow base width. In SiGe BiCMOS processes, the process flow is controlled so that the CMOS heat cycles usually occur prior to the SiGe epitaxial base growth. In SiGe:C processes, addition of carbon to the SiGe base layer further reduces the diffusion of the p-type dopant.

While most CMOS processes today support multiple gate oxide thicknesses to enable interfacing with multiple I/O-levels, there is a trend in SiGe processes to

Table 3.4 Comparison of processes of Philips' QUBiC family

	QUBiC3	QUBiC4	QUBiC4G	QUBiC4X
Year of introduction	1998	2001	2002	2004
Wafer resistivity	$20\,\Omega\cdot$cm	$20\,\Omega\cdot$cm	$20\,\Omega\cdot$cm	$\gg 20\,\Omega\cdot$cm
Lithography (μm)	0.5	0.25	0.25	0.25
BJT HBT base	Si	Si	SiGe	SiGe:C
Emitter	Implanted poly	Insitu poly	Insitu poly	Mono
f_T (GHz)	30	40	70/50	130/60
f_{max} (GHz)	60	90	100/110	140/120
f_{cross} (GHz)	15	20	30	45
f_A (GHz)	8.0	12.4	12.8	18
BV_{CEO} (V)	4.0	3.7	2.7/3.9	2.0/3.1
BV_{CBO} (V)	14	16	10/15	9/13
MIM density (fF/μm^2)	1	5	5	5/15
Metal layers	4	4/5/6	5/6	4/5/6

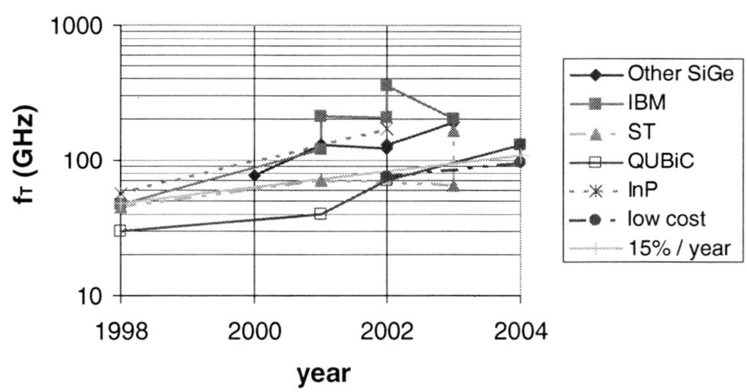

Fig. 3.27 Comparison of peak-f_T values based on data published in the time frame 1998–2004

offer transistors with different breakdown voltage levels. The reduction in breakdown voltages has enabled higher speeds, but at the same time limited the application range. In particular, Power Amplifier (PA) output functions may require high breakdown voltages, not only for normal operation but also for robustness when applying a large mismatch to the output load (ruggedness). To support implementation of PA functions, transistors with different breakdown voltages are offered in a technology. Lower collector doping reduces f_T but at the same time increases BV_{CEO}. While the QUBiC family now offers two variants, IC processes supporting three (see for example [12]) or even four (see for example [13]) different transistor breakdown voltages have already been reported.

The following graph based on several recent publications is used to analyse the trend in f_T [6, 8–27]. For reference, the 15% per year trend in f_T-increase reported in [28], starting at 4 GHz in 1980, has been included in Fig. 3.27. While the published low-cost and volume production technologies are able to continue the 15%

per year growth rate, several technologies outperform the predicted f_T trend. It is interesting to note that InP technologies remain at the forefront of the published f_T results.

It is dangerous to draw firm conclusions from these data, because the f_T figures do not relate to manufacturability, yield, cost, etc. The picture shows both experimental research results (based on leading edge lithography) and production technology results typically 1–2 generations behind the experimental research results. Whether a publication relates to a production or research technology is not always clear. Still, some remarkable trends can be highlighted.

The record published f_T is already a few years old [18]. Only recently has the focus shifted to low costs and improved manufacturability [12, 13]. Low costs are achieved mainly by omitting the costly deep trench isolation and replacing the buried sub-collector by an implanted sub-collector [13], shared with the CMOS flow.

One of the challenges involved in designing transistors with a high peak-f_T is that an increased peak-f_T is often accompanied by an increase in collector current density J_c. These high current densities need to be supported by the metal backend, where electromigration limitations are becoming a bottleneck. The trend in current density for peak-f_T is shown in Fig. 3.28. The need to operate at increased current density to achieve a higher peak-f_T is clearly supported by these data.

The increased current density is the result of vertical device scaling. A narrower base plus increased base and collector doping are applied to reduce the transit time and allow a higher maximum collector current. To support reliable operation at high current densities up to high temperatures, processes requiring operation at $J_c > 5\,\text{mA}/\mu\text{m}^2$ use Cu interconnect. Slotted contacts or double contact rows may be used to extend the current handling capabilities of interconnect [11,22].

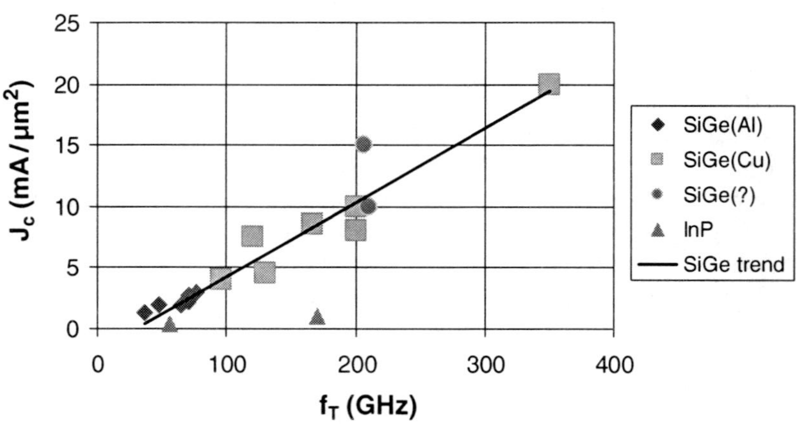

Fig. 3.28 Required current densities for peak-f_T for the processes also reported in Fig. 3.27. The type of metal used for the backend is indicated for the SiGe processes if this information is published

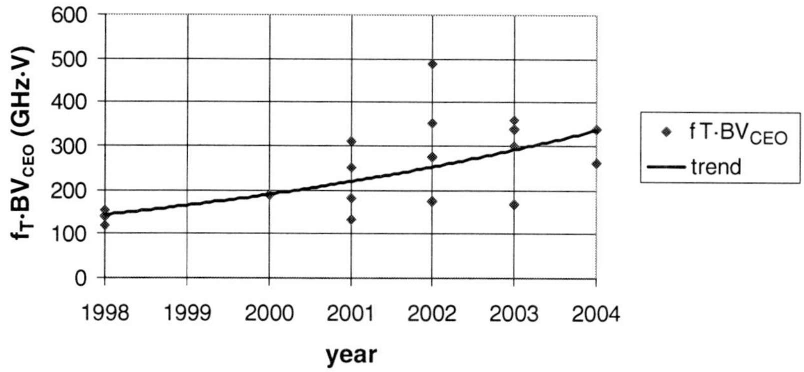

Fig. 3.29 Trend in the $f_T \cdot \mathrm{BV_{CEO}}$ product for the high-performance-style npns in Si and SiGe IC processes

The current density for InP technology is remarkably lower than required for SiGe technology. However, the relatively large minimum emitter area of state-of-the-art InP technologies relative to SiGe technologies explains why the low current density of InP technologies cannot yet be exploited to achieve lower power. For example, in the InP process reported in [27], the npn with a minimum emitter area of $1 \times 3\,\mu\mathrm{m}^2$ requires 3 mA to achieve peak-f_T. Despite the larger current density, a SiGe process with a comparable peak-f_T, e.g., [16], requires only approximately 1 mA for the minimum emitter area of $0.12 \times 1.0\,\mu\mathrm{m}^2$ to operate at peak-f_T.

Although the increase realised in f_T over the years has been accompanied by a reduction in $\mathrm{BV_{CEO}}$, the $f_T \cdot \mathrm{BV_{CEO}}$ product shows a steady growth (see Fig. 3.29). The predicted $200\,\mathrm{GHz} \cdot \mathrm{V}$ Johnson limit (for Si-based processes [37]) has recently been surpassed by several IC processes.

3.7.2 Self-Heating

The introduction of deep-trench isolation has led to an increase in the thermal resistance of devices. Since a temperature difference between devices may have a substantial impact on circuit performance (for example, it can translate into an offset in differential pairs), there is a need to include self-heating in present and future design flows. Another reason why it is important to know a device's temperature is because self-heating of the device at a high current density enhances electromigration degradation.

The increase in current density for peak-f_T demonstrated in Fig. 3.28 leads to an important increase in the self-heating of a transistor. For example, a typical temperature rise of $30\,^\circ\mathrm{C}$ has been reported for a $120\,\mathrm{GHz}$ f_T SiGe technology operating a medium-size npn (minimum emitter width; ten times minimum emitter length) at $\mathrm{V_{cb}} = 1\,\mathrm{V}$ ($\mathrm{V_{ce}}$ near $\mathrm{BV_{CEO}}$) and at a current density of $5\,\mathrm{mA}/\mu\mathrm{m}^2$ (near peak-f_T), corresponding to a power density of $0.01\,\mathrm{W}/\mu\mathrm{m}^2$ [29]. The corresponding thermal resistance is $\mathrm{R_{TH}} = 3{,}000\,\mathrm{K/W}$.

The thermal resistance R_{TH} depends on the emitter geometry. At minimum emitter width, R_{TH} will increase with a decreasing emitter length. However, smaller devices have a larger area within the deep trench per dissipated power. The trench behaves as an effective heat insulator, due to the low thermal conductivity of SiO_2 (0.014 W/cm·K at 300 K) relative to Si (1.48 W/cm·K at 300 K). So, although the smallest devices have the highest R_{TH}, they show the least self-heating [29].

In IC process design a trade-off can be made between electrical and thermal optimisation of a device. Placing the deep trench isolation further away (or reducing its depth) will reduce the thermal resistance at the cost of an increased collector-substrate capacitance.

The thermal resistance has been shown to be marginally affected by the metal interconnect at the emitter. Extensive numerical simulations showed that the maximum temperature can be reduced by 10–15% [31].

Inclusion of the oxide layer causes the thermal resistance in silicon-on-insulator (SOI)-based processes to rise dramatically [32]. SOI processes can hence only be exploited for high-speed applications when additional measures are taken to lower the thermal resistance, for example by removing the substrate and post-processing a metallization layer on the backside [32].

In its simplest form, the thermal network applied in circuit simulators is a first-order network as shown in Fig. 3.30.

In this single-exponential model, P(t) represents the instantaneous power dissipation $P(t) = I_c \cdot V_{cb} + I_e \cdot V_{be}$. The resulting voltage v(t) represents the device temperature rise due to self-heating as a function of time. Mutual heating between transistors may be included in the simulation by adding thermal networks between the voltage nodes of the transistors. The thermal resistance R_{TH} and capacitance C_{TH} of the device are usually derived via the following procedure. First, a reference measurement is obtained for V_{be} as a function of the substrate temperature while operating the device at a very low current (so that self-heating is negligible). Then, static and dynamic measurements of V_{be} are performed at a certain level of power dissipation in the device. R_{TH} and C_{TH} can be derived from the measurement results. A typical time constant for self-heating is $R_{TH} \cdot C_{TH} = 1\,\mu s$.

The single-exponential model has been shown to have a low accuracy, because the model is based not on physics but on ease of implementation in a simulator. Improved fitting between measured and simulated device temperatures can be realised using two exponential terms in series [30].

In the case of large area transistors operating at high output voltages, as for example in power amplifiers, it is more complicated to find the junction temperature.

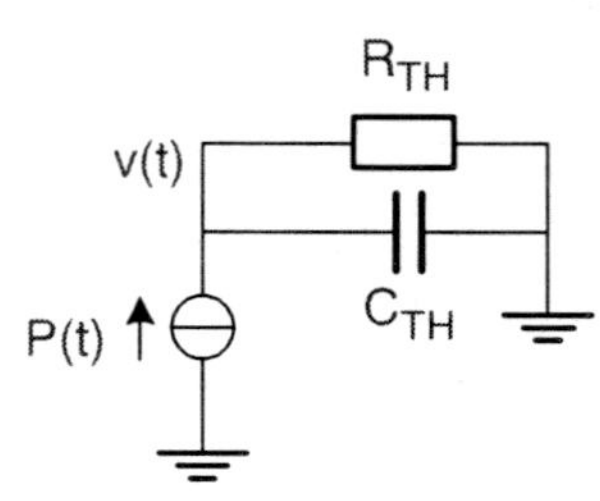

Fig. 3.30 First-order thermal network applied in most simulators

When such transistors are operated at $V_{cb} > BV_{CEO}$, both self-heating and avalanche current multiplication play a role in the current distribution of long emitter fingers, as shown in [33]. For example, when the transistors are biased at peak-f_T, interaction between avalanche current and self-heating starts to become significant at $V_{cb} > 2 \cdot BV_{CEO}$. The transistor model should include a distributed base resistance, needed to model pinch-in and distribution of the self-heating along the emitter finger and mutual heating between multiple fingers to predict the current distribution accurately. However, such models are not commonly used.

3.8 Other Trends

In this section a number of other technology trends that are important for RF circuit design will be summarised.

The first trend is the increase in the number of metal layers, combined with the introduction of metal density restrictions in the backend. In the Philips QUBiC IC process family, tiling was introduced in the first 0.25 µm generation involving five metal layers. To guarantee a sufficient yield of the metal backend, a chemical mechanical polishing procedure was used to flatten the wafer surface between the deposition of metal layers. To avoid damage to the metal due to the polishing process, sufficient coverage of metal must be ensured per unit of area. If the circuit layout does not provide sufficient metal coverage, dummy tiles are added in an automated tiling routine. Tiling may seriously affect the RF performance of the circuits due to its impact on inductors and transmission lines. In [34], the effect of tiling on the quality factor of inductors is shown to be relevant at frequencies above 10 GHz. In [5] it is shown that the way the tiles are placed with respect to each other can have an impact on the attenuation of transmission lines of approximately 0.2 dB/mm for frequencies between 10–100 GHz. Whenever allowed, tiling should be avoided in the area of RF components and interconnect.

A consequence of the increased number of metal layers is the increased height of the top metal layers above the substrate. While a typical distance of 8–10 µm is employed in today's six-metal layer processes, in [35], a distance of 16 µm is predicted for a backend with ten metal layers. The increased distance to the substrate in combination with the typically reduced inductance value required for circuits operating at increased bandwidths enables high quality factor inductor designs for GHz and Gb/s circuits. The reduced inductance value at higher frequencies reduces the area per inductor. Low power, and hence low supply voltage circuit topologies often make extensive use of inductors, for example to eliminate the dc voltage drop across load impedances. All these effects explain the rapid increase in the number of inductors per IC and the great efforts that are being made to include accurate inductor design tools in design flows, see for example [36].

A further consequence of the increased height of the top metal layers above the substrate in combination with increased operating frequencies is that the via inductance may start to play a role in circuit performance. Guidelines are needed for the via inductance to support future microwave circuit design.

A second trend is the attention that is in design flows and process optimisation being paid to passive components and interconnect. Standard transmission line configurations like GSG and GSSG, supported by verified equivalent circuit models, need to become part of the RF design flow. Very little information on transmission lines has so far been provided in publications focusing on IC processes. Sometimes a brief section is devoted to transmission lines, as in [13], but no exact line configurations or physical dimensions are given so the information is of limited use for comparisons.

Models for passive components such as the widely used π-model for resistors and capacitors are not sufficiently accurate for frequencies above 10 GHz. Distributed models will have to be introduced.

To improve on-chip isolation between circuits, it is proposed to eliminate the buried layers, also referred to as channel stoppers, outside the circuit cells [11,38]. The idea is to make islands of circuits, with maximum isolation in between. Elimination of the buried layer has been shown to be effective at frequencies up to 10 GHz; above 10 GHz the substrate behaves capacitively, irrespective of the presence of the buried layer.

3.9 Bipolar Versus RF-CMOS

For CMOS, f_A is usually dominated by f_{out} since the transistor layout can be optimized for very low gate series resistance R_g and thus high f_V (e.g., $f_V > 100$ GHz is often feasible). When comparing bipolar versus RF-CMOS, a CMOS process with comparable f_T and f_{max} may possess a relatively poor f_A, as shown in Table 3.5, where a 0.12 μm CMOS process is compared with the SiGe BiCMOS process from [8].

The device metrics for a NMOS transistor with L = 18 μm, W = 0.13 μm are shown as a function of the bias current in Fig. 3.31.

Despite the favorable f_T and f_{max}, the f_A is substantially lower in the CMOS process. This is due to the relatively low transconductance, requiring high load resistance values that reduce the output bandwidth.

Furthermore, the higher impedance level in CMOS technology makes the impact of interconnect parasitic capacitances more important, and thus it is more difficult to realize circuit bandwidths predicted by f_A. For CMOS, the favorable f_V also results in a favorable f_{cross}. Thus, the realization of microwave LC-VCOs in CMOS is usually not a problem, even in relatively outdated process generations.

Table 3.5 Comparison of typical CMOS and bipolar device metrics

	f_T (GHz)	f_{max} (GHz)	f_A (GHz)	f_{cross} (GHz)
CMOS 0.12 μm	86	138	6.7	123
QUBiC4G [8]	61	73	15.2	34

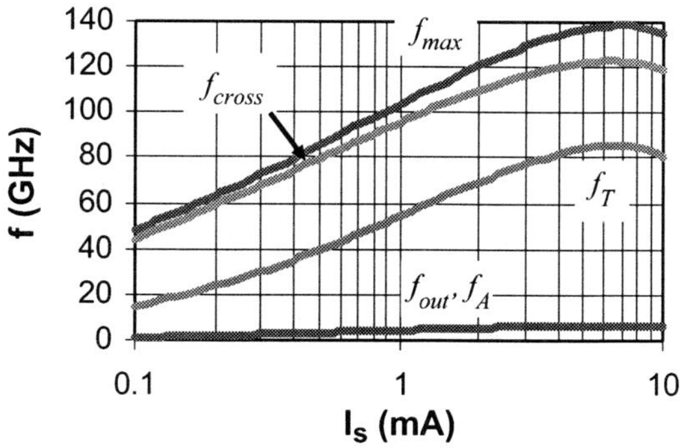

Fig. 3.31 Example CMOS device metrics as a function of the bias current

3.10 Conclusions and Outlook

A convenient methodology for deriving small-signal transistor device metrics is based on y-parameters, as analysed in Section 3.3. Measured y-parameters can be used to evaluate the device metrics without the need for parameter extraction [39]. In Section 3.4, the y-parameters were derived for a simplified transistor model. The resulting approximate equations for the device metrics provide valuable input for circuit design. However, the widely used metrics f_T and f_{max} are not directly related to the bandwidth of important circuits such as a differential pair amplifier. Circuits for high bit-rate applications make extensive use of differential pairs, and therefore the bandwidth of a differential pair amplifier is of interest. Metric f_{max} provides only limited information for the design of (broadband) circuits with high bandwidths. Despite the physical relevance of f_{max} (e.g., maximum oscillation frequency), f_{max} has no direct relation to circuit performance. In addition, the peak value of f_{max} for modern SiGe processes is typically beyond the capabilities of the measurement equipment and is derived via extrapolation from lower frequency measurements.

In contrast to f_{max}, the available bandwidth f_A introduced in Section 3.3.3, is directly relevant for circuit applications. The available bandwidth represents the bandwidth of a differential pair amplifier designed for 20 dB low-frequency voltage gain, a widely used topology in broadband circuit design. The available bandwidth can be sub-divided into two (parallel) contributions: the input bandwidth f_V and the output bandwidth f_{out}. An analysis of f_V and f_{out} across bias reveals which contribution dominates (when biasing the transistor at peak-f_T) and needs to be improved to further increase the peak-f_A. This information provides valuable feedback for further IC process optimisation. Alternatively, the impact of IC process changes on circuit performance can be evaluated on the basis of their effect on the metric f_A. To obtain a high f_A, it is not sufficient to have a high f_T. Reducing the base resistance, which has a minor impact on f_T, improves both the input bandwidth f_V and the output bandwidth f_{out}, and consequently the peak-f_A. When $gm_p \cdot \mathrm{R_b} > 1$, the output bandwidth f_{out} will saturate at a current density before the peak-f_T, and the peak-f_A will consequently occur at a current density below the current density for peak-f_T.

Reduction of the base resistance down to a level so that $gm_p \cdot R_b \leq 1$ is always very beneficial for f_A. Unfortunately, it is not common practice to evaluate f_A for an IC process. Therefore, it is not (yet) possible to benchmark many existing IC processes on the basis of f_A.

To achieve an increase in f_T, newer IC processes typically trade breakdown voltage BV_{CEO} for an increase in f_T. Reducing the breakdown voltage BV_{CEO} involves two aspects. In the first place, the modelling of avalanche currents becomes important for designing circuits with transistors operating at $V_{ce} > BV_{CEO}$. In the second place, circuits need to handle the relatively large (often $|I_b| >> I_c/\beta_0$) negative base currents that occur because of operation at $V_{ce} > BV_{CEO}$. Transistors operating at $V_{ce} > BV_{CEO}$ are typically found as output transistors of bias current sources and output driver circuits. This subject will be discussed in detail in relation to bias current circuits in Chapter 5.

In the near future, matching requirements will include matching of thermal resistances of transistors. This will depend on the layout of emitter metal and contacts, and on power dissipation in nearby components. Extraction tools will be needed to include both thermal networks and substrate networks.

In circuit design, attention will have to be paid to limiting V_{ce}, not only to reduce power dissipation and hence self-heating, but also to avoid (large) avalanche currents where possible.

The new device metrics introduced in this chapter have been published in [39] (f_A, f_V and f_{out}) and [40] (f_{cross}). The relevance of the metrics for circuit design has been highlighted in [41].

References

1. J.M. Miller, "Dependence of the input impedance of a three-electrode vacuum tube upon the load in the plate circuit," *Scientific Papers of the U.S. National Bureau of Standards*, vol. 15(351), 1920, pp. 367–386.
2. W.J. Kloosterman, J.C.J. Paasschens, D.B.M. Klaassen, "Improved extraction of base and emitter resistance from small signal high frequency admittance measurements," in *Proc. IEEE BCTM*, 1999, pp. 93–96.
3. P.A.H. Hart (ed.), *Bipolar and bipolar-MOS integration*, Section 3.10 by G.A.M. Hurkx, Elsevier, 1994, ISBN 0-444-81510-4.
4. J.M. Rollet, "Stability and power-gain invariants of linear two-ports," *IRE Trans. Circuit Theory*, CT-9:29–32, 1962.
5. P. Wennekers, R. Reuter, "SiGe technology requirements for millimeter-wave applications," in *Proc. IEEE BCTM*, 2004, pp. 79–83.
6. K. Washio, E. Ohue, et al., "A 0.2-μm 180-GHz-f_{MAX} 6.7-ps-ECL SOI/HRS self-aligned SEG SiGe HBT/CMOS technology for microwave and high-speed digital applications," *IEEE Trans. Electron Devices*, vol. 49, No. 2, February 2002, pp. 271–278.
7. T. Hashimoto, Y. Nonaka, et al., "Integration of a 0.13-μm CMOS and a high performance self-aligned SiGe HBT featuring low base resistance," in *Proc. IEDM*, 2002, pp. 779–782.
8. P. Deixler, R. Colclaser, et al., "QUBiC4G: a $f_T/f_{max} = 70/100$GHz 0.25μm low power SiGe-BiCMOS production technology with high quality passives for 12.5Gb/s optical networking and emerging wireless applications up to 20GHz," in *Proc. IEEE BCTM*, 2002, pp. 201–204.

9. A. Pruijmboom, D. Szmyd, R. Brock, R. Wall, N. Morris, K. Fong, F. Jovenin, "QUBiC3: a 0.5μm BiCMOS production technology, with $f_T = 30GHz$, $f_{max} = 60GHz$ and high-quality passive components for wireless telecommunication applications," in *Proc. IEEE BCTM*, 1998, pp. 120–123.
10. D. Szmyd, R. Brock, N. Bell, S. Harker, G. Patrizi, J. Fraser, R. Dondero, "QUBiC4: a Silicon-RF BiCMOS technology for wireless communication ICs," in *Proc. IEEE BCTM*, 2001, pp. 60–63.
11. P. Deixler, A. Rodriguez, et al., "QUBiC4X: an $f_T/f_{max} = 130/140GHz$ SiGe:C-BiCMOS manufacturing technology with elite passives for emerging microwave applications," in *Proc. IEEE BCTM*, 2004, pp. 233–236.
12. L. Lanzerotti, N. Feilchenfeld, et al., "A low complexity 0.13μm SiGe BiCMOS technology for wireless and mixed signal applications," in *Proc. IEEE BCTM* 2004, pp. 237–240.
13. D. Knoll, B. Heinemann, et al., "A modular, low-cost SiGe:C BiCMOS process featuring high-f_T and high-BV_{CEO} transistors," in *Proc. IEEE BCTM*, 2004, pp. 241–244.
14. S. Subbanna, L. Larson, et al., "Silicon-germanium BICMOS technology and a CAD environment for 2–40 GHz VLSI Mixed-Signal ICs," in *Proc. IEEE CICC*, 2001, pp. 559–566.
15. A. Joseph, D. Coolbaugh, et al., "A 0.18μm BiCMOS technology featuring 120/100 GHz (f_T/f_{MAX}) HBT and ASIC-compatible CMOS using copper interconnect," in *Proc. IEEE BCTM*, 2001, pp. 143–146.
16. B.A. Orner, Q.Z. Liu, et al., "A 0.13μm BiCMOS technology featuring a 200/280 GHz (fT/fmax) SiGe HBT," in *Proc. IEEE BCTM*, 2003, pp. 203–207.
17. B. Jagannathan, M. Khater, et al., "Self-aligned SiGe NPN transistors with 285 GHz f_{MAX} and 207 GHz f_T in a manufacturable technology," *IEEE Electron Device Lett.*, vol. 23, No. 5, May 2002, pp. 258–260.
18. J.-S. Rieh, B. Jagannathan, et al., "SiGe HBTs with cut-off frequency of 350GHz," in *Proc. IEDM*, 2002, pp. 771–774.
19. A. Chantre, M. Marty, et al., "A high performance low complexity SiGe HBT for BiCMOS integration," in *Proc. IEEE BCTM*, 1998, pp. 93–96.
20. H. Baudry, B. Martinet, et al., "High performance 0.25μm SiGe and SiGe:C HBTs using non selective epitaxy," in *Proc. IEEE BCTM*, 2001, pp. 52–55.
21. H. Baudry, B. Szelag, et al., "BiCMOS7RF: a highly-manufacturable 0.25-μm BiCMOS RF-applications-dedicated technology using non-selective SiGe:C epitaxy," in *Proc. IEEE BCTM*, 2003.
22. M. Laurens, B. Martinet, et al., "A 150GHz f_T/f_{max} 0.13μm SiGe:C BiCMOS technology," in *Proc. IEEE BCTM*, 2003.
23. D. Knoll, K.E. Ehwald, et al., "A flexible, low-cost, high performance SiGe:C BiCMOS process with a one-mask HBT module," in *Proc. IEDM*, 2002, pp. 783–786.
24. H. Rücker, B. Heinemann, et al., "SiGe:C BiCMOS technology with 3.6 ps gate delay," in *Proc. IEDM*, 2003, pp. 121–124.
25. J. Böck, H. Schäfer, K. Aufinger, et al., "SiGe bipolar technology for automotive radar applications," in *Proc. IEEE BCTM*, 2004, pp. 84–87.
26. P. Andre, J. Benchimol, et al., "InP DHBT technology and design methodology for high-bit-rate optical communications circuits," *IEEE J. Solid-State Circuits*, vol. 33, No. 9, September 1998, pp. 1328–1334.
27. N.X. Nguyen, J. Fierro, G. Peng, A. Ly and C. Nguyen, "Manufacturable commercial 4-inch InP HBT device technology," in *Proc. GaAs MANTECH*, 2002.
28. M. Sokolich, "High Speed, low power, optoelectronic InP-based HBT integrated circuits," in *Proc. CICC*, 2002, pp. 483–490.
29. J.-S. Rieh, D. Greenberg, B. Jagannathan, G. Freeman, S. Subbanna, "Measurement and modeling of thermal resistance of high speed SiGe heterojunction bipolar transistors," in *Proc. Silicon Monolithic ICs in RF Systems*, 2001, pp. 110–113.
30. D.J. Walkey, T.J. Smy, D. Marchesan, H. Tran, C. Reimer, T.C. Kleckner, M.K. Jackson, M. Schröter, J.R. Long, "Extraction and modelling of thermal behaviour in trench isolated bipolar structures," in *Proc. IEEE BCTM*, 1999, pp. 97–100.

31. D.J. Walkey, D. Celo, T.J. Smy, "A Simplified model for the effect of interfinger metal on maximum temperature rise in a multifinger bipolar transistor," *IEEE Trans. Computer-Aided Design*, vol. 22, No. 1, January 2003, pp. 15–25.
32. E. Aksen, "On-glass process option for BiCMOS technology," in *Proc. IEEE BCTM*, 2004, pp. 64–67.
33. M. Pfost, P. Brenner, R. Lachner, "Investigation of advanced SiGe heterojunction bipolar transistors at high power densities," in *Proc. IEEE BCTM*, 2004, pp. 100–103.
34. W. De Cock, M. Steyaert, A 2.5V, "10GHz fully integrated LC-VCO with integrated high-Q inductor and 30% tuning range," *Analog Integrated Circuits and Signal Processing*, vol. 33, No. 2, November 2002, pp. 137–144.
35. B. Kleveland, C.H. Diaz, et al., "Exploiting CMOS Reverse Interconnect Scaling in Multigigahertz Amplifier and Oscillator Design," *IEEE J. Solid-State Circuits*, vol. 36, No. 10, October 2001, pp. 1480–1488.
36. L.F. Tiemeijer, R.J. Havens, R. de Kort, Y. Bouttement, P. Deixler, M. Ryczek, "Predictive spiral inductor compact model for frequency and time domain," in *Proc. IEDM*, 2003.
37. E.O. Johnson, "Physical limitations on frequency and power parameters of transistors," *RCA Rev.*, vol. 26, p. 163, 1965.
38. W. Steiner, H.-M. Rein, J. Berntgen, "Experimental verification of substrate coupling in a high-gain 30 Gb/s SiGe amplifier," in *Proc. IEEE BCTM*, 2004, pp. 273–276.
39. G.A.M. Hurkx, P. Agarwal, R. Dekker, E. van der Heijden and H. Veenstra, "RF figures-of-merit for process optimisation," *IEEE Trans. Electron Devices*, vol. 51, No. 12, December 2004, pp. 2121–2128.
40. H. Veenstra, E. van der Heijden, "A 19–23 GHz integrated LC-VCO in a production 70 GHz f_T SiGe technology," in *Proc. ESSCIRC*, 2003, pp. 349–352.
41. H. Veenstra, G.A.M. Hurkx, E. v.d. Heijden, C. Vaucher, M. Apostolidou, D. Jeurissen, P. Deixler, "10–40GHz design in SiGe-BiCMOS and Si-CMOS – linking technology and circuits to maximize performance," in *Proc. European Microwave Week*, 2005.

Chapter 4
Cross-Connect Switch Design

4.1 Introduction

Many aspects that are important in high bit-rate circuit design are brought together in the design of a cross-connect switch IC for routing data in optical networks. The switch matrix, which forms the core of the cross-connect switch IC, is an excellent example showing that optimum performance can only be obtained when circuits and interconnect are optimised together. For the design and optimisation of the signal distribution inside the matrix, extensive use is made of the interconnect models described in Chapter 2 and the device metrics given in Chapter 3.

Other high bit-rate circuits are needed to support the matrix operation, such as input and output buffers and functions for built-in self-testing (e.g., a VCO and a PRBS generator). This chapter will describe the design of the RF path of a cross-connect switch IC that supports 20 differential inputs and 20 differential outputs, at data rates up to 12.5 Gb/s per input. The block diagram of the switch IC is shown in Fig. 1.9. An overview of the main specifications is provided in Table 4.1.

The main focus of this chapter is on the design of the RF signal path. The IC is designed for wire bonding, so all the bondpads must be located on the perimeter of the IC. Because of the large number of bondpads, the die area is bondpad-limited. The IC includes built-in self-test functions, implemented using a PRBS generator plus an error detector, both clocked by an on-chip tuneable LC-VCO. The VCO generates an output frequency of up to 12.5 GHz. When clocked at 12.5 GHz, the PRBS generator outputs a pseudo-random data pattern with a sequence length $2^7 - 1$ (127 bits) at 12.5 Gb/s. Design aspects of the PRBS generator and VCO will be described in Chapters 6 and 7, respectively. The CMOS configuration interface, built using standard CMOS logic, will not be described.

The basic concept of a $(N \times M)$ cross-connect switch based on a matrix architecture is shown in Fig. 4.1. Any input i(i) can be connected to any output o(i) provided that no conflicts occur (e.g., multiple inputs connected to the same output are not possible). The delay between the various signal paths in the matrix may differ. There will be a shortest possible path with minimum delay and a longest possible path with maximum delay. In the matrix of Fig. 4.1 the shortest path is from input

H. Veenstra and J.R. Long, *Circuit and Interconnect Design for RF and High Bit-Rate Applications,* 119
doi: 10.1007/978-1-4020-6884-3_4, © Springer Science + Business Media B.V. 2008

Table 4.1 Specifications for the cross-connect switch IC

Parameter	Value
Supply voltage	$2.5\,\mathrm{V} + / - 10\%$
Power dissipation	<4 W; all I/O active
Number of inputs N × outputs M	20 × 20, differential
Input sensitivity	$0.1\,V_{pp,diff}$
Output swing into 2 × 50 Ω loads	0.1–$0.6\,V_{pp,diff}$ programmable
Output rise-time (20–80%)	<30 ps
Jitter generation (RMS)	<2 ps
IC technology	SiGe BiCMOS $f_T / f_{max} = 70/100\,\mathrm{GHz}$ [1]
Support of multicast and broadcast functions	

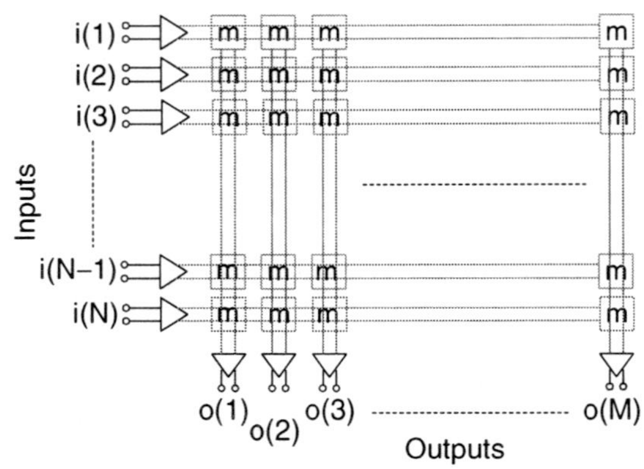

Fig. 4.1 General concept of a (N × M) matrix architecture with input buffers (left) and output buffers (bottom). The signal path is differential. The matrix node circuits 'm', at the cross-points, can be individually activated via a configuration interface (not shown). For each column, at most one matrix node circuit may be active

i(N) to output o(1), the longest path from input i(1) to output o(M). The delay difference between the shortest and longest paths defines the timing skew. The skew is not important in the switch design presented here, since no on-chip retiming functions need to be implemented. The skew is important if retiming functions are required, shared between multiple outputs, assuming that the signal paths operate at the same datarate.

The RF signal path inside the matrix is defined as the set of all possible individual signal paths between inputs i(1)–i(N) and outputs o(1)–o(M). With 20 inputs and 20 outputs, there are in total 400 possible paths through the matrix. For example, the shortest and longest paths are elements of the RF signal path. When the performances of the shortest and longest paths through the matrix are acceptable, it is reasonable to assume that the performance of all other signal paths will also be acceptable. Therefore, the analysis in this chapter will focus on the shortest and longest paths.

In Section 4.2, the design of the RF signal path will be explained. There is an important difference between the design of the signal distribution *outside* the matrix (that is, from input bondpads to the matrix inputs and from matrix outputs to the output bondpads) and the design of the signal distribution *inside* the matrix. This will be explained, and the concept of distributed capacitive loading will be introduced. The matrix core provides an excellent example to explain that circuits, interconnect and floorplan need to be designed and optimised together to achieve adequate performance. The design of the matrix node circuits will be described to detail. Design, layout and evaluation results of a test-IC, enabling detailed analysis of the signal transfer inside the matrix, will be described.

Buffer circuits are widely used across the IC. Section 4.3.1 deals with the design of intermediate buffer circuits, used inside the matrix. In Section 4.3.2 the designs of the input and output buffer circuits will be described. Simulation results obtained for the entire RF path of the switch IC will be described in Section 4.4. The on-chip supply decoupling strategy will be explained in Section 4.5. Experimental results are described in Section 4.6. Finally, a discussion of the results and the challenges involved in designing a similar cross-connect switch circuit at higher data rates will follow in Section 4.7.

The cross-connect switch IC was implemented as the very first product and process development driver for the IC process used [1]. This implied an opportunity to optimise the process features and performance for cross-connect switch applications.

4.2 Switch Matrix Design

To understand and design the RF signal path inside the matrix, it is essential to have an indication of the expected physical matrix size even before the actual design has started. On the basis of the estimated matrix size, the interconnect delay across rows and columns can be estimated and it can be assessed whether signal reflections and transmission line theory will be relevant.

The following analysis is performed to estimate the matrix size. To begin with, GSSG transmission lines are used for differential signal transfer in both rows and columns, following the preferred transmission line configuration outlined in Section 2.6 of this book. A typical transmission line width is 35 μm (5 μm line width, 4 lines, 5 μm line spacing). Since the transmission lines are placed on top of a ground shield, no circuitry is assumed to be underneath the transmission lines for rows or columns. The simplest matrix node circuit is a differential pair. This can be realised in a chip area of about 20 μm × 20 μm in the given technology. So, the minimum area per matrix node (e.g., circuit surrounded by transmission lines for row and column) is 55 μm × 55 μm. However, each matrix node needs additional area for biasing, power supply routing and local power mode decoding. It is also assumed that some power supply decoupling is implemented per matrix node circuit. So, a space as large as 100 μm × 100 μm may be required per matrix node circuit, resulting in an initial estimate of 2 mm × 2 mm for a matrix of 20 inputs by 20 outputs.

4.2.1 Transmission Lines for Rows and Columns

A 2 mm × 2 mm matrix requires signal distribution across 2 mm long rows and columns. An unloaded transmission line of 2 mm length introduces a signal delay (t_d) of approximately 2 mm × 6 ps/mm or 12 ps. With a required signal rise-time for 12.5 Gb/s signals of $t_r \leq 30$ ps, the signal delay across the 2 mm long line is significantly longer than the minimum requirement for transmission line modelling as explained in Section 2.3, equation (2.45). So, transmission line design and modelling need to be applied. Signal reflections may occur at discontinuities (for example at the ends of the line), at the points where the matrix nodes circuit inputs and outputs are connected, and at points where the transmission lines of rows and columns cross. Such reflections should be minimised because they increase the jitter generation of the switch and may introduce bit errors.

Distribution of the signals across the columns is as difficult as distribution along the rows, because the estimated lengths of the signal lines for rows and columns are the same. Signal reflections in rows and columns are of equal importance to the final performance of the IC. Therefore, it is desirable to have transmission lines of optimum performance (e.g., controlled impedance and low loss) for rows and columns.

To facilitate the optimisation of the transmission lines for rows and columns, the IC process features two thick top metal layers: 2 μm thick Metal5 and 3 μm thick Metal6.

4.2.2 The Concept of Distributed Capacitive Loading

The transmission lines of rows and columns inevitably cross inside the matrix. First, a single line crossing will be analysed. The effect of one Metal5 line crossing one Metal6 GSSG transmission line orthogonally may be included in the RLMC transmission line model by two additional capacitors C_{65} as shown in Fig. 4.2. The network between the Metal5 line and ground is not important for differential-mode signal transfer because a virtual ground exists between the two C_{65} capacitors. The value of C_{65} can be estimated on the basis of the line widths and spacing using

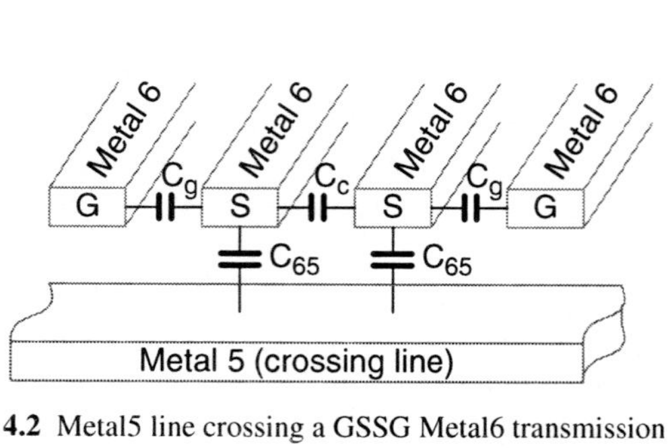

Fig. 4.2 Metal5 line crossing a GSSG Metal6 transmission line

the general equation for the capacitance between two metal plates separated by a distance d:

$$C_A = \frac{\varepsilon_0 \varepsilon_r A}{d} \tag{4.1}$$

Here, A is the overlap area between Metal5 and Metal6. In the technology used, the distance d between Metal5 and Metal6 is 1.8 μm while the relative permittivity is $\varepsilon_r \approx 4$. For 5 μm wide lines this gives $C_A = 0.5$ fF. Since C_A only refers to the capacitance due to the Metal5–Metal6 overlap area, a correction factor α needs to be included to account for fringing effects that more than double the capacitance. Using parasitic extraction for a single Metal6–Metal5 crossing of 5 μm × 5 μm, the factor α was found to be α = 3.6, so $C_{65} \approx 3.6 \cdot C_A = 1.8$ fF.

In analysing the effect of the Metal5 crossings on the signal transfer in the Metal6 transmission line, it is important to realise that the floorplan of the matrix may be designed so that the line crossings occur equally distributed across the total (2 mm) line length. So, it is sufficient to analyse a single line section for one matrix node element. Such a section has a length of 100 μm. Although multiple line crossings occur per section in the matrix, the influence of a single crossing is analysed first. A GSSG transmission line implemented in Metal6 with $Z_{0dm} = 100\ \Omega$, $t_{dm} = t_{cm} = 6$ ps/mm and $Z_{0cm} = 50\ \Omega$ is assumed as a reference. The line loss is not considered.

The lumped element values for a 100 μm long section of the reference transmission may be calculated using equations (2.85)–(2.88). This gives $C_g = 60$ fF/mm; $C_c = 30$ fF/mm; L = 0.45 nH/mm; k = 1/3. In analysing the differential-mode signal transfer it is convenient to consider the differential-mode lumped capacitance $C_{dm} = C_c + C_g/2 = 60$ fF/mm and inductance $L_{dm} = 2 \cdot L(1-k) = 0.6$ nH/mm. So, a transmission line section of 100 μm length has a lumped capacitance of $C_{dm,sec} = 6$ fF and an inductance of $L_{dm,sec} = 60$ pH. If for every section of the Metal6 transmission line a single Metal5 line crosses, the situation shown in Fig. 4.3 results.

If the Metal6 transmission line is lossless, as in Fig. 4.3, the combination of the reference transmission line plus distributed capacitive loading (capacitors $C_{65}/2$) is also lossless, provided that no series (or parallel) resistance is associated with C_{65}. For the differential-mode capacitance $C_{65}/2$, the associated loss resistance may be ignored because C_{65} is a metal plate capacitor. As long as there is no resistive term in the load impedance, the loaded transmission line will remain lossless. Since one matrix node circuit is connected per section, the input (or output) impedance of the matrix node circuit can be added in parallel to each capacitor $C_{65}/2$ in Fig. 4.3 for

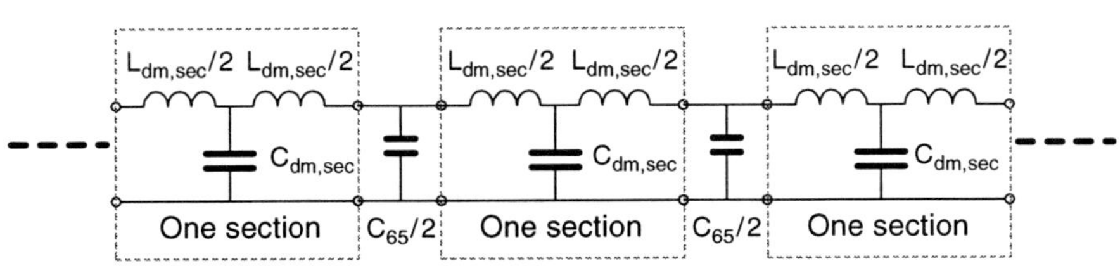

Fig. 4.3 Differential-mode equivalent circuit of a GSSG transmission line crossed by one Metal5 line per section. All Metal5 lines are identical

the transmission line of the row (or column). So, it is important that the (parallel equivalent) differential input resistance R_{pi} and output resistance R_{po} of the matrix circuits are as high as possible, preferably $|R_{pi}| \gg Z_{0dm}$ and $|R_{po}| \gg Z_{0dm}$. The situation shown in Fig. 4.3 can then be regarded as a transmission line with effective characteristic impedance and delay per section according to:

$$Z_{0dm,eff} = \sqrt{\frac{L_{dm,eff}}{C_{dm,eff}}} = \sqrt{\frac{L_{dm,\text{sec}}}{C_{dm,\text{sec}} + C_{65}/2}} \tag{4.2}$$

$$t_{dm,\text{sec},eff} = \sqrt{L_{dm,\text{sec},eff} C_{dm,\text{sec},eff}} = \sqrt{L_{dm,\text{sec}}(C_{dm,\text{sec}} + C_{65}/2)} \tag{4.3}$$

Equations (4.2) and (4.3) demonstrate the effect of distributed capacitive loading. For a single-ended transmission line, the concept of distributed capacitive loading can be summarised as follows. If a transmission line with characteristic impedance $Z_0 = \sqrt{(L/C)}$ and delay per mm $t_d = \sqrt{(LC)}$ is capacitively loaded with a total load capacitance of C_l per mm, and the load capacitance is equally distributed in multiple sections along the line, the loaded transmission line behaves as a transmission line with effective characteristic impedance $Z_{0,eff}$ and delay per mm $t_{d,eff}$:

$$Z_{0,eff} = \sqrt{\frac{L}{C + C_l}} \tag{4.4}$$

$$t_{d,eff} = \sqrt{L(C + C_l)} \tag{4.5}$$

In the case of a differential transmission line, the concept of effective characteristic impedance applies to both differential and common modes. An example of differential mode was outlined in Fig. 4.3 above.

In the example from the cross-connect matrix design, the differential-mode capacitance of the transmission line per section (100 µm) equals $C_{dm,\text{sec}} = 6\,\text{fF}$. A single Metal5 line crossing per section introduces a load capacitance equal to $C_{65}/2 = 0.9\,\text{fF}$ per section. The resulting effective differential-mode impedance and delay are $Z_{0dm,eff} = 0.93 \cdot Z_{0dm} = 93\ \Omega$; $t_{dm,eff} = 1.07 \cdot t_{dm} = 6.43\,\text{ps/mm}$. If the line is terminated at both ends by $Z_{0dm,eff}$, reflections will be insignificant. Inside the matrix, multiple lines cross per section of the transmission line: the four lines of the GSSG transmission line, supply lines and logic control lines. The multiple crossings per section can be distributed within each section to minimise signal reflections.

4.2.3 Matrix Node Circuit Design

As shown in the previous section, signal transfer inside the matrix can be implemented using the concept of distributed capacitive loading. The loaded transmission line then needs to be terminated by its effective characteristic impedance $Z_{0,eff}$. The effective characteristic impedance follows from equation (4.4). A higher load capacitance results in a lower $Z_{0,eff}$.

Current-mode logic (CML) inverters can be used for the matrix node circuits. To achieve a low power dissipation, the signal swing throughout the signal path is designed at $0.1\,V_{p,diff}$. Typical CML circuits show a small-signal gain A of 4; the operation at the reduced swing of $0.1\,V_{p,diff}$ reduces the small-signal gain to approximately $A \approx 2$.

Since the output of the matrix node circuit needs to drive the transmission line (of the column) to the output buffer, the bias current needed for the CML inverter is defined by the effective characteristic impedance of the column transmission line. The concept of distributed capacitive loading applies to the signal transfer in the column, provided that the output impedance of the matrix node circuit is mainly capacitive. This will be the case with a differential pair because the typical (parallel equivalent) output resistance of a differential pair will equal $R_{po} = 2 \cdot V_{eaf}/I_c \approx 50\,k\Omega$ (with V_{eaf} the Early voltage of typically 50 V and $2 \cdot I_c$ the bias current of the differential pair, here $I_c = 2\,mA$), so that $R_{po} \gg Z_{0dm}$.

Figure 4.4 shows the concept of signal distribution in the column using CML inverters. At most one matrix node circuit for each column may be biased at any time; all the other bias currents are then zero. For uniformity of the loaded

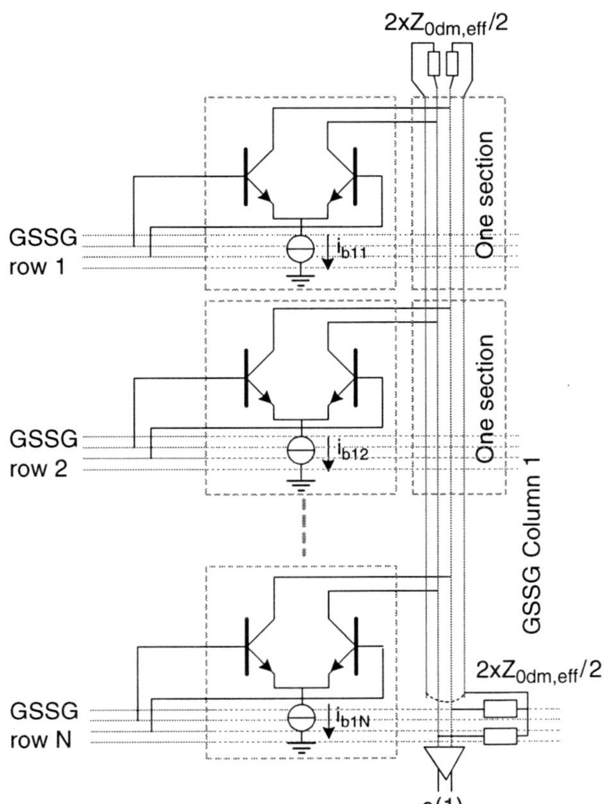

Fig. 4.4 Using differential pairs as matrix node circuits

transmission line, all sections must be of the same length and equally loaded. If the output impedance of the matrix node circuit depends on its bias state, a mismatch will be introduced at only one section. This is different in the case of row design, because in a row, multiple (in multicast mode) or even all (in broadcast mode) matrix node circuits may be active at the same time.

The parallel equivalent output resistance R_{po} of the differential pair is either infinite (in the 'off' state) or $R_{po} = 2 \cdot V_{eaf}/I_c \approx 50\,k\Omega$. In both cases the output resistance may be ignored since $R_{po} \gg Z_{0dm}$.

Assuming a low source impedance driving the differential pairs (e.g., lower than the base series resistance R_b), the parallel equivalent output capacitance C_{po} in the 'on' state equals $C_{po,on} = 0.5(C_{cs} + C_{bc}(1 + gm \cdot R_b))$ while in the 'off' state $C_{po,off} = 0.5(C_{cs} + C_{bc})$. There may be a small difference in output capacitance between the 'on' and 'off' states depending on the bias current in the 'on' state and the transistor size of the differential pair.

However, there are problems associated with the use of a differential pair as a matrix node circuit that make this solution unacceptable. In the first place, the isolation between row and column in a circuit in the 'off' state is of utmost importance for low crosstalk. The differential pair in the 'off' state has its collector-base capacitance C_{bc} connected between the inputs and outputs of the matrix node, deteriorating its isolation. In the second place, like the output capacitance, the input capacitance must also be independent of the bias state of the matrix node circuit. This does not hold for a differential pair, certainly not when biased at current densities close to peak-f_T. The diffusion capacitance has a substantial bias dependency. Moreover, the input capacitance (loading the transmission line of the row) needs to be as low as possible in order to keep the effective characteristic impedance $Z_{0dm,eff}$ of the row close to its unloaded value Z_{0dm}. A lower effective impedance requires lower termination resistance values, thereby increasing the power dissipation of the input buffers driving the transmission lines for a constant signal swing on the transmission line. In the third place, like the output resistance, the parallel equivalent input resistance must be as high as possible. Improvements in all these aspects can be realised by adding a pair of input emitter followers to each matrix node circuit. The resulting matrix node circuit is shown in Fig. 4.5.

To achieve the minimum input capacitance C_{pi}, minimum area input transistors Q_{1a}, Q_{1b} should be used. This is only possible if sufficient bandwidth ($>0.7 \cdot DR_{max}$ with $DR_{max} = 12.5\,Gb/s$ the highest supported data rate) is obtained for the entire RF signal path, requiring approximately twice this value or 16.5 GHz bandwidth for the matrix node circuit, and requiring >20 GHz bandwidth at the Q_1–Q_2 interface. The bandwidth may be limited at the interface between the emitter follower Q_1 and the differential pair Q_2. The design in the technology used is based on the following parameters.

The circuit is designed to operate at a signal swing of $0.2\,V_{pp,diff}$ at rows and columns. With a line impedance $Z_{0dm} = 100\,\Omega$ and a two-sided termination, a bias current $I_2 = 4\,mA$ is needed. Transistors Q_{2a}, Q_{2b} are operated at a peak-f_T of approximately 75 GHz. Note that the operation at $V_{cb} > 0$ is advantageous for the peak-f_T. The input capacitance of the differential pair Q_2 can be estimated using

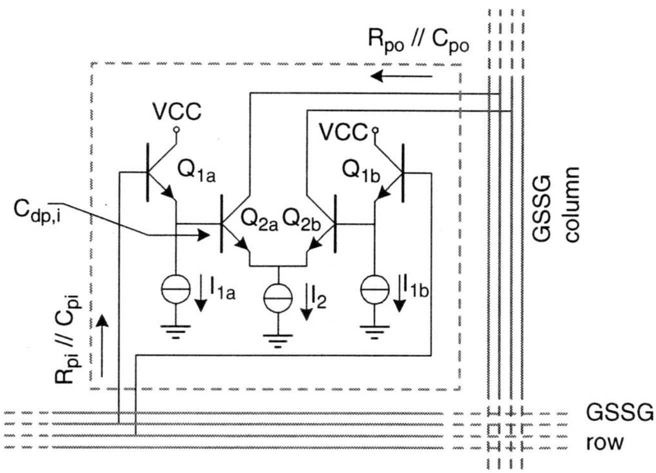

Fig. 4.5 CML matrix node circuit with input emitter followers

$$f_T = \frac{gm}{2\pi(C_{be} + C_{bc})} \tag{4.6}$$

The differential pair input capacitance is $C_{dp,i} = 0.5(C_{be} + (|A| + 1) \cdot C_{bc})$. With $|A| = Z_{0dm}/4/(1/gm + R_e) \approx 1.6$, $C_{bc} = 20\,fF$ and $f_T = 75\,GHz$, $C_{dp,i}$ becomes (in the 'on' state) approximately $C_{dp,i} \approx 100\,fF$: unacceptably large relative to the $C_{dm,sec} = 6\,fF$ lumped capacitance of a 0.1 mm section of the line. To lower the input capacitance, emitter followers Q_{1a}, Q_{1b} are added. If the emitter series resistance R_e of the emitter followers is ignored, the transistors need to be biased at $I_1 = 0.7\,mA$ each to drive the 100 fF differential load capacitance with sufficient bandwidth, that is larger than 20 GHz. For current density reasons, transistors Q_{1a}, Q_{1b} need to be at least twice the minimum area to handle 0.7 mA. The emitter series resistance is however relatively high for the technology used. A transistor biased at peak-f_T has approximately $R_e = 1.6/(40 \cdot I_c)$, thereby more than doubling the emitter follower output resistance R_{out} relative to the situation in which $R_e = 0$, since $R_{out} \approx 1/gm + R_e$. Therefore, transistors Q_1 are chosen to be six times the minimum area with $C_{bc} = 8\,fF$, and are biased at 1.7 mA each.

The addition of emitter followers causes the differential input capacitance (in the 'on' state) to decrease to $C_{pi} = 15\,fF$. This is a major improvement relative to the 100 fF input capacitance of the differential pair, but still implies a significant load to the transmission line. To achieve a further reduction in the capacitive load to the line, the circuit was extended with a second differential pair Q_{4a}, Q_{4b} plus emitter followers Q_{3a}, Q_{3b} (see Fig. 4.6).

The impedance levels for the input emitter followers Q_{3a}, Q_{3b} plus differential pair Q_{4a}, Q_{4b} are a factor of 3 higher than those for the circuit around Q_1 and Q_2. This was realised by choosing the emitter lengths for Q_3 and Q_4 a factor of 3 smaller than those for Q_1 and Q_2, using bias current $I_4 = I_2/3 = 1.3\,mA$ and $I_3 = I_1/3 = 0.56\,mA$, and using $R_{4a} = R_{4b} = 0.75 \cdot Z_{0dm} = 75\ \Omega$. With these design parameters, small input transistors Q_3 can be used, realising a low input capacitance $C_{pi} = 5\,\text{fF}$ (in the 'on' state). Since all impedance levels at the Q_3–Q_4 interface have been scaled by a factor of 3 with respect to the Q_1–Q_2 interface, the bandwidth at the Q_3–Q_4 interface is identical to that at the Q_1–Q_2 interface. The bandwidth at the R_4–Q_1 interface is sufficiently high; with $R_{4a} + R_{4b} = 150\,\Omega$ and a total capacitance $C_{pi,Q1} + C_{po,Q4} = 15\,\text{fF} + 8\,\text{fF}$, the bandwidth at the R_4–Q_1 interface is 46 GHz.

The output capacitance C_{po} of the circuit shown in Fig. 4.6 can be derived using the analysis based on Fig. 4.7. In the 'off' state, all bias currents in the matrix node

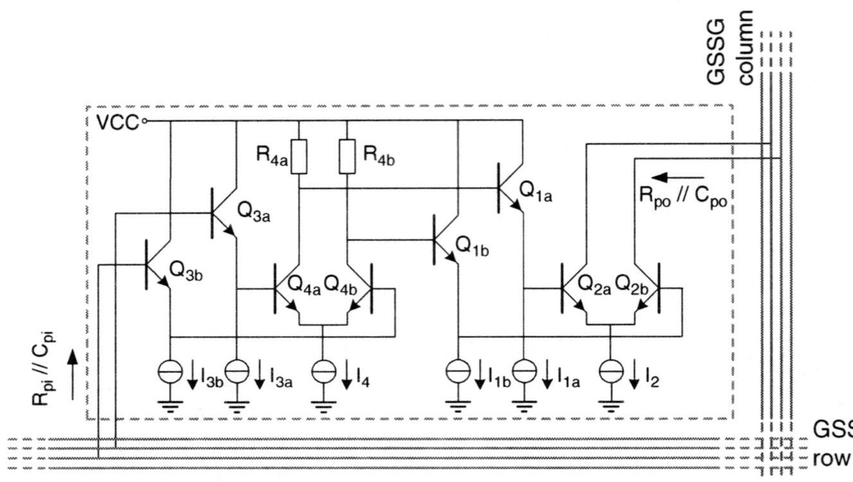

Fig. 4.6 Final design of the matrix node circuit

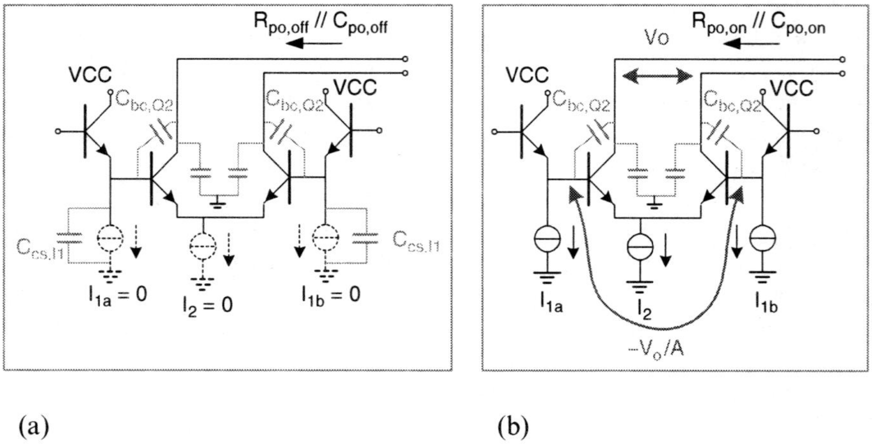

Fig. 4.7 Output part of the matrix node circuit showing the most relevant capacitances in the 'off' state (a) and the 'on' state (b)

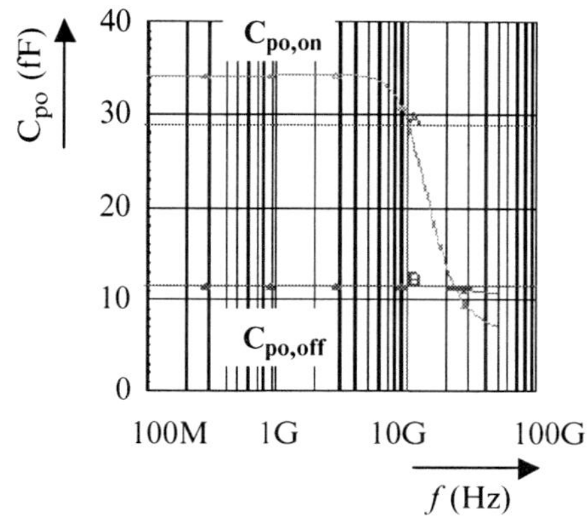

Fig. 4.8 Simulated differential output capacitance of the matrix node circuit

circuit are zero. When the circuit is in the 'off' state, the differential output capacitance is

$$C_{po,off} = 0.5 \cdot \left(C_{cs,Q2} + \frac{C_{bc,Q2} \cdot C_{cs,I1}}{C_{bc,Q2} + C_{cs,I1}} \right) \tag{4.7}$$

When the circuit is in the 'on' state, the output Miller effect amplifies the contribution of $C_{bc,Q2}$. The output capacitance in the 'on' state is

$$C_{po,on} = 0.5 \cdot (C_{cs,Q2} + C_{bc,Q2}(1 + 1/|A|)) \tag{4.8}$$

With $C_{cs,Q2} = 14\,\text{fF}$, $C_{bc,Q2} = 20\,\text{fF}$, $C_{cs,I1} = 5\,\text{fF}$ and $|A| = 1.4$ the resulting output capacitances are $C_{po,off} = 9\,\text{fF}$ and $C_{po,on} = 24\,\text{fF}$. In Fig. 4.8, the output capacitance obtained in a Spectre circuit simulation is shown for both states. The simulation results are in reasonable agreement (within 20% accurate) with the calculated values. Inaccuracies are mainly due to small differences in the dc operating points of the circuit and the ignoring of $C_{be,Q2}$ and $R_{b,Q2}$ in the calculations.

The transmission line of the column is loaded by the distributed output capacitance of the matrix node circuits. In total, 19 circuits (in the 'off' state) provide a load capacitance of $C_{po,off}$ each; one circuit (in the 'on' state) provides a load capacitance $C_{po,on}$. Using equation (4.4), the effective characteristic impedance of the column transmission line can be calculated. If the total load per section were to be only the capacitive load from the matrix node circuit in the 'off' state, and assuming a line length of 100 μm with lumped capacitance $C_{dm,sec} = 6\,\text{fF}$, the effective impedance would decrease to $Z_{0dm,eff} = 63\ \Omega$. The section with the active matrix node circuit introduces a mismatch due to the somewhat higher capacitive load; this mismatch results in a negligible reflection (as will be shown below), from analysis of the signal distribution of the longest path through the matrix.

The following analysis is used to estimate the input capacitance C_{pi} of the matrix node circuit in more detail.

The input emitter followers Q_{3a}, Q_{3b} are capacitively loaded by the differential pair Q_{4a}, Q_{4b} (see Fig. 4.6). The capacitive term in the emitter current of the emitter followers results in a negative real part of the input impedance, due to the phase shift in the current gain β of Q_3. In general, the input impedance of a capacitively loaded

Fig. 4.9 Analysis of the input impedance of a capacitively loaded emitter follower

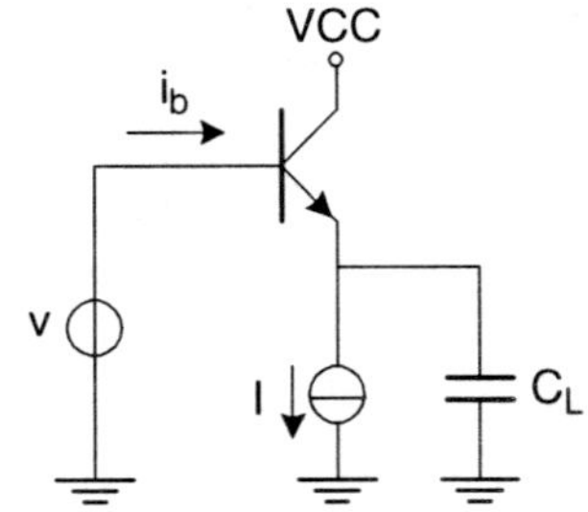

emitter follower can (in a certain frequency range, as will be explained below) be mapped onto a parallel network of a capacitor C_p plus a frequency-dependent negative resistor R_p. This will be explained using the schematic of Fig. 4.9.

The following relation holds for the input admittance Y_i:

$$Y_i = \frac{i_b}{v} = \frac{i_e}{v(\beta+1)} \approx \frac{j\omega C_L}{1+\frac{\beta_0}{1+j\beta_0\omega/\omega_T}}$$
$$= \frac{j\omega C_L}{(\beta_0+1)^2+\left(\frac{\beta_0\omega}{\omega_T}\right)^2}\left(\beta_0+1+\left(\frac{\beta_0\omega}{\omega_T}\right)^2+j\beta_0^2\frac{\omega}{\omega_T}\right) \tag{4.9}$$

For $\omega < \omega_T$, the resulting input admittance can be approximated by

$$Y_i \approx -\frac{\omega^2 C_L}{\omega_T} + j\frac{\omega C_L}{\beta_0+1} \tag{4.10}$$

The real part $Re(Y_i)$ represents the frequency-dependent parallel equivalent input resistance $R_p \approx -\omega_T/\omega^2 C_L$; the imaginary part $Im(Y_i)$ represents the parallel equivalent input capacitance $C_p \approx C_L/\beta_0$. From this analysis it follows that the parallel equivalent input resistance R_p is negative with a frequency dependence of −40 dB/decade. The input capacitance of the emitter follower is approximately equal to C_{bc}, provided that $C_L/\beta_0 \ll C_{bc}$ and provided that the output resistance of the current source is sufficiently high, so that the low-frequency voltage gain of the emitter follower equals unity. If the voltage gain is less than unity, part of the base-emitter capacitance will add to the input capacitance.

The simulated differential input resistance R_{pi} of the matrix node circuit of Fig. 4.6 is shown in Fig. 4.10. The single-ended load capacitance C_L is the input capacitance of the differential pair Q_4 in Fig. 4.6 and equals $C_L = 66\,fF$ (e.g., a factor of 3 lower than the input capacitance of Q_2 as calculated for Fig. 4.5). For example, the absolute value of R_{pi} at $f = 1\,GHz$ can be verified using equation (4.10): $R_{pi} = -2\omega_T/\omega^2 C_L$; using $f_T = 75\,GHz$ and $C_L = 66\,fF$ this gives $|R_{pi}| = 362\,k\Omega$ (versus 460 kΩ in the simulation). The slope of −40 dB/decade is found for frequencies between $f_L < f < f_H$.

At $f < f_L = f_T/\beta_0$, the current gain of the input emitter follower provides less than 45° phase shift, and R_{pi} is consequently positive.

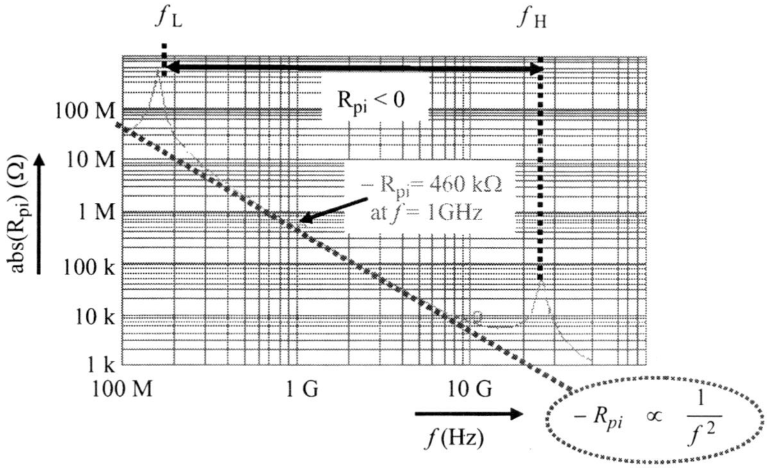

Fig. 4.10 Simulated parallel equivalent input resistance for the matrix node circuit

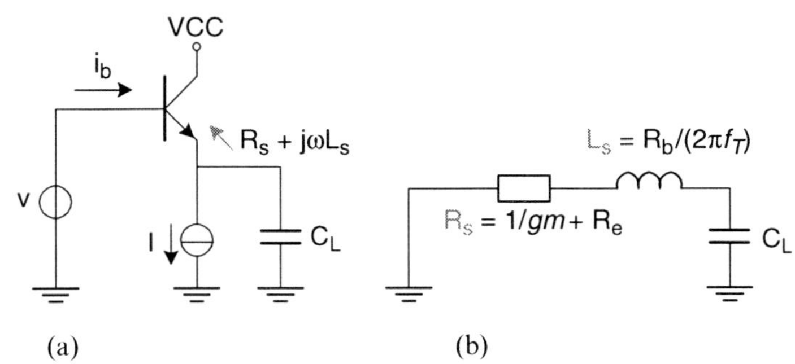

Fig. 4.11 Capacitively loaded emitter follower (a) and equivalent series-resonance circuit at the output (b). Together with the load capacitance, the output impedance of the emitter follower constitutes a series resonance circuit with resonance frequency f_H. At $f > f_H$, the emitter follower is not capacitively loaded

At $f = f_H$, a series resonance from the input capacitance of differential pair Q_{4a}, Q_{4b} plus output inductance of emitter followers Q_{3a}, Q_{3b} occurs, as illustrated in Fig. 4.11.

With (single-ended values) $C_L = 66\,\text{fF}$ and $L_s = R_b/2\pi f_T = 0.53\,\text{nH}$ (since for Q_3, $R_b = 250\ \Omega$ and $f_T = 75\,\text{GHz}$), the calculated resonant frequency is $f_H = 1/2\pi\sqrt{(L_s \cdot C_L)} = 27\,\text{GHz}$. At $f > f_H$, $\omega L_s > 1/\omega C_L$ and consequently the emitter follower is no longer capacitively loaded. This explains why the input resistance becomes positive at $f > f_H$.

In the frequency range of interest for this design (up to 20 GHz), the input impedance of the active matrix node circuit behaves as a parallel network of $C_{pi,on} = 5\,\text{fF}$ and $|R_{pi}| > 6\,\text{k}\Omega$, R_{pi} negative. Since $|R_{pi}| \gg Z_{0dm}$, the signal amplitude on the transmission line is almost independent of the matrix configuration, even in multicast and broadcast modes. The input impedance of an inactive matrix node circuit can be represented by a single differential capacitance with value $C_{pi,off} = C_{bc}/2 = 2\,fF$. Note that the interconnect parasitic capacitance between the input transistors and GSSG transmission line adds to the input capacitance of the matrix node circuit C_{pi}.

4.2.4 Cross-Connect Switch IC Floorplan

Since the cross-connect switch IC is bondpad-limited, all four sides of the IC are used to distribute the bondpads across the perimeter. To simplify the RF signal distribution, the floorplan is designed so that the inputs are divided between two opposite sides, and the outputs are divided between the two sides orthogonally to the inputs (see Fig. 4.12).

Inside the matrix, signals need to be distributed from the input buffers to the matrix node circuits (in rows) and from the matrix node circuits to the output buffers (in columns). To minimise the number of logic control signals inside the matrix, each matrix node circuit has a dedicated logic decoding circuit. Implementing the rows and columns in pairs ensures that one wire is saved per transmission line pair (by merging two GSSG lines into a single GSSGSSG line), resulting in a size reduction of 10 µm per pair, corresponding to an overall 90 µm size reduction for the total matrix width and total height.

The floorplan is shown in detail in Fig. 4.13.

Multiple matrix node circuits are grouped together, sharing supply (column wise) and ground (row wise) paths.

During the development of the cross-connect switch IC, a test-IC was also developed enabling evaluation of the signal transfer across a single Metal6 GSSG transmission line (row) inside the matrix [2]. The test-IC allows verification of the concept of distributed capacitive loading by means of on-wafer measurements.

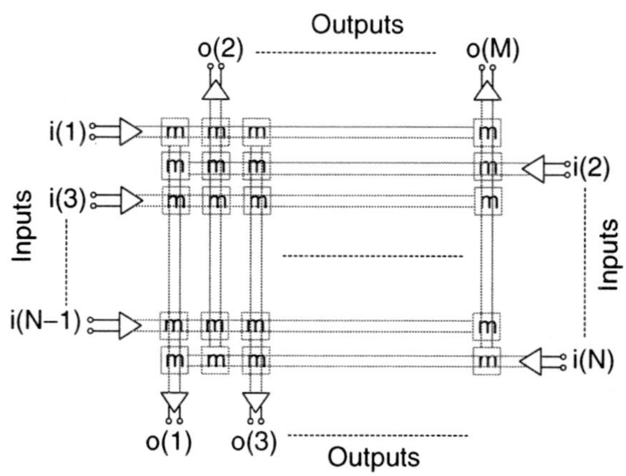

Fig. 4.12 Matrix with RF inputs and outputs equally distributed across all four sides of the IC

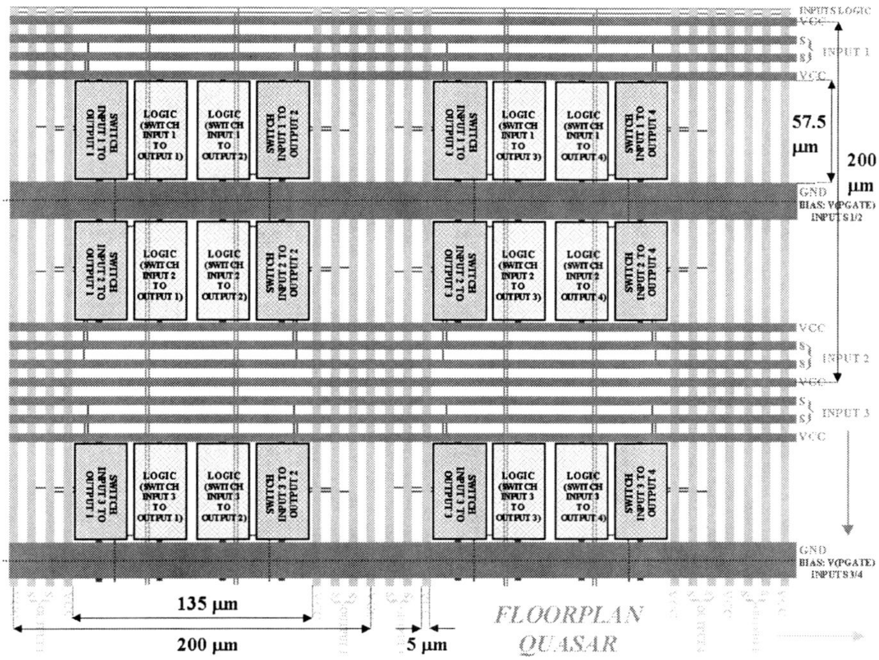

Fig. 4.13 Floorplan detail of the matrix, zoomed-in to a 3-input, 4-output section

In total, 20 matrix node circuits according to the design shown in Fig. 4.6 are distributed across the 2-mm-long transmission line. The block diagram of the test-IC is shown in Fig. 4.14. One row (horizontal) is crossed by in total 20 columns (vertical). Except for the outer 2 columns, the columns cross the row in pairs. By implementing the columns in pairs, one wire can be saved per transmission line pair (merging two GSSG lines into a single GSSGSSG line), resulting in a width reduction of 10 μm for each column pair. The output signals of the columns cannot be monitored; only the signal transfer across the row is studied. The output signal of each matrix node circuit is dumped into load resistors at each end of the column transmission line. A 2-bit wide programming bus (p2, p1) controls the bias state of the matrix node circuits between all the circuits in the 'off' state, one circuit in the 'on' state, half of the circuits in the 'on' state or all the circuits in the 'on' state (broadcast mode).

To help find the effective differential-mode characteristic impedance, an overview of all load capacitances per two sections is presented in Table 4.2. The values are based on calculations and/or simulations. Note that only the matrix input capacitance depends on the state of the matrix, all other contributions are passive.

Due to the distributed capacitive loading, the effective line impedance has decreased from $Z_{0dm} = 100\ \Omega$ (unloaded) to $Z_{0dm,eff} = 52\ \Omega$ with one circuit active or $Z_{0dm,eff} = 48\ \Omega$ with all circuits active.

A photomicrograph of the test-IC with connected wafer probes is shown in Fig. 4.15.

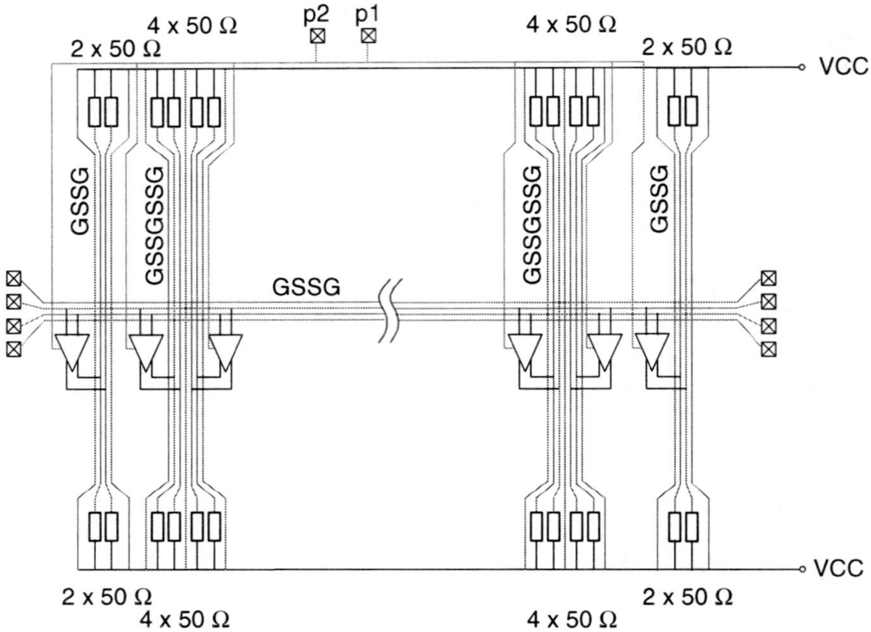

Fig. 4.14 Block diagram of the test-IC for studying the signal transfer using the concept of distributed capacitive loading. Probe pads are indicated by symbols ⊠. GSSG RF probe pads are used to evaluate the signal transfer across the GSSG transmission line (row); dc probe pads p1 and p2 control the state of the matrix node circuits at the cross-points

Table 4.2 Differential-mode capacitances per two sections

	Active/passive	C ('off' state)	C ('on' state)
2 matrix circuits	active	4 fF	10 fF
200 μm unloaded line section (GSSG row)	passive	12 fF	12 fF
Interconnect GSSG row to 2 matrix circuits	passive	4 fF	4 fF
Interconnect inside matrix circuits	passive	2 fF	2 fF
Crossing 4 control lines	passive	1 fF	1 fF
Crossing GSSGSSG column	passive	6.3 fF	6.3 fF
Crossing supply lines	passive	4 fF	4 fF
Total		**33.3 fF**	**39.3 fF**

In addition to the Metal6 transmission line with distributed matrix node circuits, an unloaded transmission line has also been placed on the same wafer for reference. Also, an unloaded transmission line implemented in Metal5 has been included. The Metal5 transmission line is needed for the columns of the matrix. On-wafer evaluation of the transmission lines is performed using the procedure described in Chapter 2. The evaluation results are summarised in Table 4.3. The second and third

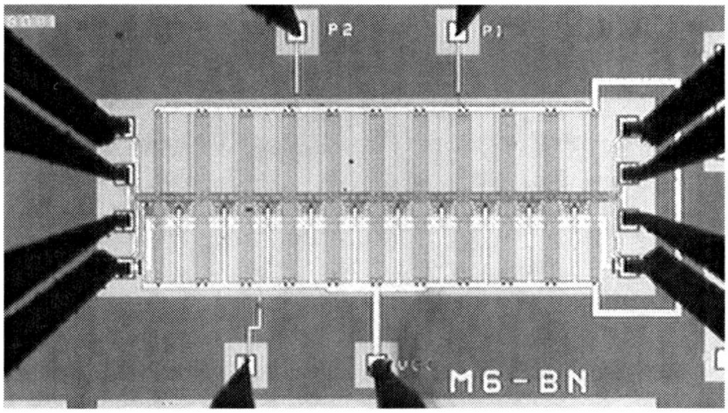

Fig. 4.15 Chip photomicrograph of the test-IC, studying the effect of distributed capacitive loading on the signal transfer across the 2-mm-long transmission line. In total, 20 (dummy) matrix node circuits are connected to the transmission line

Table 4.3 Characteristic impedance and delay derived from data measured for transmission lines with and without matrix node circuits

	Z_{0dm} (Ω)	t_{dm} (ps/mm)	Z_{0cm} (Ω)	t_{cm} (ps/mm)
Unloaded, Metal6	90	5.7	45	6.7
Loaded; inactive	45	7.6	20	12.6
Loaded; all active	40	8.1	20	12.6
Unloaded, Metal5	80	6.7	35	8.1

rows refer to the block diagram shown in Fig. 4.14, with matrix node circuits as shown in Fig. 4.6.

The unloaded Metal6 line, designed for a differential-mode characteristic impedance of $Z_{0dm} = 100\ \Omega$ and with an expected delay of $t_{dm} = 6\,\text{ps/mm}$, has a measured characteristic impedance $Z_{0dm} = 90\ \Omega$ and delay $t_{dm} = 5.7\,\text{ps/mm}$. These results are considered to be reasonably accurate, given the simplified model that was used for the simulations. In order to establish a measured characteristic impedance of 100 Ω, the line geometry may be adjusted on the basis of the evaluation results. The Metal5 transmission line has a somewhat lower characteristic impedance because the line is closer to the ground shield and further away from the air above the IC. Since Metal5 is further away from air, the effective dielectric permittivity $\varepsilon_{r,eff}$ is larger for Metal5, resulting in a greater delay for the unloaded line than the unloaded Metal6 line.

Of most importance for the matrix design are the second and third rows in Table 4.3. Due to the distributed capacitive loading, the measured line impedance has decreased from $Z_{0dm} = 90\ \Omega$ (unloaded) to $Z_{0dm,eff} = 40\ \Omega$ with all circuits active. The line impedance reduction realised in the simulation is comparable (from 100 Ω to 48 Ω).

4.3 Buffer Circuits

Buffers are used at several locations on the IC. Section 4.3.1 describes the intermediate buffers used in the switch matrix, Section 4.3.2 describes the input and output buffers.

4.3.1 Intermediate Buffer Circuits

In designing the signal path of the complete cross-connect IC, sufficient margin should be available to cope with inaccurate line and parasitic models. Robust signal transfer inside the matrix can be achieved by reducing the length of the transmission lines for both rows and columns. The line lengths may be halved by introducing intermediate signal buffers. As can be seen in Fig. 4.16, in the case of an unloaded 2-mm-long line, the highest sensitivity to wrong termination of the line occurs at a 20 GHz signal. This is because at 20 GHz, the 2 mm line length corresponds to $\lambda/4$, or $f_{\lambda/4} = 20\,\text{GHz}$.

In the case of a transmission line with a distributed capacitive load (as in the matrix), the line delay will increase and the frequency $f_{\lambda/4}$ will decrease correspondingly. The data spectrum of 12.5 Gb/s signals extends to roughly 12.5 GHz, and the sensitivity to incorrect line termination must hence be low up to 12.5 GHz. The introduction of intermediate buffers halving the transmission line length will double $f_{\lambda/4}$ and hence introduce sufficient margin to cope with modelling inaccuracies plus processing and temperature variations.

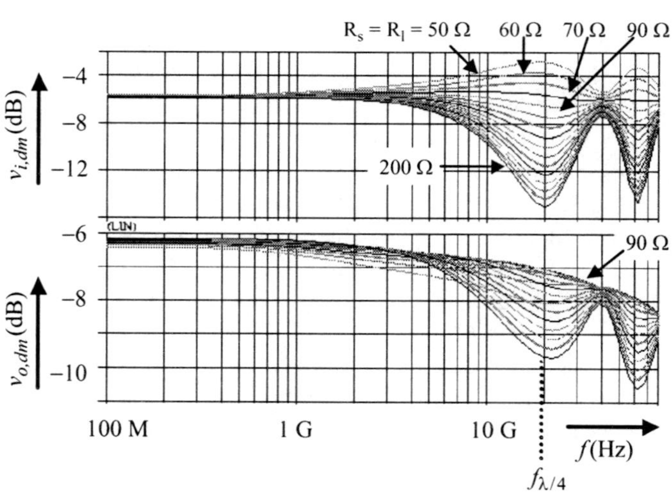

Fig. 4.16 Fasterix simulation result obtained for a 2-mm-long Metal6 GSSG transmission line on a Metal1 ground shield, differential mode, for source and load impedance values $R_s = R_l$ from 50 Ω to 200 Ω in 10 Ω increments. The top graph shows the signal at the input of the line, the bottom graph shows the signal at the output of the line

Fig. 4.17 Intermediate buffer circuit

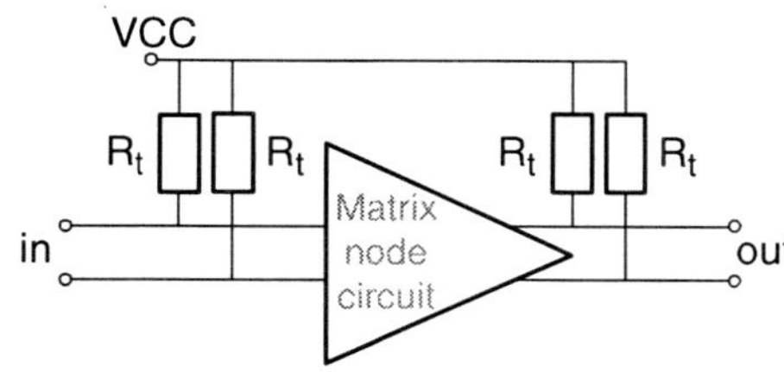

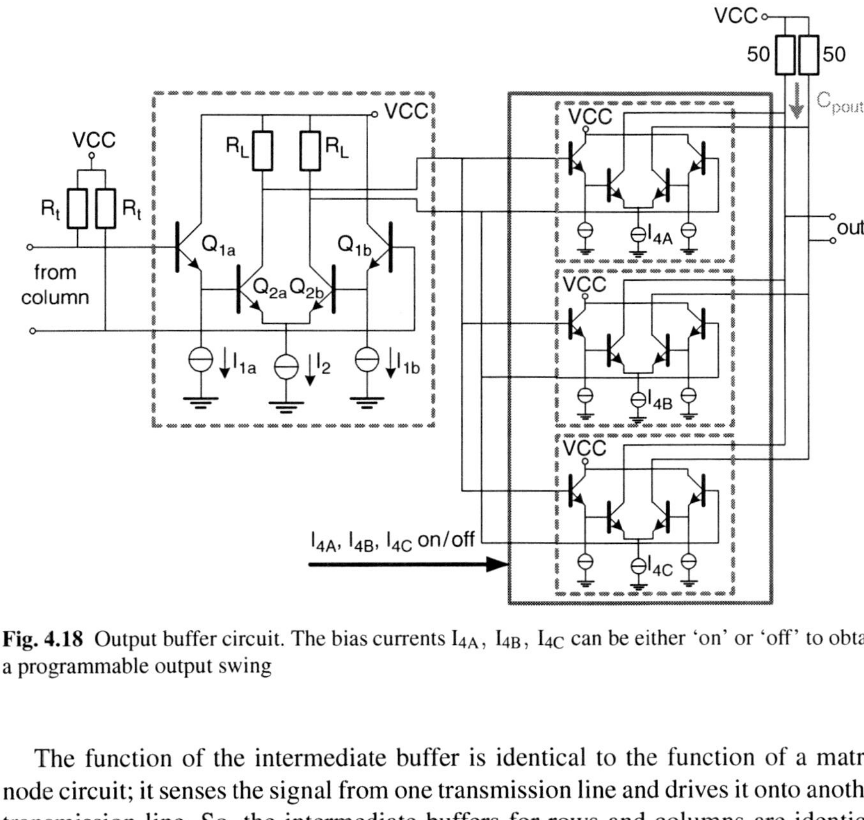

Fig. 4.18 Output buffer circuit. The bias currents I_{4A}, I_{4B}, I_{4C} can be either 'on' or 'off' to obtain a programmable output swing

The function of the intermediate buffer is identical to the function of a matrix node circuit; it senses the signal from one transmission line and drives it onto another transmission line. So, the intermediate buffers for rows and columns are identical and are built from a matrix node circuit surrounded by four line termination resistors $R_t = Z_{0dm,eff}/2$ (see Fig. 4.17).

4.3.2 Input and Output Buffer Circuits

The output signal swing of the IC needs to be programmable up to $0.6\,V_{pp,diff}$. Since the signal swing throughout the matrix is $0.2\,V_{pp,diff}$, the bias current required for the output differential pair must be three times higher than that of the matrix node circuit. This is realised by connecting three differential pairs in parallel. The output buffer circuit is based on emitter followers and differential pair circuits, as shown in Fig. 4.18.

Each differential pair is identical to the matrix node circuit output differential pair (Q_{2a}, Q_{2b} in Fig. 4.6) and biased at the same current. As in the matrix node circuits, the differential input capacitance of the output buffer with n differential pairs in the 'on' state, $n \in [1..3]$, is $n \cdot 15\,fF$. With a single-ended load impedance $R_L = 25\,\Omega$, the input differential pair (Q_{2a}, Q_{2b}) can drive the three parallel output differential pairs with sufficient bandwidth (>20GHz), even at the maximum output signal swing. The output buffer output capacitance C_{pout} depends on the state of the three output differential pairs, and equals the sum $C_{pout} = n \cdot C_{po,on} + (3-n) \cdot C_{po,off}$ with $C_{po,on} \approx 30\,fF$ and $C_{po,off} \approx 10\,fF$ (see Fig. 4.8). The buffer output is terminated twice: by the on-chip 50 Ω load resistors in parallel to the $Z_{0dm} = 100\,\Omega$ transmission line towards the output bondpads. Hence, the differential-mode load resistance for the output buffer is $100\,\Omega // 100\,\Omega$ or $50\,\Omega$. The worst-case output bandwidth is $1/(2\pi \cdot 50 \cdot 90f) = 36\,GHz$. This leaves sufficient margin for adding electro-static discharge (ESD) protection circuitry and bondpads at the buffer output nodes.

Additional circuitry has been added in front of the buffer to allow output signal polarity programming and to add an output to the internal bit-error rate test circuit, needed to support built-in self-testing. These circuits have a minor impact on the signal integrity.

The input buffer is almost identical to the intermediate buffer circuit. To protect the input transistors against excessive reverse base-emitter junction voltage, anti-parallel diodes are added to the base-emitter junctions of the first differential pair. In addition, ESD protection diodes are added between the input bondpads and supply plus ground nodes. Additional circuitry has been added to allow input signal polarity programming and to implement an additional input for the internally generated pseudo-random data signal, needed to support built-in self-testing. These circuits have a minor impact on the signal integrity.

4.4 Complete RF Signal Path

A detailed block diagram of the cross-connect switch IC is shown in Fig. 4.19. The signal path, although shown single-ended, is fully differential. The (20×20) matrix has been split into four identical (10×10) sub-matrices. Intermediate buffers have been inserted between the (10×10) sub-matrices.

Spectre simulation results obtained for the shortest and longest signal paths through the matrix will be presented below. The simulation results are based on full circuit simulations including transmission line models and extracted layout parasitic capacitances. MEXTRAM transistor models were used.

4.4.1 Small-Signal Simulations

Small-signal simulations were performed for the longest and shortest signal paths in the matrix. There were four paths of maximum length through the matrix (for

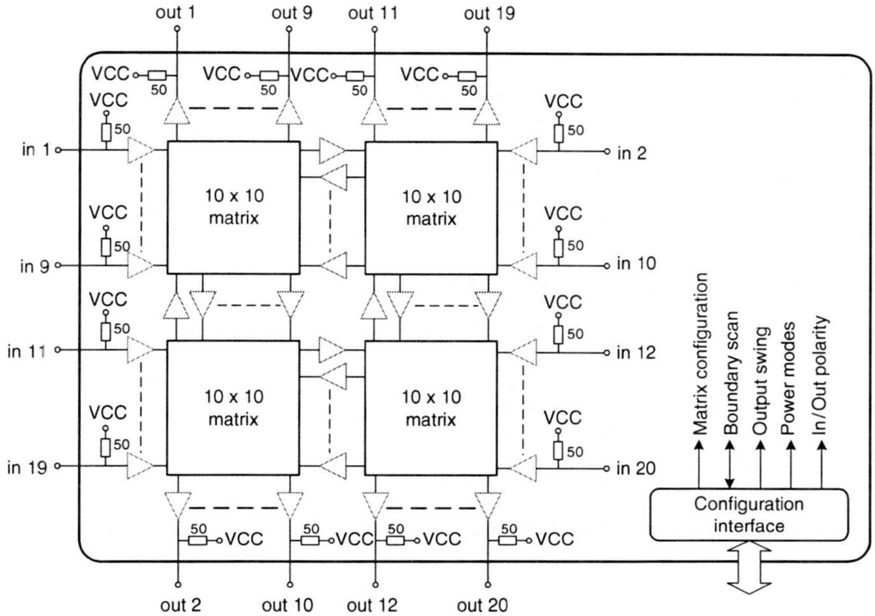

Fig. 4.19 Block diagram of the cross-connect switch IC in more detail

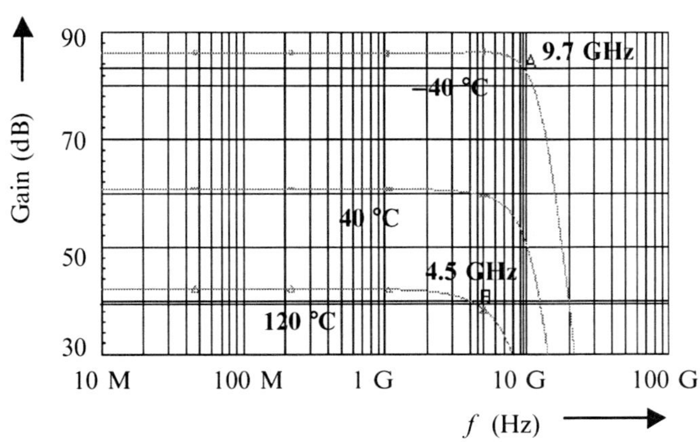

Fig. 4.20 Small-signal simulation results obtained for the longest signal path at 2.3 V supply at minimum, nominal and maximum junction temperatures

example between input 1 and output 20 in Fig. 4.19). In the simulation of the longest path, the intermediate buffers were part of the signal path. No intermediate buffers were included in the simulation of the shortest path (for example between input 19 and output 2).

Figure 4.20 shows the ac simulation results obtained for the longest signal path at three different temperatures and a 2.3 V supply voltage (worst-case) in nominal processing. The minimum bandwidth for the longest path in the matrix was 4.5 GHz at 120°C.

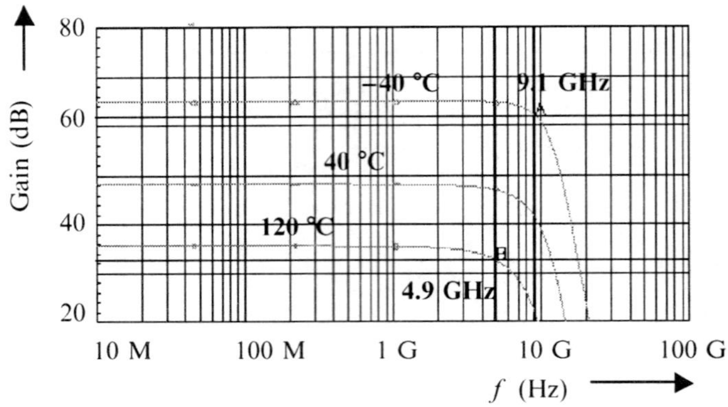

Fig. 4.21 Small-signal simulation results obtained for the shortest signal path at 2.3 V supply at minimum, nominal and maximum junction temperatures

Figure 4.21 shows the results obtained for the shortest signal path in the matrix. The minimum bandwidth for the shortest path in the matrix was 4.9 GHz at 120 °C.

The gain realised for the longest signal path is higher than that realised for the shortest path due to the two additional intermediate buffers. Although the worst-case bandwidth in the case of both the shortest and longest paths seems insufficient to support 12.5 Gb/s operation, in large-signal simulation this proved not to be the case. This is because the small-signal simulation results are only valid when operating at a very low input signal amplitude, well below the required sensitivity level. In practice, the circuits will be operated at large-signal amplitudes. At large-signal amplitudes, the average input capacitance of a differential pair will be smaller due to the non-linearity of the base-emitter capacitance C_{be}. At non-zero differential input signal amplitudes, one base-emitter junction will be forward biased (increasing its C_{be}) while the other base-emitter junction will be reverse biased (reducing its C_{be}). Since the two base-emitter junctions are connected in series, the differential input capacitance will reach its maximum of $C_{be}/2$ for zero input signal amplitude, and will be smaller than $C_{be}/2$ at non-zero input signals. In the small-signal simulations, however, the value of C_{be} was independent of the signal amplitude, and the small-signal simulation results may hence be regarded as too pessimistic for large-signal operation.

It is interesting to verify that the signal path has insufficient bandwidth for small-signal operation. This can be done by applying a low input signal amplitude, well below the specified sensitivity level. Figure 4.22 shows transient results (eye diagrams) obtained for the longest signal path at VCC = 2.3 V, 120 °C and $0.2\,mV_{p,diff}$ input signal amplitude. A pseudo-random binary sequence signal was applied to the input. A bit-rate of 6.5 Gb/s revealed a near-perfect eye diagram, while the eye diagram showed poor jitter performance at higher bit-rates (9.5 and 12.5 Gb/s). A rule of thumb for small-signal analysis is that the overall bandwidth must be at least 70% of the bit-rate. From the 4.5 GHz worst-case bandwidth found in ac simulations (Fig. 4.20) it follows that the bit-rate must remain below approximately 6.4 Gb/s.

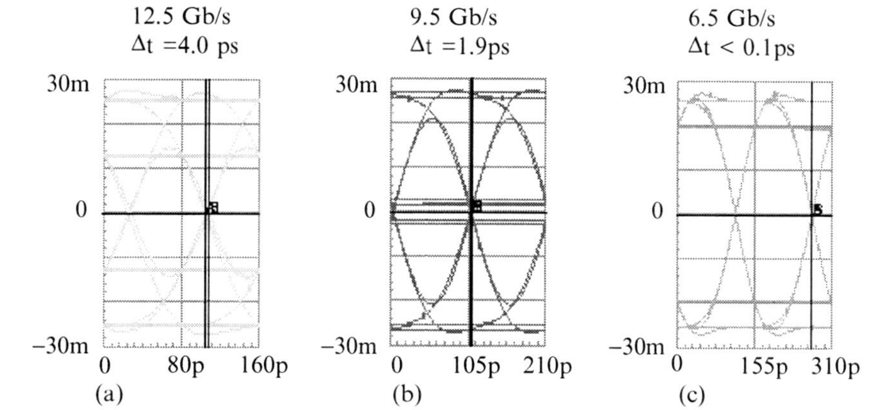

Fig. 4.22 Transient simulation results (eye diagrams) obtained for the longest path at VCC = 2.3 V, 120°C, at an input signal amplitude of $0.2\,mV_{p,diff}$ and three different bit-rates: 12.5 Gb/s (a), 9.5 Gb/s (b) and 6.5 Gb/s (c). Δt represents the peak-peak jitter at the zero-crossings

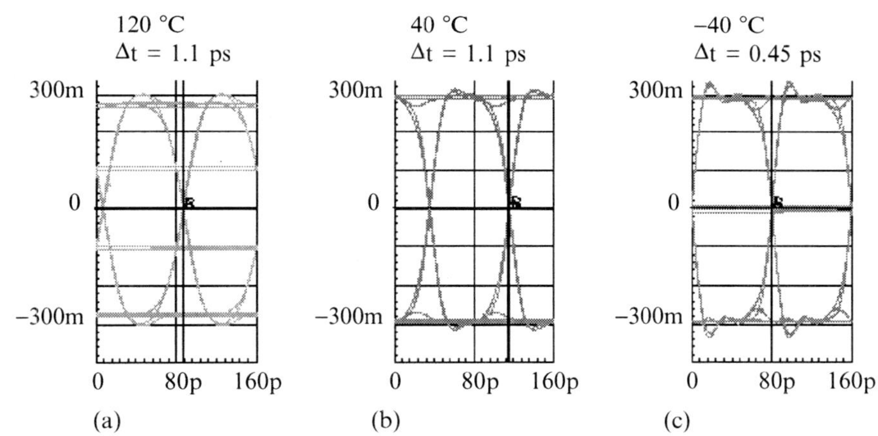

Fig. 4.23 Transient simulation results (eye diagrams) obtained for the longest path at 2.3 V supply and 12.5 Gb/s at three different temperatures: 120°C (a), 40°C (b) and −40°C (c)

At input signal amplitudes larger than $100\,mV_{pp,diff}$, the overall bandwidths of 4.5 GHz (longest path) and 4.9 GHz (shortest path) are not applicable. The maximum speed of the circuits for large-signal operation will depend more on the slew rate, determined by currents charging and discharging capacitive loads.

4.4.2 Large-Signal Simulations

Large-signal simulations were also performed for the longest and shortest signal paths in the matrix. Figure 4.23 shows the transient results (eye diagrams) obtained

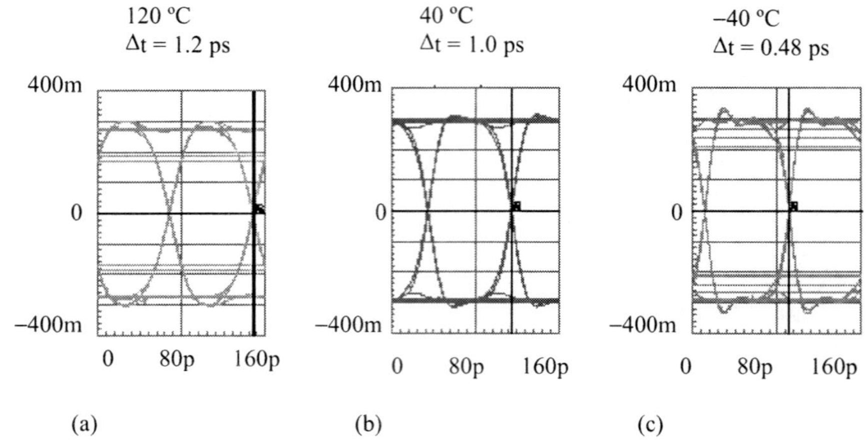

Fig. 4.24 Transient simulation results (eye diagrams) obtained for the shortest path at 2.3 V supply and 12.5 Gb/s at three different temperatures: 120 °C (a), 40 °C (b) and −40 °C (c)

for the longest path at three different temperatures and a 2.3 V supply voltage, with the output buffer programmed to the maximum swing of 0.6 $V_{pp,diff}$, corresponding to worst-case operating conditions. The highest jitter measured for the longest path in the matrix was 1.1 ps at 40 °C and 120 °C.

Figure 4.24 shows the results obtained for the shortest path in the matrix. The highest jitter measured for the shortest path in the matrix was 1.2 ps at 120 °C.

The transient results show good performance, despite the marginal overall small-signal bandwidth.

4.5 Supply Decoupling

The IC has dedicated supply and ground pins per RF input and RF output. Additional supply domains are used for the digital circuits and the built-in self-test circuits. The supply decoupling is distributed across the IC. Small (1 pF) decoupling capacitors are included per matrix node circuit. Close to the input and output buffer circuits, larger decoupling networks (with 100 pF total capacitance per buffer) are used. It is important to have a low-ohmic supply network close to the input and output signal line termination resistors, because the supply network is part of the common-mode termination impedance. The low-ohmic supply decoupling network avoids reflections of common-mode input signals.

To analyse the effectiveness of the supply decoupling network, Spectre circuit simulations were performed of the longest RF signal path. In the simulations, transmission line models were used for the supply and ground paths. Part of the supply line model used for the simulation of the RF path, including the supply line network to the input buffer and matrix, is shown in Fig. 4.25.

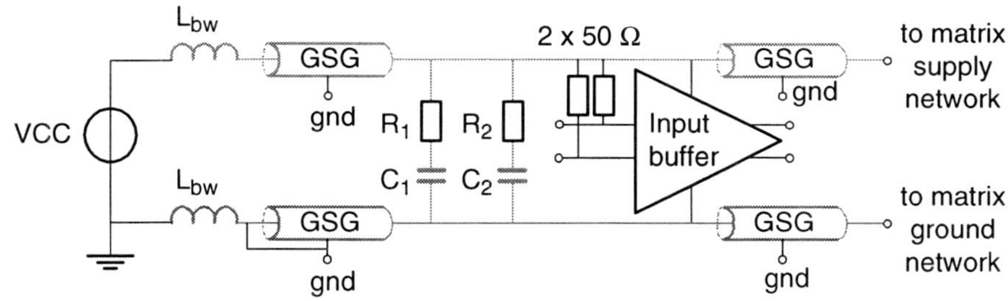

Fig. 4.25 Supply and ground path network, including supply decoupling, of an input buffer

A single-ended GSG transmission line model was used for the supply and ground paths. The series resistance of the supply lines was included in the transmission line model in order to obtain an accurate estimate of the supply and ground line voltage drops. An inductance $L_{bw} = 1\,nH$ modelled each bondwire. It is important to include such a bondwire inductance because the supply network inductance forms a high-Q resonant circuit in combination with the on-chip supply decoupling capacitors (C_1 and C_2 in Fig. 4.25). Some damping of the resonance occurred due to the series resistance of the supply lines. If, however, a resonance of the supply signal is found, the Q-factor of the resonant circuit may effectively be reduced by inserting a resistor in series with the supply decoupling capacitor (R_1 and R_2 in Fig. 4.25). Seen from the input buffer circuit side, a parallel resonant circuit is formed by the two bondwires with the decoupling capacitor. Since a parallel resonant circuit behaves as an open circuit at its resonance frequency, large-signal amplitudes may occur at the supply lines at the resonant frequency. This makes it important to reduce the quality factor of the supply network at the self-resonant frequency.

The supply decoupling network was implemented as two parallel branches $R_1 - C_1$ and $R_2 - C_2$, with $C_2 = C_1/10$. The resonant frequency f_{r2} of the supply network with C_2 was therefore at $f_{r2} = \sqrt{10} \cdot f_{r1}$. The damping resistor R_2 was scaled in accordance with the increased resonant frequency: if $C_i^{+1} = C_i/a$, then $R_i^{+1} = R_i \cdot \sqrt{(a)}$. A distributed decoupling network is more effective than a single lumped decoupling capacitor because it reduces resonance in the supply. If the supply decoupling is equally distributed along the supply line, the supply network may be considered a low-ohmic transmission line.

A 1 pF supply decoupling capacitor is included in every matrix node circuit. Again a series resistor is used to avoid ringing of the local supply voltage.

Simulation results showing the effect of the supply decoupling are shown in Fig. 4.26. To find possible supply line ringing, the circuits were switched from sleep mode to active mode at $t = 1\,ns$; at $t = 15\,ns$ the output signal amplitude was reprogrammed. This caused increments in the current consumption of the circuits, thereby stimulating potential instabilities. As can be seen, ringing occurred at a frequency $f_0 \approx 300\,MHz$ at $R_s = 1\ \Omega$. This is the result of the 2nH bondwire inductance (supply plus ground) plus on-chip supply network inductance and 100 pF decoupling capacitor. The self-resonant frequency at $L = 2\,nH$ and $C = 100\,pF$ was

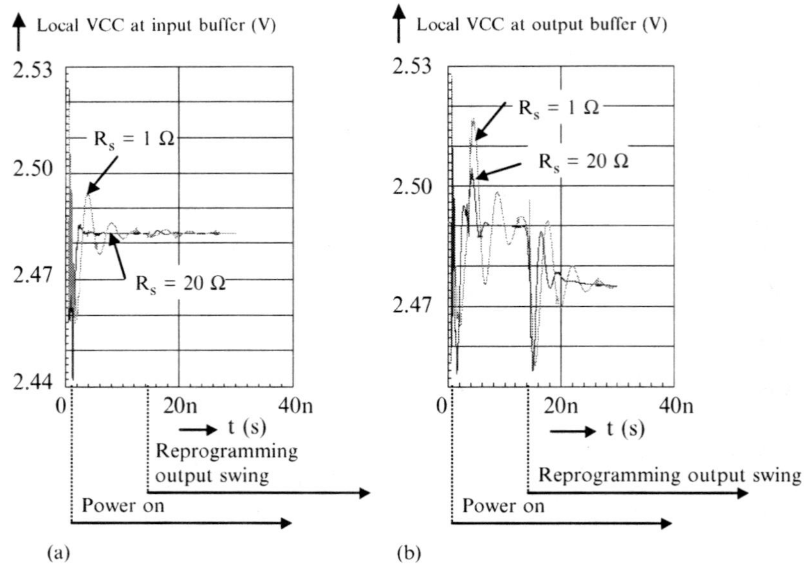

Fig. 4.26 Local supply voltage, at the input buffer (a) and output buffer (b) using an on-chip decoupling capacitor of 100 pF. A series resistor R_s was inserted in series with the decoupling capacitors of 1 Ω and 20 Ω, respectively. At t = 15 ns, the swing of the output buffer was reprogrammed, causing an increase in the supply current of the output buffer

$f_0 = 1/(2\pi\sqrt{(LC)}) = 355\,\text{MHz}$, while the quality factor of the supply network at the self-resonant frequency (ignoring the load impedance from the circuits connected to the supply) was $Q_0 = 1/(2\pi f_0 \cdot R_s \cdot C_s) = 4.5$. Increasing the value of the series resistance effectively prevents ringing, as shown in Fig. 4.26 by the $R_s = 20\,\Omega$ curves. Note that the quality factor dropped to $Q_0 \ll 1$ in this case; lower resistance values are feasible while still avoiding ringing.

To avoid ringing and at the same time provide decoupling up to the highest possible frequency, a value of $Q_0 = 0.7$ should be used. In the example supply decoupling network this can be realised by using $R_s = 3.7\,\Omega$.

More severe noise on the supply lines can occur when several cross-point circuits are reprogrammed simultaneously. This has however no impact on the potential ringing of the supply lines. To study ringing of the supply lines, it is sufficient to study the supply network, e.g., as shown in Fig. 4.25, and optimise the decoupling network as suggested in this section.

The value for the bondwire inductance $L_{bw} = 1\,\text{nH}$ may be rather pessimistic in practice. To optimise the on-chip supply decoupling, the analyses for the ringing from this section may be repeated when more accurate off-chip supply network models are available.

4.6 Experimental Results

The chip photomicrograph of the entire cross-connect switch IC is shown in Fig. 4.27. The locations of the large circuit blocks are indicated in the photo. The on-chip transmission lines are clearly visible, both outside and inside the matrix. Five bondpads were used per differential input or output signal: four for the GSSG transmission line plus one for a dedicated power supply line. The GSSG transmission line continued between the IC and the ball grid array (BGA) package via the bondwires. The IC measured $6 \times 6\,\mathrm{mm}^2$.

A photo of the IC mounted in its HBGA475 package is shown in Fig. 4.28. The $35\,\mathrm{mm} \times 35\,\mathrm{mm}$ BGA package is dedicated for this cross-connect IC. The differential transmission lines for the 20 RF inputs plus 20 RF outputs on the package are clearly visible, as are the wirebonds.

An evaluation board has been developed in which only a sub-set of the 20 inputs and 20 outputs are made accessible at connectors. The inaccessible inputs and outputs are terminated into dummy 50 Ω resistors on the board. The paths that can be evaluated include the shortest and longest paths through the matrix. A photograph of the evaluation board connected to a Tektronix communication analyser, with the IC operating at 12.5 Gb/s, is shown in Fig. 4.29.

The IC has been evaluated using an externally applied PRBS input signal up to 14.3 Gb/s, the highest data rate supported by the Advantest PRBS generator. The eye-diagram obtained for the longest path, measured single-ended, is shown in Fig. 4.30.

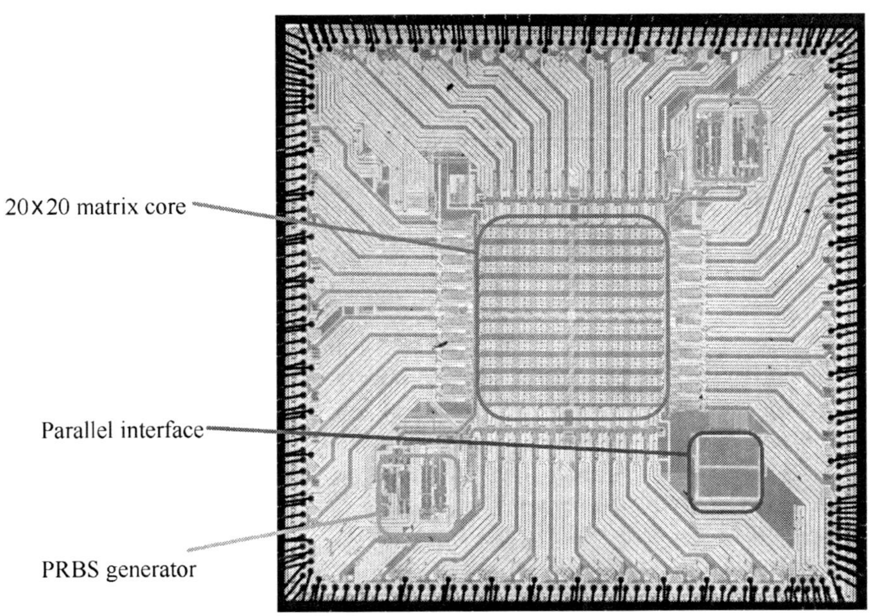

Fig. 4.27 Chip photomicrograph of the cross-connect switch IC

Fig. 4.28 Cross-connect switch IC mounted in its BGA package

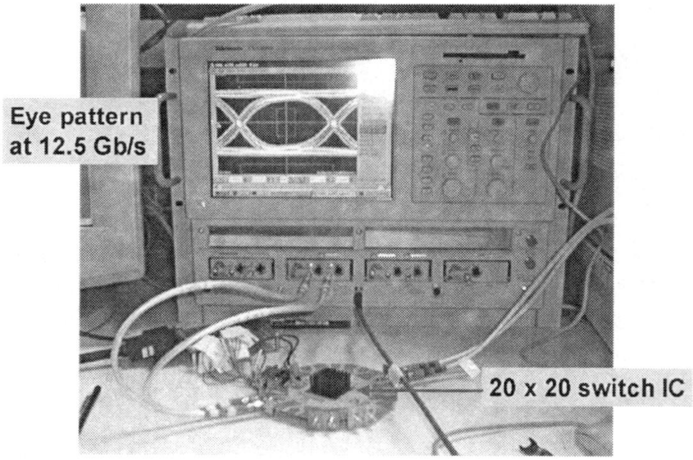

Fig. 4.29 Cross-connect switch IC under test. A pseudo-random input signal is applied to the IC at 12.5 Gb/s. The output eye diagram is shown on the communication analyser

The measured jitter at 14.3 Gb/s was 2.3 ps RMS. At the specified maximum bit-rate of 12.5 Gb/s, the RMS output jitter remained below 2 ps. The measured performance of the IC meets the specifications given in Table 4.1.

The crosstalk inside the switch has been analysed in the following way. A sinusoidal input signal of nominal amplitude is applied to the input of the longest path, and the output signals of the neighbouring channels are analysed and compared with the desired output signal using a spectrum analyser. The path with input signal is from input 1 to output 0 (see Fig. 4.31).

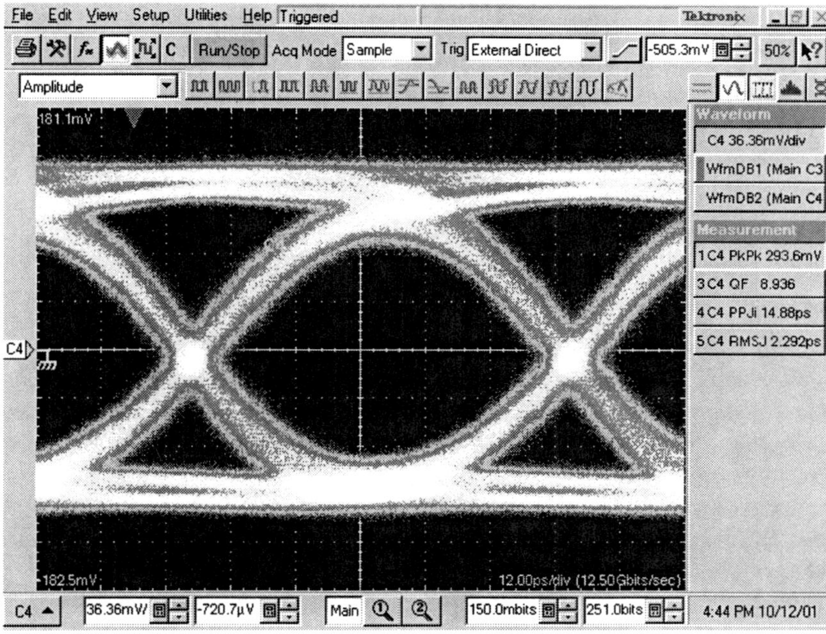

Fig. 4.30 Typical output eye diagram obtained at 14.3 Gb/s at an input swing of 0.3 $V_{pp,diff}$. The output swing was programmed to 0.4 $V_{pp,diff}$

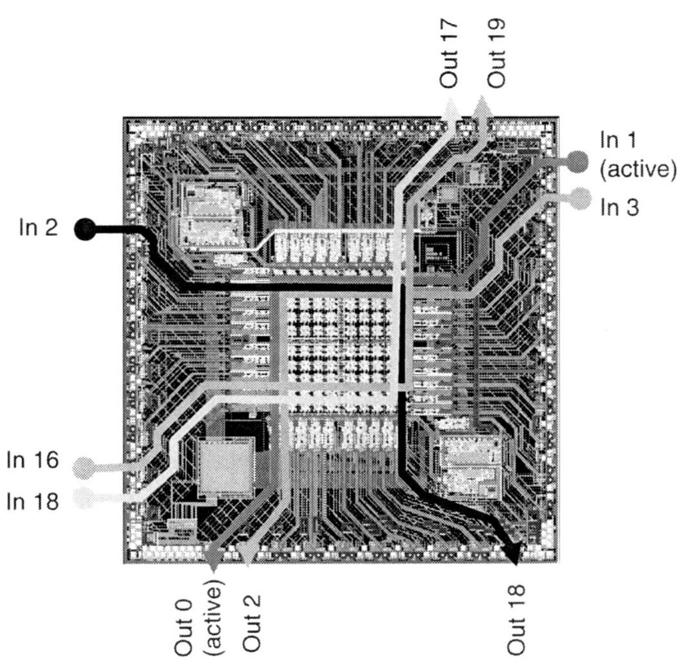

Fig. 4.31 Evaluation of crosstalk. A sinusoidal signal is applied to input 1; the matrix is configured for connecting input 1 to output 0. The crosstalk to several neighbouring channels is measured

Using this approach, the crosstalk can be expressed in dB as a function of frequency. The worst-case crosstalk is found at output out2 and is 35 dB below the output level at output out0 for frequencies up to 10 GHz, which demonstrates that low crosstalk is feasible despite the small distance between the transmission line interconnects inside the matrix.

4.7 Conclusions and Outlook

This chapter describes the design and realization of a cross-connect switch IC with 20 inputs and 20 outputs in a SiGe IC technology. The cross-connect switch IC applies signal distribution in a matrix architecture. On-chip transmission lines are used for the signal distribution between bondpads and input and output buffers and for the rows and columns inside the matrix. To facilitate transmission line design for rows and columns inside the matrix, the IC technology provides two thick top metal layers. The cross-connect function provides an excellent example of an application in which circuits and interconnect need to be designed and optimised together.

The concept of distributed capacitive loading is applied inside the matrix. The matrix node circuit was designed for minimum input capacitance and high input resistance, that is a differential input resistance $R_i \ll Z_{0dm}$ with Z_{0dm} being the differential-mode characteristic impedance of the unloaded transmission line. The parasitic capacitances due to crossing interconnects are distributed across the matrix node transmission line section. The loaded transmission line consequently has a reduced characteristic impedance $Z_{0dm,eff}$ and increased delay $t_{dm,eff}$ with respect to the unloaded transmission line. By terminating the loaded transmission line with its effective characteristic impedance, reflections are minimized and broadband signal transfer across relatively long interconnect is made possible.

The design procedure for the matrix node circuit starts at the output and ends at the input and can be summarised as follows. The simplest matrix node circuit is a differential pair. The CML signal swing together with the characteristic impedance of the double-terminated column transmission line define the required bias current of the differential pair. The transistor size of the differential pair follows from the bias current. The differential pair is operated at peak-f_T. Although biasing at peak-f_T is not optimum for f_A, it results in a small transistor size and hence low output capacitance. The input capacitance of the differential pair is calculated from the f_T. Input emitter followers are added to reduce the capacitive load to the transmission line. The bias current from the input emitter followers is calculated from the required bandwidth at the interface between the emitter followers and the differential pair. Since the bias current of the emitter followers requires transistors of more than minimum size, a second emitter follower plus differential pair is added at the input of the matrix node circuit, at an increased impedance level (and hence a reduced bias current and transistor size).

The signal distributions in rows and columns are of equal importance for the performance of the matrix. At most one matrix node circuit is active in each column, whereas multiple or even all circuits may be active in a row. So a low output capacitance of the matrix node circuit in the 'off' state is important in designing the column. A relatively high output capacitance in the 'on' state introduces mismatch at only a single location. In this design, this was shown to be acceptable.

The design procedure for input and output buffers is almost identical to that for the matrix node circuit. The maximum output swing required for the output buffer defines a relatively high bias current. The sensitivity required for the input buffer in combination with the CML signal swing inside the matrix defines the minimum small-signal gain. The requirement to use minimum-size input transistors does not hold for the input and output buffers. For the input buffer, sufficient input bandwidth is required in combination with the specified source impedance level, resulting in a maximum allowed input capacitance. ESD requirements must be taken into account at the input of the input buffer and the output of the output buffer. When ESD protection circuitry is included, sufficient bandwidth is still possible in the longest path through the matrix.

As follows from the design procedure, both the input and the output bandwidths are of the same importance for the bandwidth of the signal path. Therefore, f_A is a good FOM for the design of a cross-connect switch IC. The IC technology used has $f_A = 12\,\text{GHz}$ at $10\times$ low-frequency gain and turns out to be adequate for 12.5 Gb/s. Although the small-signal low-frequency gain in the differential pairs in the matrix node circuits is only about 2, the loading of the outputs of the differential pairs puts extra emphasis on the output bandwidth. In addition, many circuits are cascaded in the longest path of the matrix.

Maximum sensitivity to an incorrect transmission line termination resistance value occurs at a line length corresponding to $\lambda/4$ (and $\lambda/4 + n \cdot \lambda/2$, at integer n). In the case of an unloaded on-chip transmission line and a 12.5 Gb/s data rate, maximum sensitivity to mismatch will occur (at $n = 0$) at a line length of 2 mm. Intermediate buffers are added after 1 mm of line to reduce the sensitivity to incorrect termination of the loaded transmission line inside the matrix. The intermediate buffer circuits are identical to the matrix node circuits surrounded by termination resistors.

Small-signal simulations showed a relatively poor worst-case bandwidth for the longest path through the matrix. Indeed, a poor eye diagram was obtained when a very small input signal was applied in simulations, so that all circuits in the longest path operated in the small-signal regime. At typical input signal levels, however, the dynamic input capacitance of the differential pairs results in sufficient bandwidth and a satisfactory eye diagram.

The power supply decoupling can best be distributed across the supply network. A resistor is inserted in series with each decoupling capacitor to prevent potential ringing from the decoupling capacitors in combination with the supply line inductance. Transmission line models are applied to obtain a realistic model of the impedance of the supply network over frequency. It is essential to include a realistic model for the supply bondwires in designing the decoupling network.

Since transmission line termination resistors are placed at the end of the transmission lines, that is, close to the input and output buffers, supply decoupling is needed close to these RF I/O termination resistors. Since the termination resistors are connected to the positive supply, the supply network forms part of the termination network for common mode. This may seem of little importance, but it is actually essential, because evaluation is often performed using a single-ended input signal source or a single-ended output signal (e.g., in analysing the output eye diagram using a communications analyser).

The cross-connect switch presented here is designed for a maximum supply voltage VCC_{max} below BV_{CEO}. This means that breakdown will not affect the circuit. If a higher nominal supply voltage were allowed (e.g., 3.3 V), simpler circuits would have been possible as more base-emitter voltages may then be stacked. For example, a second pair of input emitter followers could be added to the matrix node circuit, replacing the differential amplifier (Q_{4a}, Q_{4b}, R_{4a}, R_{4b}) plus the input emitter followers (Q_{3a}, Q_{3b}) in Fig. 4.6. The resulting matrix node circuit would then be based on double emitter-coupled logic (EECL). EECL has been successfully applied to other high bit-rate circuit designs [4]. On the basis of EECL, the matrix size can be reduced, although no chip area reduction will follow because the design is bondpad-limited. The reduced current consumption in the matrix would not lead to a power reduction due to the increased supply voltage. Instead, the bandwidth of the matrix may be further enhanced, supporting a higher maximum data rate. So, a higher supply voltage enables a higher maximum data rate.

For a 40 Gb/s switch design, the same concepts for signal distribution may be applied as described in this chapter. Extra bandwidth may be obtained by increasing the supply voltage (and hence using EECL-based matrix node and input and output buffer circuits) in combination with an improved IC technology. The required factor of 3–4 speed improvement may then come partly from an improved IC technology (approximately one half) and the remainder from improved circuit concepts. The IC technology should thus provide an f_A of at least 22 GHz.

An increased supply voltage results in a design operating at a supply voltage $VCC > BV_{CEO}$. Aspects of circuit design at supply voltages above BV_{CEO} will be discussed in Chapter 5.

The cross-connect switch IC described in this chapter was presented at the International Solid-State Circuits Conference (ISSCC) in 2003 [3]. The IC was introduced on the market in July 2002 as Philips Semiconductors' TZA2060 (now NXP).

References

1. P. Deixler, R. Colclaser, et al., "QUBiC4G: a $f_T/f_{max} = 70/100$GHz 0.25 μm low power SiGe-BiCMOS production technology with high quality passives for 12.5Gb/s optical networking and emerging wireless applications up to 20GHz," in *Proc. IEEE BCTM*, 2002, pp. 201–204.
2. H. Veenstra, E. van der Heijden, D. van Goor, "Optimising broadband signal transfer across long on-chip interconnect," in *Proc. ESSCIRC*, 2002, pp. 763–766.

3. H. Veenstra, P. Barré, E. v.d. Heijden, D. van Goor, N. Lecacheur, B. Fahs, G. Gloaguen, S. Clamagirand, O. Burg, "A 20-input 20-output 12.5 Gb/s SiGe crosspoint switch for optical networking with <2ps RMS jitter," *ISSCC Dig. Tech. Papers*, 2003, pp. 174–175.
4. H.-M. Rein, M. Moller, "Design considerations for very-high-speed Si-bipolar IC's operating up to 50 Gb/s," *IEEE J. Solid-State Circuits*, vol. 17, No. 8, August 1996, pp. 1076–1090.

Chapter 5
Bias Circuits Tolerating Output Voltages Above BV_{CEO}

5.1 Introduction

Emitter followers are frequently used to obtain the highest possible speed of bipolar circuits for high bit-rate applications. For example, double emitter-coupled logic (EECL) is usually faster than emitter-coupled logic (ECL). The maximum bit-rate of the cross-connect switch described in Chapter 4 can be increased by employing double emitter followers in the RF signal path. However, to be able to do so, the supply voltage must be increased to a value above the collector-emitter breakdown voltage (BV_{CEO}) of the high-speed transistors of the technology.

Since the base-emitter voltage V_{be} for transistors operating at peak-f_T has remained approximately constant at about 0.9 V over the last decade, the supply voltage required for a given circuit topology has also remained constant. For example, ECL circuits require a minimum supply voltage of approximately $2.5 \cdot V_{be}$ and usually operate from a 3.3 V nominal supply; EECL circuits require a minimum supply voltage of $3.5 \cdot V_{be}$ and are typically operated from a 5 V nominal supply. The notations V_{be}, V_{bc}, I_c, etc. used in this chapter refer to dc quantities.

Circuits operating at supply voltages above BV_{CEO} are also found in applications in which large voltage swings and high efficiency are required, such as laser modulator drivers for high bit-rate systems and radio frequency (RF) power amplifiers. Supply voltages as high as two to three times BV_{CEO} have been reported for such applications [1]. This chapter deals with the consequences of circuit operation at supply voltages above the breakdown voltage BV_{CEO} of the transistor.

The evolution of SiGe devices with transit frequencies above 70 GHz was accompanied by a decrease in breakdown voltages below 3 V. Operation from a standard supply voltage (e.g., 5 V) may consequently exceed the transistor BV_{CEO}. In particular, the biasing sub-circuitry must be compliant enough to absorb the balance given a fixed supply voltage. A transistor's current source may be forced into breakdown by such topological and supply constraints or when operating at extremes such as elevated temperature.

H. Veenstra and J.R. Long, *Circuit and Interconnect Design for RF and High Bit-Rate Applications,* 153
doi: 10.1007/978-1-4020-6884-3_5, © Springer Science + Business Media B.V. 2008

The breakdown voltages of transistors therefore play a role in the selection of circuit topologies, bias conditions and voltage swings. Two relevant breakdown parameters for the bipolar transistor are the collector-emitter breakdown voltage with an open-circuited base terminal (BV_{CEO}) and the collector-base breakdown voltage with an open-circuited emitter (BV_{CBO}), as illustrated in Fig. 5.1.

Due to the trade-off between transit time and breakdown voltage [2, 3], the speed improvements realized in modern SiGe and SiGe:C processes were (partially) achieved by reducing breakdown voltages. The evolution of a high-performance production BiCMOS process over the time frame from 1998 until 2004 is shown as an example in Table 5.1 [4–7].

Note that many modern bipolar and BiCMOS technologies now offer multiple transistor designs tailored for either high speed or high breakdown (e.g., the 60 GHz f_T/3.1 V breakdown versus 130 GHz f_T/2.0 V breakdown devices listed in Table 5.1 for QUBiC4X).

In modern SiGe and SiGe:C processes, device metrics such as the unity current gain bandwidth f_T and unity power gain frequency f_{max} increase when the collector-base voltage V_{cb} (and collector-emitter voltage V_{ce}) is increased (e.g., $V_{cb} = 0$ V to $V_{cb} = 1$ V). However, designing for the highest speed or bandwidth causes transistors to operate in the signal path at $V_{cb} > 0$ V, bringing them closer to breakdown. Above the collector-base breakdown voltage BV_{CBO}, the collector-base junction breaks down, regardless of the impedance connected between the base and emitter terminals. Therefore, BV_{CBO} is an absolute maximum for V_{cb}, and circuits should be designed to operate at $V_{cb} < BV_{CBO}$ under all operating conditions. However, depending on the network connected between the base and the emitter, circuits may be operated at V_{ce} above BV_{CEO}. For example, the output stage of an RF power

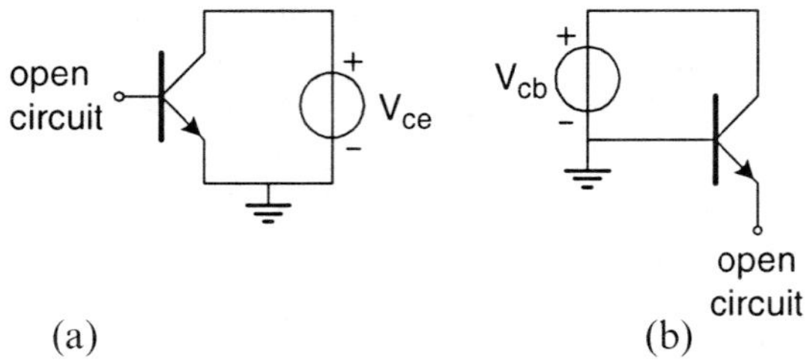

Fig. 5.1 Transistor in the open base (a) and open emitter (b) configurations

Table 5.1 Evolution of Philips' high-performance BiCMOS production IC processes

	QUBiC3 [4]	QUBiC4 [5]	QUBiC4G [6]	QUBiC4X [7]
Year introduced	1998	2001	2002	2004
BJT HBT base	Si	Si	SiGe	SiGe:C
f_T (GHz)	30	40	70/50	130/60
f_{max} (GHz)	60	90	100/110	140/120
BV_{CEO} (V)	4.0	3.7	2.7/3.9	2.0/3.1
BV_{CBO} (V)	14	16	10/15	9/13

amplifier can be designed to tolerate collector-emitter voltages greater than BV_{CEO} by driving the base terminal with a relatively low impedance.

Bias transistors in high-speed and high-power applications also need to operate across a wide range of collector-emitter voltages and to handle $V_{ce} > BV_{CEO}$. The design of bias circuits for operation at collector-emitter voltages continuously above BV_{CEO} will be considered in this chapter. A number of widely used and newly proposed bias current circuits will be analysed. In Section 5.2 a brief review of the relationship between collector-base avalanche current and breakdown will be presented. Using avalanche current multiplication theory, the operation of the widely used simple 2-transistor current mirror will be analysed in Section 5.3, focusing on its operation at output voltages above BV_{CEO}. To increase the accuracy of the simple 2-transistor mirror, a buffer transistor is often added to supply base current to all the devices. The output characteristics of the current mirror with buffer will be analysed in Section 5.4.

The breakdown voltage of a current mirror operating at an output voltage above BV_{CEO} may be further increased by employing improved biasing circuits for the internal buffer circuit. In addition, avalanche current compensation can be applied to improve the accuracy of the current mirror ratio at output voltages above BV_{CEO}. A feedforward avalanche current compensation technique will be proposed and demonstrated in Section 5.5.1. Avalanche current compensation via feedback will then be demonstrated in Section 5.5.2. The results will be summarised and discussed in Section 5.6.

The current mirror circuits described in this chapter are implemented in the QUBiC4G SiGe IC technology (see [6] and Table 5.1), and are designed for a nominal 1:10 current mirror ratio and 10 mA output current.

5.2 Principle of Collector-Base Avalanche Current

The collector-base avalanche current and a simplified model for collector-base breakdown in a bipolar transistor will be described in this section. The following analysis is valid for a transistor of arbitrary emitter area. A model for the transistor that includes the avalanche current effect is shown in Fig. 5.2. Avalanche current between the collector and the base is modelled by the current source I_{avl}, and this model is valid in the forward active region of operation (i.e., at $V_{be} > 0$ and $V_{cb} > 0$).

On the basis of Fig. 5.2, the base (I_b) and collector (I_c) currents including avalanche current from the base-collector junction can be expressed as follows:

$$I_b = \frac{I_{c0}}{\beta_0} \cdot e^{\frac{V_{be}}{V_T}} - (M-1)I_{c0} \cdot e^{\frac{V_{be}}{V_T}} \tag{5.1}$$

$$I_c = M \cdot I_{c0} \cdot e^{\frac{V_{be}}{V_T}} \tag{5.2}$$

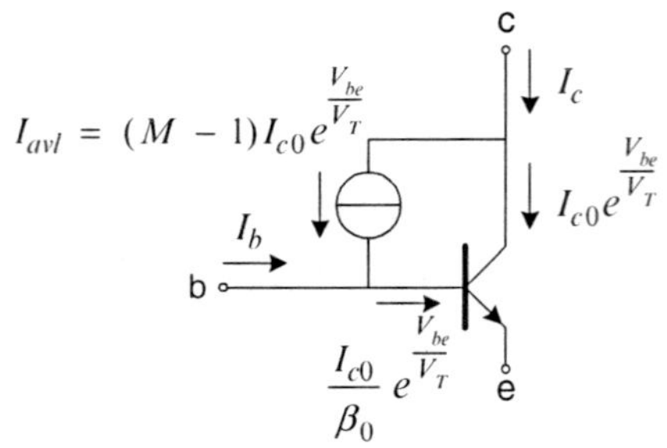

Fig. 5.2 Transistor with collector-base avalanche current source

where $V_T = kT/q$ is the thermal voltage, β_0 is the dc current gain and V_{be} is the voltage across the base-emitter junction. M is the avalanche current multiplication factor [8], which is defined as:

$$M = \frac{1}{1 - \left(\frac{V_{cb}}{BV_{CBO}}\right)^{\eta}} \tag{5.3}$$

A typical value for η in equation (5.3) is 3. Note that as V_{cb} approaches BV_{CBO}, the avalanche multiplication factor $M \rightarrow \infty$ and the collector-base junction breaks down. Therefore, the maximum useable reverse collector-base voltage is BV_{CBO}, which is independent of the circuit topology.

The collector-emitter breakdown voltage in the open-base configuration (BV_{CEO}) will occur when the avalanche current (last term in equation (5.1)) equals the recombination current (i.e., base current without avalanche generation) so that the net base current becomes zero (i.e., $I_b = 0$). The collector current can be written as a function of the base current by eliminating $e^{(Vbe/VT)}$ from equations (5.1) and (5.2):

$$I_c = I_b \frac{M}{1 + \frac{1}{\beta_0} - M} \tag{5.4}$$

and breakdown will occur when $M = 1 + 1/\beta_0$. Substituting this value for M in equation (5.3) gives V_{cb} at BV_{CEO} as:

$$V_{cb}|_{BVCEO} = \frac{BV_{CBO}}{\sqrt[\eta]{\beta_0 + 1}} \tag{5.5}$$

and since $V_{ce} = V_{cb} + V_{be}$, BV_{CEO} can be written as:

$$BV_{CEO} = V_{be} + \frac{BV_{CBO}}{\sqrt[\eta]{\beta_0 + 1}} \tag{5.6}$$

BV_{CBO} is typically a few times larger than BV_{CEO}. Typical values for the technology used in this study [6] are $BV_{CBO} = 10\,V$, $BV_{CEO} = 2.7\,V$ and $\beta_0 = 220$.

From equation (5.3) it follows that when V_{cb} is zero there will be no avalanche current, because M will be unity. As V_{cb} increases, M becomes slightly larger than

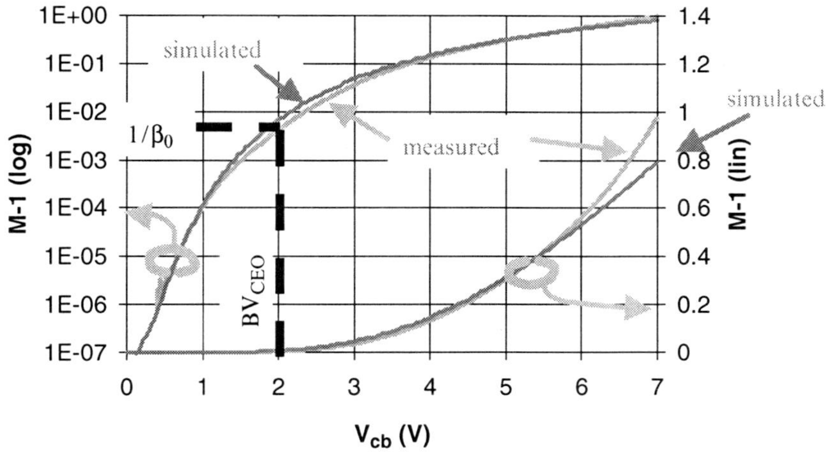

Fig. 5.3 Avalanche multiplication factor (M − 1) versus collector-base voltage V_{cb} at $V_{be} = 0.7$ V. Breakdown voltage BV_{CEO} will occur when (M − 1) = $1/\beta_0$, as indicated

unity. It is therefore common practice to evaluate (M − 1) rather than M as a function of V_{cb}. In this chapter, the avalanche multiplication factor may refer to either M or M − 1, depending on the context. The measured and simulated curves obtained for the parameter (M − 1) on both logarithmic (left ordinate) and linear scales (right ordinate) are shown in Fig. 5.3. The compact transistor model MEXTRAM was used in the simulations [9, 10]. The results were obtained using the procedure described in [11], based on equations (5.1)–(5.3).

As V_{cb} is increased starting from $V_{cb} = 0$, the resulting decrease in base current and increase in collector current are used to determine M. At a constant V_{be}, the variation in base current caused by a change in V_{cb} (i.e., ΔI_b) is

$$\Delta I_b = -I_b|_{V_{cb}} + I_b|_{V_{cb}=0} = (M-1)I_{c0} \cdot e^{\frac{V_{be}}{V_T}} \tag{5.7}$$

from equation (5.1). From equations (5.2) and (5.7) it then follows that

$$\frac{I_c}{\Delta I_b} = \frac{M}{M-1} \tag{5.8}$$

or

$$M - 1 = \frac{\frac{\Delta I_b}{I_c}}{1 - \frac{\Delta I_b}{I_c}} \tag{5.9}$$

A procedure based on this result follows to extract M via either measurement or simulation. The factor (M − 1) of equation (5.9) can be determined by measuring the base and collector currents as a function of V_{cb} while keeping V_{be} constant. As can be seen in Fig. 5.3, the simulation model agrees with the measurement up to $(M-1) \approx 0.5$, or up to $V_{cb} \approx 2.5 \cdot BV_{CEO}$. The collector-base junction breakdown at BV_{CBO} was not included in the MEXTRAM simulation model.

Equation (5.3) predicts that the avalanche current multiplication factor M does not depend on the actual bias condition (V_{be}) of the transistor, which is true for low to moderate values of V_{be}. The voltage drop across the extrinsic collector resistance offsets the base-collector terminal voltage from the internal junction voltage. This offset causes a shift in M versus V_{cb} as the collector (and base) currents increase. In addition to this effect, the collector current affects the electric-field profile in the collector-base depletion layer (also known as the Kirk effect) at high values of V_{be}. This also makes M a function of V_{be}.

The avalanche current multiplication factor (M) for transistors operating at $V_{cb} = 0$ is unity, and there is no avalanche current. At $V_{cb} \ll BV_{CEO}$, the avalanche current multiplication factor is still close to one (i.e., $M \approx 1$), and avalanche currents may therefore be ignored in circuits operating at supply voltages VCC well below BV_{CEO}. As V_{cb} approaches BV_{CEO}, the base current is significantly influenced by avalanche multiplication because $(M-1)$ is of the same order of magnitude as $1/\beta_0$. However, the collector current is only slightly affected, as $M \approx 1$. At V_{cb} values above the breakdown voltage BV_{CEO} (i.e., $V_{cb} > BV_{CEO}$), the increase in collector current will become significant and current will now flow out of the base terminal. For example, at $(M-1) = 0.5$, the collector current will increase by 50% with respect to nominal, while the base current will equal 50% of the nominal collector current and the transistor dc current gain will be $I_c/I_b \approx -3$. This large dc current flowing out of the base terminal must be dissipated by the network connected to the base terminal.

5.3 Analysis of Simple 2-Transistor Current Mirrors

A simple current mirror biased using a diode-connected transistor is shown in Fig. 5.4. This circuit topology is often referred to as simple 2-transistor current mirror [12].

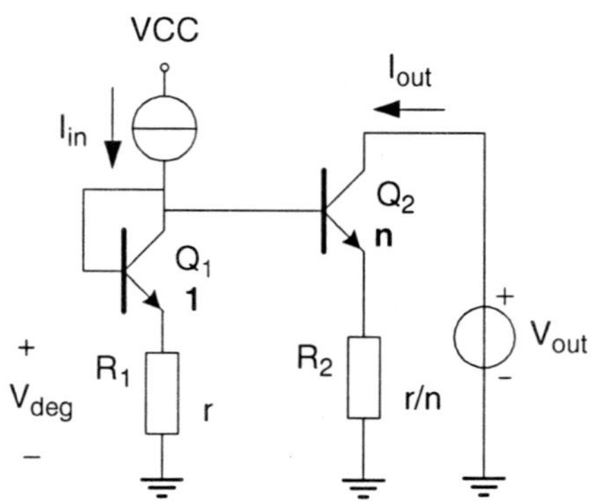

Fig. 5.4 1 : n degenerated simple 2-transistor current mirror

The input current I_{in} is mirrored to the output as current I_{out}. In many cases I_{out} is set n times larger than I_{in} by making the emitter area of the output transistor Q_2 n times larger than the area of the input transistor Q_1. In the circuit shown in Fig. 5.4, $n = 10$. Degeneration resistors R_1 and R_2 (where $R_2 = R_1/n$) are added to reduce the sensitivity of the output current to mismatch between transistors Q_1 and Q_2. At a degeneration voltage V_{deg} larger than the thermal voltage $V_T \approx 25\,\text{mV}$, the accuracy of the current mirror is determined mainly by the matching between the degeneration resistors R_1 and R_2. In the following example (from Fig. 5.4), $V_{deg} = 0.2\,\text{V}$ and ideal matching is assumed between the components. In addition, the output resistance in the absence of avalanche multiplication is ignored (so an infinite Early voltage is assumed).

If the output voltage is low enough so that $V_{ce} < BV_{CEO}$ across the output transistor Q_2, avalanche currents will be insignificant and the current mirror inaccuracy will be mainly due to the finite dc current gain (β_0) of the npn transistors. On the condition that V_{ce} is less than BV_{CEO}, the ratio of the output and input currents is:

$$\frac{I_{out}}{I_{in}} = \frac{n}{1 + (n+1)/\beta_0} \tag{5.10}$$

At output voltages V_{out} less than $(BV_{CEO} + V_{deg})$, inaccuracy in the mirror ratio defined by equation (5.10) will arise from the $(n + 1)$ base currents that are subtracted from the input reference current I_{in}. When the output voltage exceeds $(BV_{CEO} + V_{deg})$, the base current due to avalanche breakdown of Q_2 will become significant. When accounting for both the finite current gain β_0 and the avalanche current based on the model shown in Fig. 5.2, the simple 2-transistor current mirror generates the following ratio of output and input currents:

$$\frac{I_{out}}{I_{in}} = \frac{nM}{1 + \frac{n+1}{\beta_0} - n(M-1)} \tag{5.11}$$

Here, factor M is the avalanche multiplication factor for the output transistor Q_2. If $M = 1$, equation (5.11) simplifies to equation (5.10). With increasing M, the mirror ratio increases because of an increase in the output current. Accurate modelling of M for V_{cb} above BV_{CEO} is required in order to determine the output current at these higher output voltages.

In equation (5.11) the denominator is zero at

$$M - 1 = \frac{1}{n} + \frac{1}{\beta_0} + \frac{1}{n\beta_0} \approx \frac{1}{n} \tag{5.12}$$

From equation (5.12) it follows that breakdown will occur at $(M - 1) \approx 1/n$ in the case of the simple 2-transistor current mirror, with n being the emitter area ratio of the output transistor and the diode. This is a much higher value than the breakdown condition for a single transistor with an open-circuited base (i.e., where $(M - 1) \approx 1/\beta_0$). So, the output breakdown voltage BV_{CED} for the simple 2-transistor current mirror of Fig. 5.4 is defined as the collector-emitter voltage at which $(M - 1) = 1/n$ and $BV_{CED} > BV_{CEO}$. A higher BV_{CED} can be obtained by using a lower ratio of emitter areas, n. In Fig. 5.5, output breakdown voltages BV_{CEO} of approximately 2.7 V (at $V_{cb} = 2.0$ V, where $(M - 1) = 1/\beta_0$) and BV_{CED} of approximately 4.3 V (at $V_{cb} = 3.6$ V, where $(M - 1) = 1/n$; $n = 10$) are indicated in the plot showing avalanche multiplication factor measurements. The same data points for BV_{CEO} and BV_{CED} can be found from the $I_c - V_{ce}$ curves for a transistor with an open-base terminal (BV_{CEO}) and the simple 2-transistor current mirror of Fig. 5.4 (BV_{CED}), as shown in Fig. 5.6. The finite output impedance for voltages above breakdown is determined by the extrinsic collector and emitter resistances of the output transistor Q_2.

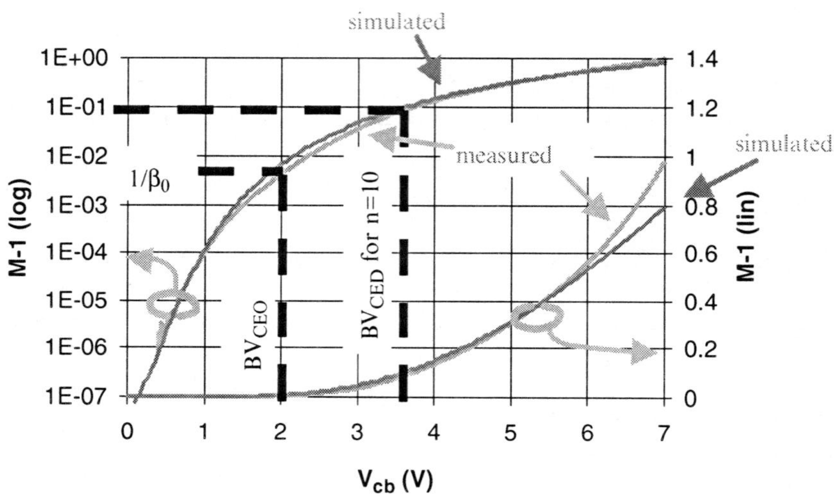

Fig. 5.5 Avalanche multiplication factor curves shown in Fig. 5.3, indicating BV_{CEO} and BV_{CED} for $n = 10$

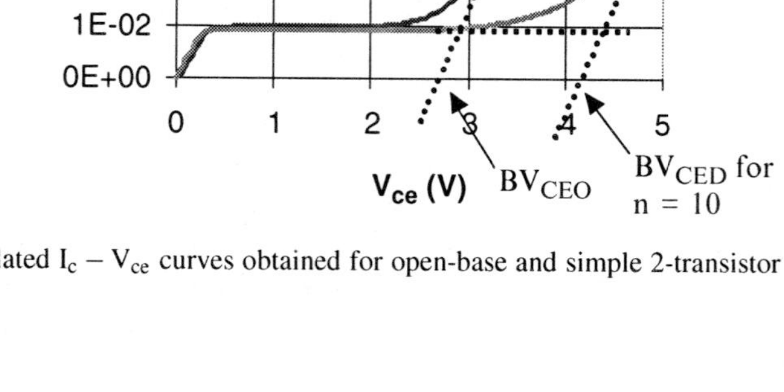

Fig. 5.6 Simulated $I_c - V_{ce}$ curves obtained for open-base and simple 2-transistor current mirror configurations

5.4 Analysis of Current Mirrors with Internal Buffer

From equation (5.11) and Fig. 5.6 it follows that the output current for the simple 2-transistor current mirror increases well before V_{out} reaches breakdown voltage BV_{CED}. This is mainly caused by the avalanche current I_{avl}, which adds to the collector current of the input transistor Q_1, thereby increasing the voltage at the base terminal of Q_2. The flow of the avalanche current to ground is indicated by arrows in Fig. 5.7.

The output impedance of the current source is improved by adding buffer transistor Q_3, as shown in Fig. 5.7b. The buffer transistor supplies base currents to transistors Q_1 and Q_2, thereby reducing the current drawn by mirror transistor (Q_2) from the reference current (I_{in}) by a factor β_0.

When the emitter current of Q_4 is minimised (i.e., $R_3 \rightarrow \infty$), the mirror ratio for the circuit of Fig. 5.7b at output voltages $V_{out} < (BV_{CEO} + V_{deg})$ will equal

$$\frac{I_{out}}{I_{in}} = \frac{n}{1 + \frac{n+1}{\beta_0^2}} \tag{5.13}$$

Network Q_4 and R_3 in Fig. 5.7b provides a path for an additional current (I_{R3}) which will bias buffer transistor Q_3. The extra bias current flowing in Q_3 will increase its base current somewhat, and will reduce the accuracy of the current mirror relative to the prediction of equation (5.13). Nevertheless, the accuracy of the modified mirror will be significantly better than that of the original circuit of Fig. 5.7a. Q_4 and R_3

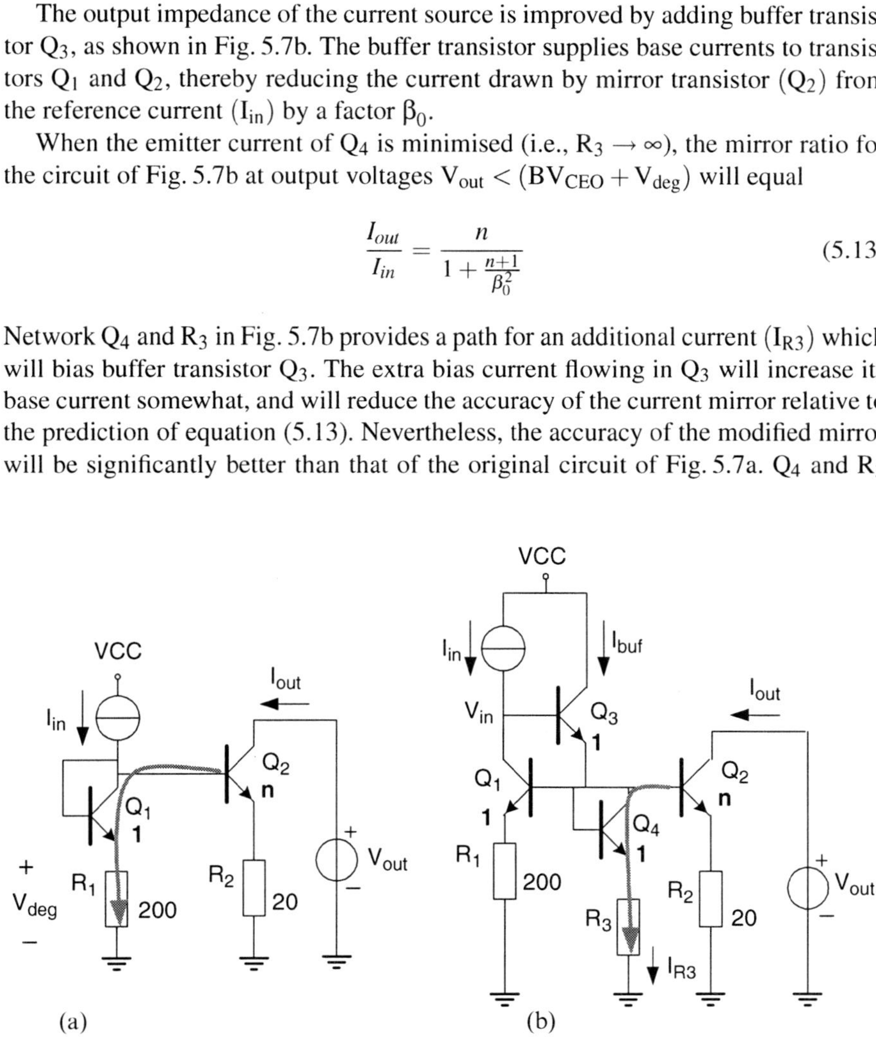

Fig. 5.7 Simple 2-transistor current mirror (a) and current mirror with buffer transistor Q_3 (b). The emitter area scaling 1 : n is indicated in bold. The arrows indicate the main path to ground of the avalanche current of Q_2

are also needed for high-frequency stability and provide a low-ohmic path for current to flow from the base terminal of Q_2 to ground, thereby reducing the tendency of the base voltage to increase due to the flow of an avalanche current. The total bias current for buffer transistor Q_3 is $I_{buf} = I_{R3} + I_{b,Q1} + I_{b,Q2}$. At output voltages V_{out} approaching $(BV_{CEO} + V_{deg})$, the base current needed to bias the output transistor $(I_{b,Q2})$ will decrease, thereby reducing I_{buf}. When avalanche breakdown does occur, the base current in the output transistor reverses, and it supplies a current which biases both Q_1 and R_3 and reduces the current flowing in the buffer transistor Q_3.

The bias current in Q_3 will become zero when $I_{b,Q2} = -(I_{b,Q1} + I_{R3}) \approx -I_{R3}$, which will occur at an output voltage at which

$$M - 1 = I_{R3}/nI_{in} \tag{5.14}$$

In this condition, buffer transistor Q_3 will be biased off, and its output impedance will become large (i.e., it will theoretically approach infinity). Output transistor Q_2 will then no longer be driven by a low impedance and the collector current will begin to rise sharply. The actual breakdown voltage for the current mirror with buffer can be derived from the $(M-1)$-curve (of Fig. 5.3), given the parameter $(M-1)$ defined by equation (5.14).

With further increases in the output voltage, the avalanche current multiplication factor $(M-1)$ will exceed the $I_{R3}/n \cdot I_{in}$ ratio. The current flowing through resistor R_3 will then be supplied entirely by the current flowing out of the base terminal of output transistor Q_2. The base-emitter junction of Q_3 will now be reverse biased. When buffer transistor Q_3 turns off, the base voltage of transistor Q_1 will begin to rise and Q_1 will be quickly driven into saturation. This will result in a $2 \cdot V_{be}$-drop in the voltage V_{in} (see Fig. 5.7b) between the condition in which Q_3 is conducting current (and V_{in} is defined by the base-emitter voltage drops across Q_3 and the other transistors in the mirror) and that in which Q_3 is turned off and transistor Q_1 is driven into saturation by the reverse base current flowing out of Q_2.

For example, at $I_{R3}/n \cdot I_{in} = 0.1$, breakdown will occur at $(M-1) = 0.1$ or $V_{cb} \approx 3.6\,V$ (from Fig. 5.5). This corresponds to $V_{out} = V_{cb} + V_{be} + V_{deg} \approx 4.6\,V$. Simulation and measurement results obtained for a current mirror fabricated with these design parameters are shown in Fig. 5.8.

The predictions resulting from the simulation using the MEXTRAM model and the experimental measurements are in excellent agreement, as can be seen in the figure. The output current remains accurate up to output breakdown, due to the path for the avalanche current I_{avl} via transistor Q_4 and resistor R_3 to ground. A slight increase in output current occurs before breakdown. In the example circuit, $I_{R3} = 0.1 \cdot n \cdot I_{in}$ so that breakdown occurs at $M - 1 = 0.1$, and therefore the output current increases by a factor 1.1 at breakdown. Also clearly visible is the steep drop in V_{in} caused by turn-off of the buffer transistor at breakdown.

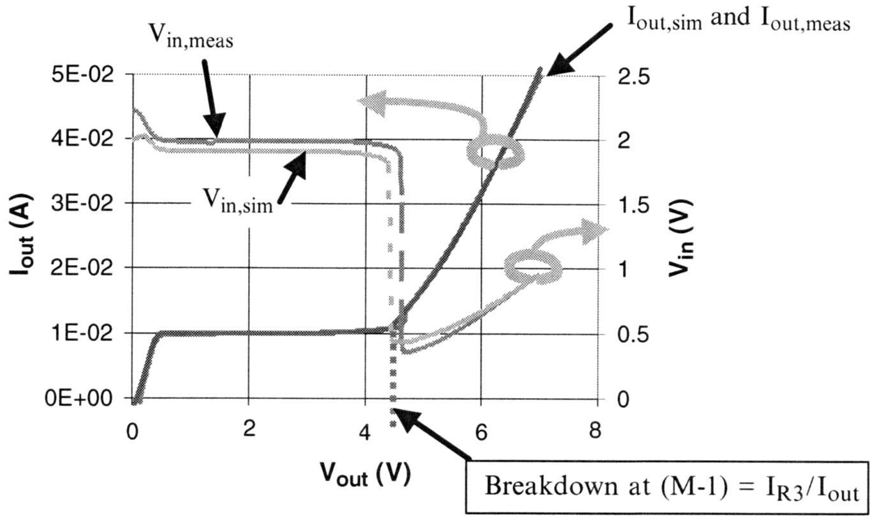

Fig. 5.8 Measured and simulated input voltage V_{in} and output current I_{out} obtained for the current mirror with buffer

5.5 Avalanche Current Compensation

To compensate for avalanche currents, feedforward and feedback techniques are described in this section.

5.5.1 Feedforward Techniques

The results of the study presented in the previous section show that adding the capability to sink reverse base current flowing from the output transistor to the bias circuit increases the breakdown voltage of the output transistor. Therefore, the output breakdown voltage for the current mirror with buffer described in the previous section could be improved further by lowering the value of R_3 in order to increase the buffer bias current (i.e., increase I_{R3} as shown in Fig. 5.7b). The resulting improvement in the output breakdown voltage can be predicted on the basis of the $(M-1)$-curve of Fig. 5.5 if the $I_{R3}/n \cdot I_{in}$-ratio is known (i.e., from equation (5.14)). However, increasing the nominal buffer bias current increases the power consumed in the circuit under all operating conditions. It should be noted that it is more efficient to sink base current from Q_2 only when necessary (i.e., when $V_{out} > BV_{CEO}$). The circuit shown in Fig. 5.9 is designed to do this.

Current I_{ff} is intended to sink avalanche current I_{avl} flowing out of the base terminal from transistor Q_2. The current mirror formed by transistors Q_5/Q_6 generates an additional bias current I_{ff} only when the current source output voltage (V_{out}) rises above a predefined threshold voltage. Below breakdown, I_{ff} is close to zero, and the

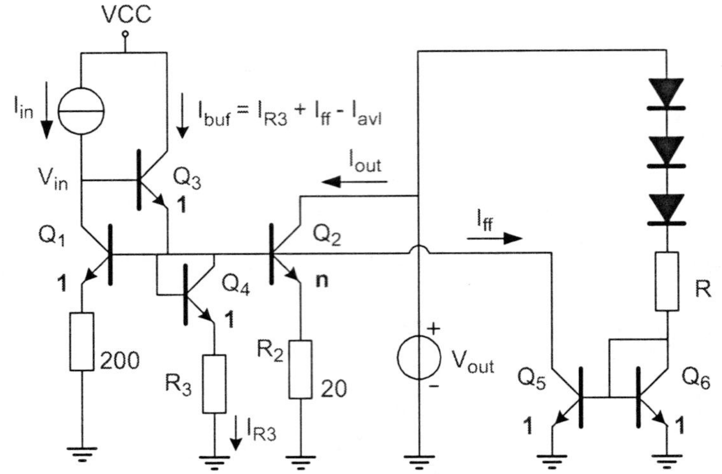

Fig. 5.9 Current mirror with buffer and feedforward avalanche current compensation

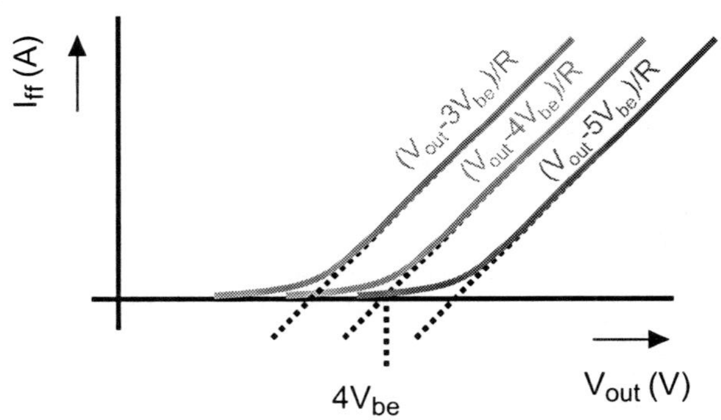

Fig. 5.10 Output current I_{ff} of the avalanche current compensation circuit as a function of the current mirror output voltage V_{out}. Curves are shown for different threshold voltages, realised by connecting a different number of diodes in series with resistor R

circuit's behaviour is identical to that of the current mirror with buffer circuit shown in Fig. 5.7b. When current mirror Q_5/Q_6 is active, the current I_{ff} adds to the current I_{buf} that is biasing the buffer transistor Q_3 in Fig. 5.9:

$$I_{buf} = I_{R3} + I_{b,Q1} + I_{b,Q2} + I_{ff} \approx I_{R3} + I_{ff} - I_{avl} \tag{5.15}$$

Transistor Q_5 supplies an additional buffer bias current I_{ff} according to

$$I_{ff} = (V_{out} - x \cdot V_{be})/R \tag{5.16}$$

In equation (5.16), the threshold voltage $x \cdot V_{be}$ for I_{ff} is defined by the number of diodes connected in series with resistor R. The example curves shown in Fig. 5.10 illustrate the relationship between current I_{ff} and the voltage V_{out} as the number of diodes connected in series increases. At least one diode forward voltage drop is required to bias the diode-connected transistor Q_6. The circuit shown in Fig. 5.9 was implemented using four diodes connected in series (i.e., $x = 4$), resulting in a threshold voltage of $4 \cdot V_{be}$ (approximately 3.2 V), which is about 1 V below the output breakdown voltage when the compensation current I_{ff} is zero.

Resistor R was chosen so that $I_{ff} = I_{R3}$ (doubling I_{buf}) near the output breakdown condition for a 1 mA input current. In the circuit in Fig. 5.9, output breakdown will occur when

$$M - 1 = \frac{I_{R3} + I_{ff}}{nI_{in}} \tag{5.17}$$

Since I_{ff} is independent of I_{in}, breakdown will occur at a higher output voltage at reduced input currents (as follows from equation (5.17)). At the nominal 1 mA input current, breakdown of the example current mirror with feedforward avalanche compensation occurs at an $(M - 1)$ of 0.2. In Fig. 5.3 it is shown that an $(M - 1)$ of 0.2 will occur at a V_{cb} of approximately 4.3 V, which will result in an expected output breakdown voltage of $V_{cb} + V_{deg} + V_{be}$ of approximately 5.3 V. The measured output breakdown voltage is 5.1 V, as shown in Fig. 5.11.

So, using the feedforward avalanche current compensation circuit, the output breakdown voltage of the current mirror is increased by about 0.5 V. As shown in the previous section, if breakdown occurs at $(M - 1) = y$, then the output current at breakdown will be a factor of $(1 + y)$ higher. Since the avalanche current adds to the collector current, the current mirror output current will increase by a factor 1.2 at breakdown (given that $(M - 1) = 0.2$), as can be seen from the measurement results presented in Fig. 5.11.

Reducing the value of R (see Fig. 5.9) increases the compensation current I_{ff} and provides an opportunity for a further increase in the circuit breakdown voltage.

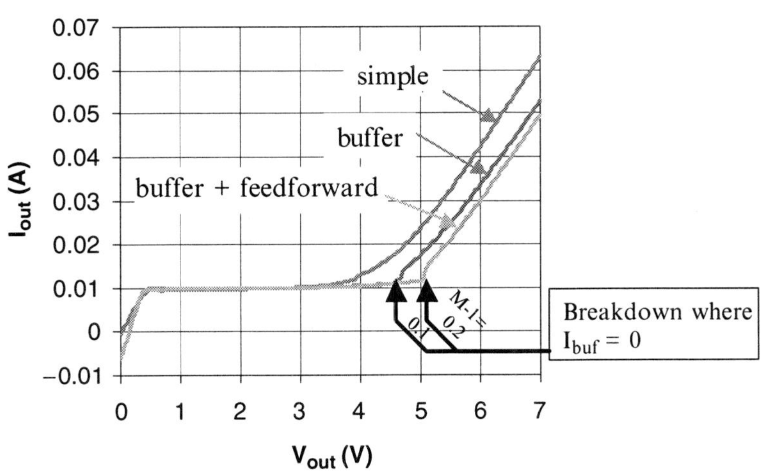

Fig. 5.11 Measured output current versus output voltage of the simple 2-transistor current mirror, current mirror with buffer and current mirror with buffer and feedforward avalanche current compensation

However, exact compensation of the avalanche current flowing from the output transistor is not possible with this circuit, because the feedforward current I_{ff} does not track the avalanche current I_{avl}. The compensation current is fixed by the choice of resistor R and the threshold voltage. It depends linearly on the output voltage, while the avalanche current is a non-linear function of the output voltage as defined by the avalanche current multiplication factor M.

5.5.2 Feedback Techniques

This section describes a circuit technique which further enhances the output breakdown voltage and improves the accuracy of the current mirror. The objective is to develop a bias circuit which sinks avalanche current flowing out of the output transistor's base terminal only when necessary, and can compensate for inaccuracy in the mirror output current caused by avalanche current flowing into the collector terminal of the output device.

It was shown in the previous sections that the output breakdown voltage of the current mirror is substantially improved by modifying the buffer circuit to prevent the situation in which $I_{buf} = 0$. A bias circuit that uses negative feedback to dynamically bias the output transistor and sink the avalanche current only as required is shown in Fig. 5.12.

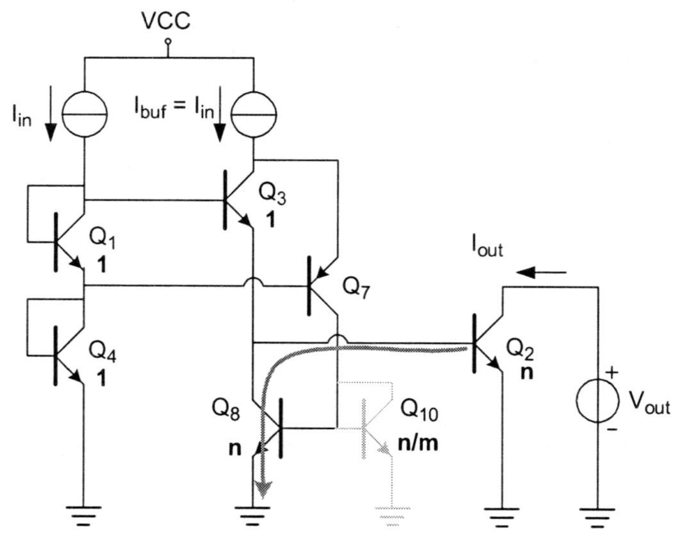

Fig. 5.12 Current mirror with modified buffer. Buffer transistor Q_3 is biased from its collector side. The emitter area scaling 1 : n is indicated in bold. The large arrow indicates the flow of the avalanche current of Q_2. Transistor Q_{10} is optional and may be applied to improve the bandwidth of the circuit

A second input current I_{buf} equal to I_{in} has been added to this circuit. The output current I_{out} is defined by a V_{be}-loop: $V_{be,Q2} = V_{be,Q4} + V_{be,Q1} - V_{be,Q3}$. The emitter area ratios for transistors Q_1–Q_4 indicated in Fig. 5.12 implement a mirror ratio 1:n. Note that the bias current of transistor Q_3 does not depend on the avalanche current produced by Q_2 in this circuit. The bias current for Q_3 is supplied by forcing the buffer bias current I_{buf} from the collector side. A feedback loop is used to define the collector voltage for Q_3, and this is implemented using transistors Q_7 and Q_8. The collector voltage of Q_3 is fixed at $2 \cdot V_{be}$ so that Q_3 operates at a collector-base voltage of 0 V. Note that a p-type device (Q_7) is needed for the circuit shown in Fig. 5.12. A pnp or PMOS transistor can be used.

In this new bias circuit implementation, avalanche current produced by transistor Q_2 is sunk by the collector current of Q_8, as indicated by the large arrow in Fig. 5.12:

$$I_{c,Q8} = I_{buf} - I_{b,Q2} \approx I_{buf} + I_{avl,Q2} \tag{5.18}$$

Although the nominal collector current for Q_8 is 1/n times the output current (e.g., 1 mA for a 1:10 mirror with an output current of 10 mA, as in previous examples), avalanche current increases the collector current of Q_8 significantly. For example, when operating at an output voltage V_{out} of 7 V, the collector-base voltage of Q_2 is approximately 6 V, and $M - 1$ is therefore about 0.55 (as can be seen in Fig. 5.3) and I_{avl} is $0.55 \cdot I_{out}$. To handle these relatively large avalanche currents, transistor Q_8 should have about the same emitter area as output transistor Q_2. In comparison with the circuits of Figs. 5.7b and 5.9, the buffer output impedance Z_{buf} seen from the base of transistor Q_2 is substantially reduced by the buffer topology shown in Fig. 5.12. For the circuits shown in Figs. 5.7b and 5.9, the low-frequency driving impedance at the base of Q_2 can be approximated by $Z_{buf} \approx 1/gm_3$ with $gm_3 \approx qI_{buf}/kT$. In the buffer topology used in Fig. 5.12, the additional loop gain introduced by transistor Q_8 reduces Z_{buf} to $Z_{buf} \approx 1/(gm_3 \cdot (\beta + 1))$ with β being the current gain of transistor Q_8. The dc bias current for pnp transistor Q_7 is relatively low (e.g., $I_{c,Q7} = I_{buf}/(\beta + 1)$). This low current results in a poor bandwidth of the internal buffer, and hence a limited capability to sink high-frequency dynamic avalanche currents of transistor Q_2. The addition of transistor Q_{10}, with emitter area n/m, effectively increases the dc bias current for transistor Q_7 to $I_{c,Q7} = I_{buf}/(m + 1)$), thereby increasing the bandwidth of the buffer and hence the circuit's capability to track dynamic avalanche currents of output transistor Q_2. With transistor Q_{10} present, the collector current of Q_3 scales by a factor of $m/(m + 1)$. Since Q_3 is part of the V_{be}-loop defining the current mirror ratio, the buffer bias current needs to be increased by a factor of $(m + 1)/m$ to restore the overall current mirror ratio to n. For high-frequency stability of the internal buffer bias circuit, a capacitor can be connected in parallel to the collector-base junction of transistor Q_8, thereby reducing the high-frequency loop-gain. At high frequencies, transistor Q_8 will then act as a diode.

Since the avalanche current produced by Q_2 forms (part of the) collector current flowing in Q_8, it is also possible to implement avalanche current compensation using a feedback technique. The intention of avalanche current compensation is to

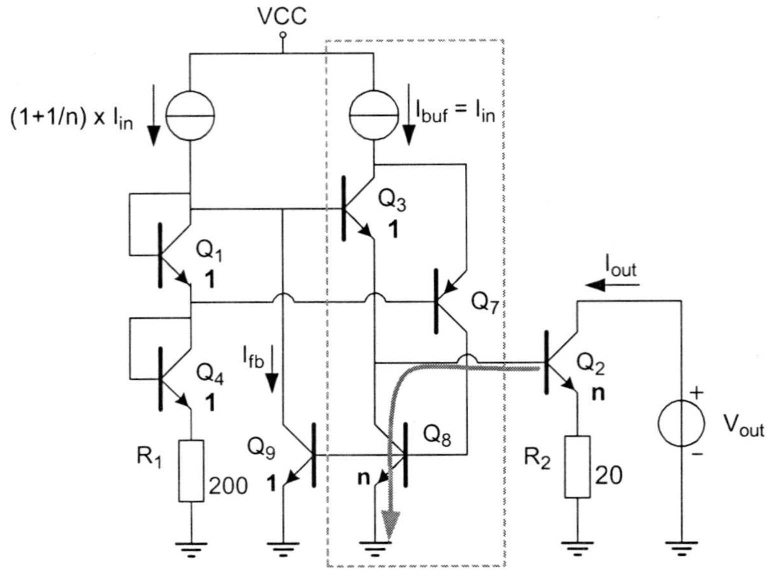

Fig. 5.13 Current mirror with buffer and feedback avalanche current compensation. The emitter area scaling 1 : n is indicated in bold. The buffer is surrounded by the dotted box. The large arrow indicates the flow of the avalanche current of Q_2

improve the accuracy of the current mirror up to output breakdown. The proposed circuit is shown in Fig. 5.13.

The compensation is implemented using an additional n : 1 ration current mirror (Q_8/Q_9 in Fig. 5.13) to generate the feedback current I_{fb}, where:

$$I_{fb} = \frac{I_{buf} + I_{avl,Q2}}{n} \tag{5.19}$$

Current I_{fb} is then subtracted from the current mirror input reference current (I_{in}) to reduce the current flowing through transistors Q_1 and Q_4. Reducing the reference current by I_{avl}/n restores the output current of the mirror to the desired value. Subtraction of I_{buf}/n (which equals I_{in}/n) from the input current is not desired, but it can be corrected by simply increasing the input current from I_{in} to $(1 + 1/n) \cdot I_{in}$, as indicated in Fig. 5.13.

As follows from equation (5.19), the increase in output current caused by avalanche current from Q_2 flowing into the output of the mirror is effectively counteracted by reducing the collector currents of Q_1 and Q_4. Simulation results obtained for this mirror designed for an output current of 10 mA are shown in Fig. 5.14.

The simulation results demonstrate the effectiveness of the avalanche current compensation circuit, as the output current remains close to the desired value over a wide range of output voltages and well above BV_{CEO}.

The base current $I_{b,Q2}$ of output transistor Q_2 is small and positive at output voltages V_{out} less than BV_{CEO}. At V_{out} values above BV_{CEO}, base current $I_{b,Q2}$ is

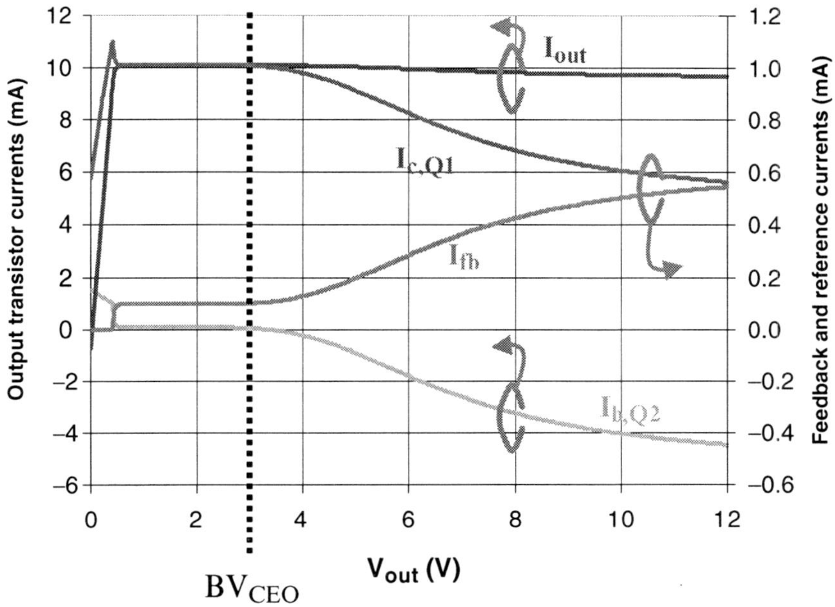

Fig. 5.14 Simulated output current I_{out}, base current of the output transistor $I_{b,Q2}$, feedback current I_{fb} and collector current $I_{c,Q1}$ in the reference transistor Q_1 as a function of the mirror output voltage V_{out} for the circuit of Fig. 5.13 with $n = 10$

negative and can reach relatively large values (as can be seen in Fig. 5.14). Subtraction of I_{avl}/n from the input reference current causes $I_{c,Q1}$ to decrease with increasing avalanche current flow. Since junction breakdown is not included in the MEXTRAM transistor model used for the simulations, the current mirror functions in simulation even at V_{out} above BV_{CBO}. However, in practice, operation will be limited to a maximum V_{out} equal to the collector-base breakdown voltage BV_{CBO}.

As can be seen from the measurements obtained for this mirror fabricated in the QUBiC4G process (shown in Fig. 5.15), feedback provides an effective means for compensating for avalanche current, thereby increasing the useable range of the circuit up to breakdown voltage BV_{CBO}.

In the measurements, a small output current increase is obtained for voltages V_{out} greater than approximately 5 V. This is believed to be due to self-heating of the output transistor, which causes an increase in collector current at a given base-emitter bias voltage. Self-heating was not modelled in the simulations whose results are shown in Fig. 5.14. Note that at a V_{out} of 7 V, the collector-base voltage of transistor Q_2 is approximately 6 V. This corresponds to an $M - 1$ of about 0.55 (from Fig. 5.3) and an I_{avl} of approximately $0.55 \cdot I_{out}$. So, the base current flowing from transistor Q_2 is 0.55 times its nominal collector current. Without avalanche current compensation (and ignoring self-heating), the output current would have increased to $I_{out,nom} + I_{avl} = 15.5\,\text{mA}$.

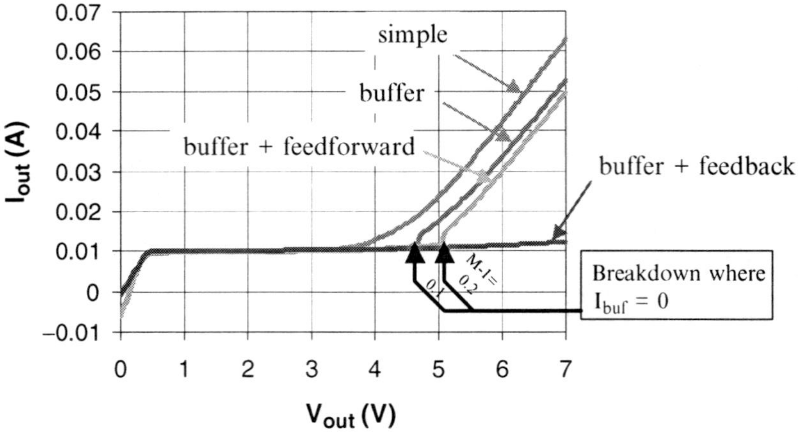

Fig. 5.15 Output characteristics measured for all four current mirror prototypes. The lowest curve corresponds to the circuit with avalanche current compensation using feedback

5.6 Conclusions and Outlook

Several bias current circuits have been proposed and their output characteristics have been analysed. In the case of the simple 2-transistor current mirror, the collector-base avalanche current causes a sharp increase in the output current error at output voltages near and above BV_{CEO}. The output breakdown voltage for this mirror (i.e., BV_{CED}) depends on the output/reference current ratio (n) and will occur when the avalanche multiplication factor $(M-1)$ is equal to 1/n.

Addition of a buffer transistor to reduce the loading of the output transistor on the reference diode will result in output transistor breakdown when the current flow in the buffer transistor decreases to zero. So, the output breakdown voltage of the current mirror depends on the impedance of the buffer bias circuit and its ability to sink the avalanche current flowing out of the base terminal of the output transistor at voltages above (approximately) BV_{CEO}. The results of the study presented in this chapter have shown that the output breakdown voltage increases as the nominal buffer bias current (i.e., I_{R3} in Fig. 5.7b) is increased. A feedforward technique was also presented, which allows operation at elevated output voltages without an excessive increase in the nominal power consumed by the output transistor biasing circuitry.

Moreover, it was shown that an effective way of increasing the output breakdown voltage for a mirror is to bias the buffer transistor using a feedback regulator. Potentially large avalanche currents from the output transistor can be sunk by a large npn without disturbing the bias of the buffer transistor using the circuit shown in Fig. 5.12. This scheme also enables regulation of the mirror output current in the avalanche regime, by subtracting a scaled copy of the avalanche current from the input reference current. Accurate modelling of the avalanche current multiplication

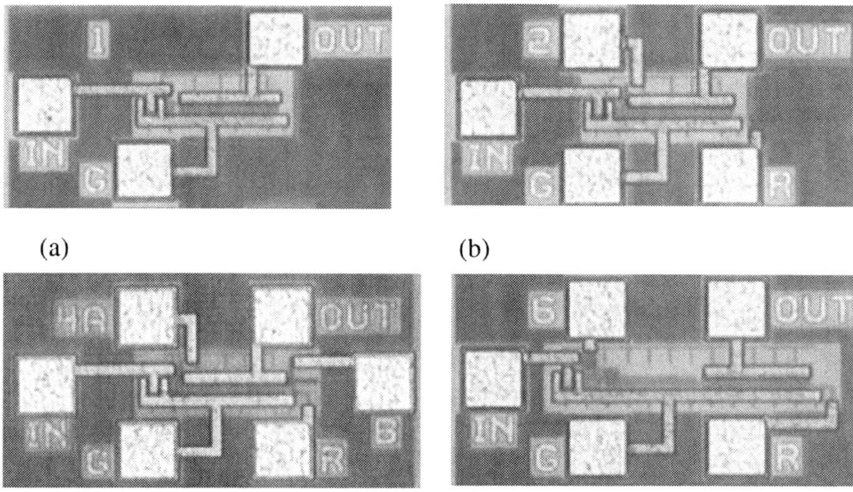

(a) (b)

(c) (d)

Fig. 5.16 Chip photomicrographs of the four current mirrors: simple 2-transistor current mirror (a); current mirror with buffer (b); current mirror with buffer and with feedforward avalanche current compensation (c); current mirror with buffer and avalanche current compensation using feedback (d)

factor up to M of approximately 2 is required for accurate circuit simulations of a practical current mirror operating in the avalanche regime.

Photomicrographs of the four current mirrors described in this chapter are shown in Fig. 5.16. The die area of the current mirror with feedback avalanche current compensation is approximately double the area consumed by the simple 2-transistor current mirror when implemented in the QUBiC4G technology.

To avoid junction breakdown, it is necessary to limit the collector-base voltages of all the transistors in the circuit to less than BV_{CBO}. As the improvement in gain-bandwidth product of sub-micron SiGe and SiGe:C IC technologies has been accompanied by a reduction in the collector-base breakdown voltage (as highlighted in Table 5.1), it is now common practice among most manufacturers to offer npn transistors with different collector dopant concentrations, thereby implementing different breakdown voltages. For example, the QUBiC4G and QUBiC4X technologies listed in Table 5.1 offer 2 npn designs with different breakdown voltages (and different f_T's). The current sources demonstrated in this chapter may alleviate the need for high breakdown voltage devices, which will simplify the fabrication process and reduce production costs.

For every transistor in a given circuit, the maximum (negative) base current can be predicted on the basis of the collector-base bias voltage. The avalanche current multiplication curve for each transistor design (note that each transistor design has a unique curve as shown in Fig. 5.3, for example) can be used to determine whether potentially large negative base currents are affecting circuit behaviour

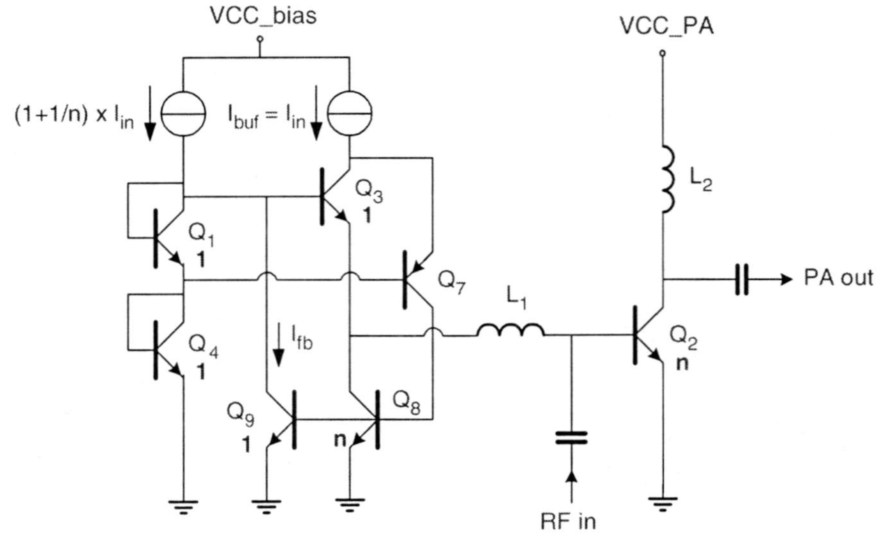

Fig. 5.17 Proposed PA circuit, applying a biasing circuit with feedback avalanche current compensation

and/or performance. Extensive simulation work is usually performed to verify that a circuit will perform according to the desired specifications across all anticipated processing, supply and temperature variations. Circuits may fail as the collector-emitter voltage approaches BV_{CEO} because of avalanche current flows. If the collector-emitter voltage across the output transistor of a biasing current source exceeds BV_{CEO}, the techniques described in this chapter provide an effective means of extending the output voltage range and ensuring proper operation of the circuit.

The avalanche current compensation techniques presented in this chapter are not limited to biasing circuits either. For example, an RF power amplifier (PA) circuit may be designed as shown in Fig. 5.17.

The PA output transistor (Q_2) may be operated at collector-emitter voltages beyond BV_{CEO} because low-frequency avalanche currents are effectively compensated by the biasing circuit. The quiescent current of the PA is set by the input reference current source I_{in}. Reference diode Q_4 and PA transistor Q_2 should be of identical transistor type for current matching purposes. At the PA RF-input, inductor L_1 isolates the dc biasing circuitry from the RF signal path. The avalanche current compensation is isolated from the PA circuit at high frequencies via inductor L_1. Therefore, only low-frequency components of the avalanche currents are compensated. Dynamic avalanche currents in the output stage and their effect on the linearity and reliability of the PA stage would have to be carefully considered in any practical design. Inductors L_1 and L_2 may be implemented using transmission lines with an electrical length of $\lambda/4$, to translate the low impedance dc bias and power supply nodes to high impedance circuits at the base and collector terminals of transistor Q_2.

Whenever a transistor is operated at collector-emitter bias voltages larger than BV_{CEO} so that $(M-1)$ is greater than about 0.1, relatively large currents will flow out of the base terminal. For example, when operating at $(M-1)=0.5$, the base current will be 50% of the nominal collector current. Consequently, the physical layout of a circuit must be designed so that relatively large negative base currents can be handled. The base terminal is often connected using relatively narrow wires and a small number of vias relative to the collector and emitter terminals in a typical circuit. There must also be a sufficient number of base ohmic contacts in order to handle electromigration limitations of metal interconnections at the base terminal when potentially large avalanche currents are anticipated in a circuit.

To support the design of circuits with transistors operating above BV_{CEO}, accurate modelling of the avalanche multiplication factor M is essential. In fact, the avalanche current multiplication factor curve (similar to the one shown in Fig. 5.3) provides essential information for designers of circuits intended for operation at a supply voltage greater than BV_{CEO}. So, the avalanche current multiplication factor curves (representative of each transistor style available in the technology) should be available to circuit designers. This could be facilitated by publishing measured and simulated curves in technology design manuals. Comparisons of measurements and simulation results are useful for circuit design, because junction breakdown is not usually modelled in circuit simulators. Differences between measurements and simulation results are hence likely at collector-emitter voltages beyond $2 \cdot BV_{CEO}$. To predict and analyse breakdown behaviour of bias circuits as a function of temperature, it is necessary that the avalanche multiplication factor curve is available as a function of temperature.

The high-frequency output impedance of the new bias circuits proposed in this chapter has not yet been analysed. However, the output impedance is important for most bias circuits that are used in high bit-rate circuits. Therefore, analysing the output impedance requires attention in future work.

All bias current circuits discussed in this chapter are designed for a nominal output current of 10 mA. The transistors in the bias circuits are relatively small. A further study using transistor models with self-heating and with thermal networks between the transistors is interesting for bias circuits designed for higher output currents (e.g., $\gg 10\,\text{mA}$).

In the theoretical analyses of the relationship between collector-base avalanche current and breakdown presented in Section 5.2, the emitter and base series resistances of the transistor were ignored. These series resistances can however be important for breakdown behaviour. When the transistor is operated in the forward active region at $V_{cb} \gg BV_{CEO}$, the base current becomes relatively large and a significant voltage drop can exist across the base and emitter series resistances. The voltage across the base-emitter junction is then different from the base-emitter terminal voltage: it is increased by the voltage drop across the base series resistance, but reduced by the voltage drop across the emitter series resistance. An increased base series resistance thus leads to a reduced breakdown voltage when operating at high current densities. Similarly, the emitter series resistance increases the breakdown voltage for high current densities.

References

1. G. Freeman, M. Meghelli, et al., "40-Gb/s circuits built from a 120-GHz f_T SiGe technology," *IEEE J. Solid-State Circuits*, vol. 37, No. 9, September 2002, pp. 1106–1114.
2. E.O. Johnson, "Physical limitations on frequency and power parameters of transistors," *RCA Rev.*, vol. 26, p. 163, 1965.
3. K.K. Ng, M.R. Frei, C.A. King, "Reevaluation of the $f_t BV_{ceo}$ limit on Si bipolar transistors," *IEEE Trans. Electron Devices*, vol. 45, No. 8, August 1998, pp. 1854–1855.
4. A. Pruijmboom, D. Szmyd, R. Brock, R. Wall, N. Morris, K. Fong, F. Jovenin, "QUBiC3: a 0.5 μm BiCMOS production technology, with $f_T = 30$GHz, $f_{max} = 60$GHz and high-quality passive components for wireless telecommunication applications," in *Proc. IEEE BCTM*, 1998, pp. 120–123.
5. D. Szmyd, R. Brock, N. Bell, S. Harker, G. Patrizi, J. Fraser, R. Dondero, "QUBiC4: a silicon-RF BiCMOS technology for wireless communication ICs," in *Proc. IEEE BCTM*, 2001, pp. 60–63.
6. P. Deixler, R. Colclaser, et al., "QUBiC4G: a $f_T/f_{max} = 70/100$GHz 0.25um low power SiGe-BiCMOS production technology with high quality passives for 12.5Gb/s optical networking and emerging wireless applications up to 20GHz," in *Proc. IEEE BCTM*, 2002, pp. 201–204.
7. P. Deixler, A. Rodriguez, et al., "QUBiC4X: An $f_T/f_{max} = 130/140$GHz SiGe:C-BiCMOS manufacturing technology with elite passives for emerging microwave applications," in *Proc. IEEE BCTM*, 2004, pp. 233–236.
8. S.M. Sze, *Semiconductor devices, physics and technology*, Section 4.2: Static Characteristics of Bipolar Transistors, Wiley, New York, 1985.
9. H.C. de Graaff, W.J. Kloosterman, "The MEXTRAM bipolar transistor model," *Philips Research Unclassified Report* NL-UR 006/94, Eindhoven, 1994.
10. J.C.J. Paasschens, W.J. Kloosterman, and R. v.d. Toorn, "Model derivation of mextram 504, the physics behind the model," *Philips Research Unclassified Report* NL-UR 2002/806, Eindhoven, 2002.
11. P.F. Lu, T. Chen, "Collector-base junction avalanche effects in advanced double-poly self-aligned bipolar transistors," *IEEE Trans. Electron Devices*, vol. 36, No. 6, June 1989, pp. 1182–1188.
12. P.R. Gray, P.J. Hurst, S.H. Lewis, R.G. Meyer, *Analyses and design of analog integrated circuits*, 4th edn., Section 4.2.2, Wiley, New York, 2001, pp. 255–257.

Chapter 6
Design of Synchronous High-Speed CML Circuits, a PRBS Generator

6.1 Introduction

In this chapter, the design of a high-speed pseudo-random binary sequence (PRBS) generator requiring only a single clock input will be described. The target is to achieve an output bit-rate of at least 40 Gb/s. Detailed circuit simulations using the SiGe technology also used for the 12.5 Gb/s cross-connect switch described in Chapter 4 revealed that a clock to data delay of approximately 15 ps per latch allows the design of a half-rate PRBS core. However, it is not feasible to design the output data multiplexer and output buffer for 40 Gb/s operation. In the simulations, adequate performance was obtained up to approximately 30 Gb/s. To achieve the target 40 Gb/s, an InP HBT technology was selected [1] which results in an improvement in f_T, f_{max} and f_A over the available SiGe technology by a factor of approximately 2 [2]. The design techniques for high bit-rate circuits in SiGe and InP HBT technology are very similar, as will be demonstrated in this chapter.

The maximum speed of digital circuits in a given IC process is often benchmarked on the basis of minimum gate delays, obtained from a ring oscillator. Such a ring oscillator can be built from simple inverters and therefore provides an indication of the maximum achievable speed in a process. However, the design of the ring oscillator tells us little about how more complex gates and latches need to be designed for optimum speed.

In synchronous digital functions, latches are clocked by a common signal. In a synchronous design, the maximum speed may be limited by the delay in either the data or the clock paths. Designing the latch for the lowest propagation delay does not mean that the digital function is optimised for speed, because the clock input of each latch may provide a significant load for the clock line, thereby complicating the clock distribution. Since the propagation delay across the on-chip interconnect may become a significant portion of a bit-time, both data and clock signal distribution play a role in the overall performance. Accurate interconnect models are needed in order to predict the delay across the clock interconnect, to determine the impact on performance of connecting the latches to the clock line, and hence to optimise the latch design and the clock distribution simultaneously. The interconnect design and

H. Veenstra and J.R. Long, *Circuit and Interconnect Design for RF and High Bit-Rate Applications*, 175
doi: 10.1007/978-1-4020-6884-3_6,

models proposed in Chapter 2 of this book will be used in the design of the PRBS generator described in this chapter.

In Section 6.2, more background on PRBS generators will be presented. A brief analysis of the npn device metrics for the InP technology will be given in Section 6.3 of this chapter. In Section 6.4.1, the block diagram of the PRBS generator will be described in greater detail. The all-zero detection and correction circuit will be discussed in Section 6.4.2. The interconnect design and modelling strategies presented in Chapter 2 will be applied to the clock distribution inside the PRBS generator. The design and optimisation of the latch and clock distribution will be described in Section 6.4.3. On-wafer evaluation results are presented in Section 6.5. The concept of distributed capacitive loading will be reviewed in Section 6.6. A discussion of the experimental results and conclusions will be given in Section 6.7.

6.2 PRBS Background

A PRBS generator is an excellent example of a circuit of significant complexity whose performance is substantially improved when both the CML gate delay and the signal distribution are optimised simultaneously. The maximum output bit-rate obtained with a PRBS generator provides a good indication of the maximum speed of other digital functions of comparable complexity in a given technology. A PRBS generator uses at least one data feedback path. In the feedback path, the timing alignment between clock and data signals severely affects the highest speed of the generator. Timing is influenced by factors such as the circuit design of the logic gates and clock drivers, interconnect design and IC floorplan.

In general, if the total delay across the clock line for a digital function is a significant portion of a bit-time, the clock distribution needs to be analysed and optimised using transmission line models. The interconnect models and approach for clock distribution used in a PRBS generator also apply to other digital functions. Simpler functions such as frequency divide-by-2 circuits may reach higher speeds, especially if the latches are optimised for the specific frequency range (for example, by using narrow-band dynamic dividers).

Aside from its use as a technology benchmark, a PRBS generator is widely used for testing digital communication functions. A suitable test configuration for broadband communication systems involves applying pseudo-random data to the communication system under test and measuring the bit-error rate at the output. This configuration is shown in Fig. 6.1. If all functions are implemented on-chip, a built-in self-test (BIST) function can be realised, enabling on-wafer testing at full speed.

Using pseudo-random data, eye patterns can be generated and analysed. For example, the timing jitter generation from a cross-connect switch is measured by comparing jitter from the input signal versus jitter from the output signal. PRBS sequences with various lengths can be generated, but pattern lengths of $2^7 - 1$ or $2^{31} - 1$ bit are often used. The shorter lengths are typically used as vehicles for IC technology demonstrators. Longer sequence lengths are of practical use in the

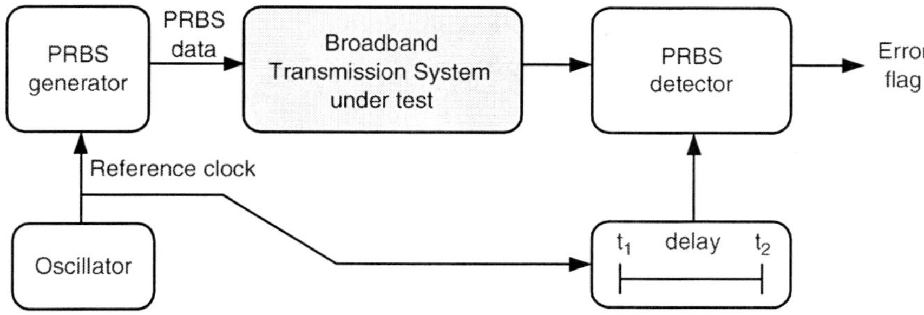

Fig. 6.1 Testing a communication system using pseudo-random data

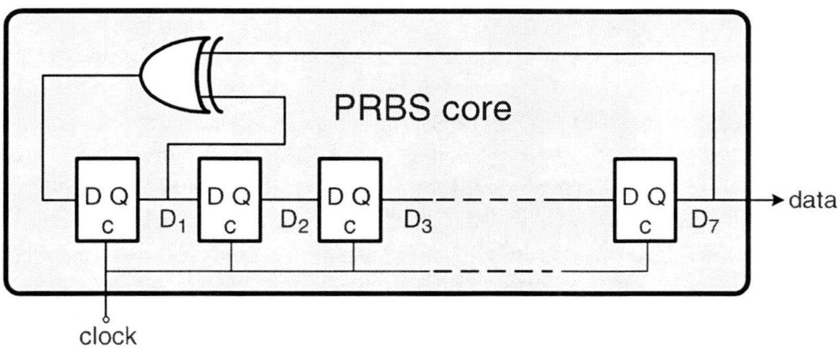

Fig. 6.2 Block diagram of a full-rate PRBS generator with sequence length $2^7 - 1$ bit

generation of ultra wide-band (UWB) signals using pseudo-noise code biphase modulation, for example. The modulated UWB signal has a discrete spectrum with lines spaced at the PRBS data rate divided by the sequence length. With a longer sequence length, more spectral lines per MHz of (on average) lower power are generated, making it easier to comply with for example governmental rules for spectral emissions.

The output signal of a PRBS generator is not random but pseudo-random, as can be seen from the autocorrelation function R of the PRBS sequence. For a PRBS sequence of $2^{\mathrm{N}} - 1$ bit, the autocorrelation function R is

$$R(i) = \sum_{k=1}^{2^N-1} Q(k)Q(k+i) = \begin{cases} 2^N - 1 & \text{for } \mathrm{i} = 0,\ 2^{\mathrm{N}} - 1,\ 2(2^{\mathrm{N}} - 1), \ldots \\ -1 & \text{otherwise.} \end{cases} \qquad (6.1)$$

The pattern length defines the number of latches that are needed in the core of the PRBS generator. For example, for a pattern length of $2^7 - 1$ bit, 14 latches (implementing 7 D-type flip-flops) are needed to generate a 7 bit delay in the generator core. The basic block diagram of a PRBS generator is shown in Fig. 6.2. The construction of the feedback path for generating a maximum-length PRBS sequence is described in [3]. The figure shows a full-rate design in which the output bit-rate in bit/s is equal to the clock frequency.

The PRBS data is available at any point in the data loop. With a 7 bit delay in the loop, a total of 2^7 or 128 states exist. However, the all-zero state is forbidden, because otherwise the latches would remain in this state forever.

As can be seen in Fig. 6.2, the clock drives all 14 latches. At the 40 Gb/s bit-rate, the delay between data input and output for each latch is 12.5 ps. The delay across an unloaded 1-mm-long on-chip clock line is typically 6 ps. Since the typical size of a CML latch in the technology considered in this work [1] is $0.1 \times 0.1\,\text{mm}^2$, placing all 14 latches in a row results in a minimum clock delay of 8.4 ps along the clock line. Capacitive loading further increases the delay in the clock line, while resistive loading may force the need for additional clock buffering, which will also increase the total clock line delay. So, the clock line delay corresponds to a significant portion of a bit-time in this design.

Since the data inside the PRBS generator circles around in a loop, the (mis-) alignment of data and clock at the point where the loop is closed (e.g., the output of the exclusive-OR gate in Fig. 6.2) is important for realising the maximum achievable bit-rate. So, accurate modelling of the clock and data signal distribution is required. Transmission line models are needed for the clock and data lines to study the line delay and possible signal reflections. Using such models, the impact of the latch clock input impedance on the clock distribution can be analysed, and optimisation of the floorplan of the IC can be exploited to obtain low delay and jitter generation in the loop.

High bit-rate PRBS generators are typically not based on full-rate architectures. PRBS data may be generated by multiplexing two identical but time-shifted PRBS sequences, each at half the bit-rate, as shown in Fig. 6.3. When the time-shift between the patterns is correct, the multiplexer output data will be an exact copy of the original PRBS signal, but at twice the bit-rate. This concept, referred to as the 'cycle-and-add' property of PRBS sequences [4], may be repeated, resulting in a quarter-rate architecture. Half-rate and quarter-rate architectures relax the requirements for the PRBS core. The maximum bit-rate in sub-rate PRBS designs is typically limited by the output data multiplexer, clock distribution and buffer circuits.

Table 6.1 gives an overview of previously published single-chip PRBS generators. The designs are based on half-rate or quarter-rate architectures. The sequence length for the generators with output bit-rates above 20 Gb/s is 127 bits in all cases.

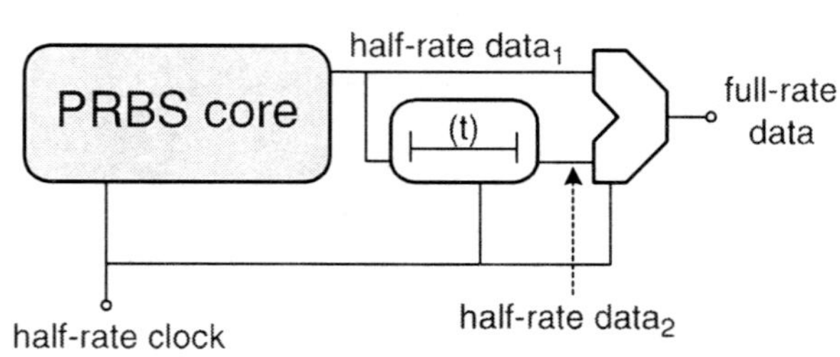

Fig. 6.3 Half-rate PRBS generator block diagram

Table 6.1 Benchmarking previously published PRBS generators

Reference	Kromat [5]	Chen [6]	Schumann [7]	Knapp [8]
Max. bit-rate	11.5 Gbs	21 Gb/s	25 Gb/s	40 Gb/s
Core bit-rate	2.875 Gb/s	10.5 Gb/s	12.5 Gb/s	20 Gb/s
Sequence	$2^{15}-1$, $2^{23}-1$	2^7-1	2^7-1	2^7-1
Auto-start	Yes	No	No	Yes
Trigger output	Yes	Yes	No	Yes
# clock inputs	2	2	2	2
Technology	Si	GaAs HBT	Si	SiGe:C
f_T	25 GHz	40 GHz	50 GHz	106 GHz
Bit-rate/f_T	0.46	0.53	0.50	0.38
Size (mm^2)	4×8	3.2×3.2	1.1×0.86	0.86×0.7
Power dissipation	6.2 W	1.1 W	2.3 W	1.2 W

A longer sequence length requires more latches, and hence increases the power dissipation.

A trigger output, providing a signal with a period which is an integer times the sequence length, is convenient for evaluation of the output data pattern. The automatic start function requires detection plus correction of the all-zero state.

Note that all of the designs listed in Table 6.1 use two clock inputs of identical frequency, whose phase relationship requires accurate external alignment to obtain the reported maximum bit-rate. One clock drives the PRBS generator core at half (or one-quarter) of the desired bit-rate, while the other clock is used for the 2:1 (or 4:1) multiplexer which interleaves bit streams to realize the serial Gb/s data output. Having two clock inputs requiring external phase alignment makes the circuits unsuitable for BIST applications, and therefore the need for two clock inputs must be eliminated.

6.3 InP Technology

While most BiCMOS technologies are very suitable for large-scale integration, InP technologies are intended for small-scale to medium-scale integration. The typical complexity in InP technology is in the range between 10 and 10,000 components per IC. In the case of ICs with more than 1,000 components, yield may drop considerably. The power dissipation in a typical digital circuit with more than 1,000 transistors will readily exceed 1 W, introducing challenges to get rid of the heat. Still, the InP technology is very suitable for high bit-rate circuits of moderate complexity such as clock and data recovery circuits and clock conversion circuits for optical networking.

The minimum emitter area of the npn transistor in the InP technology described in [1] is $1\,\mu m \times 3\,\mu m$, which is relatively large compared with most state-of-the-art silicon BiCMOS technologies. The device metrics for a $1\,\mu m \times 5\,\mu m$ emitter area npn transistor in this technology are shown in Fig. 6.4. The results were obtained using a level-1 (SPICE Gummel-Poon) device model.

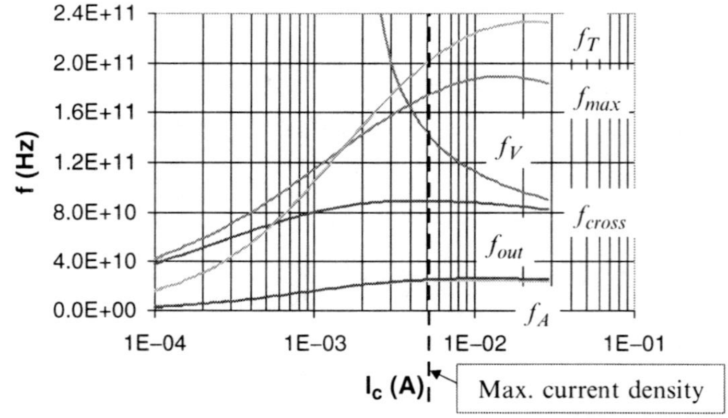

Fig. 6.4 Device metrics for a $1\,\mu m \times 5\,\mu m$ InP npn at $V_{cb} = 0\,V$, $25°C$, obtained using a level-1 npn model

To avoid electromigration problems, the maximum current density allowed is $1\,mA/\mu m^2$. So, the peak–f_T/f_{max} cannot be obtained with a production IC, unless measures are taken to limit the maximum junction temperature. It is interesting to note that in this InP technology the available bandwidth f_A is fully determined by the output bandwidth f_{out}. This is because the input bandwidth f_V is relatively high. To achieve a further increase in f_A in this technology, the effect of the base resistance R_b on the output bandwidth is determined as follows. The output Miller effect plays a role in the output bandwidth at collector currents where $gm \cdot R_b > 1$. The point at which $gm \cdot R_b = 1$ is found at the crossing of the f_V and f_T curves; in Fig. 6.4 this is at $I_c = 3.4\,mA$. So, with the example transistor size and operating conditions, at $I_c >$ $3.4\,mA$ the output capacitance C_{22} $(= C_{cs} + C_{bc}(1 + gm \cdot R_b)$, see also Section 3.3.3) will start to increase and f_{out} will saturate (as can indeed be observed in Fig. 6.4).

To conclude this discussion of the InP technology, it is necessary to increase the maximum allowable current in the npn to enable biasing of the transistor at peak-f_T, for example by increasing the number of contacts or the size of the contact holes to the emitter. Then it becomes interesting to lower the base resistance R_b in order to increase f_{out} at high bias currents. A good target is to achieve $gm \cdot R_b = 1$ at peak-f_T; this implies reducing R_b by a factor of 2–4. An increase in f_A is also feasible by reducing C_{bc} as this will also lead to an increase in f_{out}. Note that more accurate directions for technology improvements are obtained when the device metrics are evaluated using measured y-parameters instead of device model simulations. Without the proposed technology improvements, so using the InP technology as it is, the device metrics are already favourable in comparison with the available SiGe technology. Table 6.2 compares the main device metrics for the npn used in the two technologies considered for the PRBS generator. Two columns are shown for the InP technology: one with the peak values and one giving the maximum values when the current density is limited to avoid electromigration problems.

Table 6.2 Comparison of simulated npn device metrics at $V_{cb} = 0\,V$, $25\,°C$

	InP [1]	InP [1], restricting I_c	SiGe [2]
Peak-f_T (GHz)	233	200	61
Peak-f_{max} (GHz)	189	173	73
Peak-f_A (GHz)	25	24	15
Peak-f_{cross} (GHz)	89	89	34

6.4 PRBS Generator Design

In this section, the PRBS generator design is described. The functions of the different building blocks are discussed in Section 6.4.1. Circuit-level details are given for the all-zero detection and correction function in Section 6.4.2. A detailed discussion on the clock distribution, including latch circuit design, is presented in Section 6.4.3.

6.4.1 PRBS Generator Block Diagram

In this section, the concept of the PRBS generator will first be explained on the basis of a full-rate architecture. The actual implementation is based on a half-rate architecture. The extra hardware needed to transform the full-rate concept to a half-rate architecture will be described. The block diagram will then be extended with a trigger function and all-zero detection and correction circuitry.

The concept of the PRBS generator is shown in Fig. 6.5. In this figure, the PRBS core is based on a full-rate architecture. The block diagram of Fig. 6.5 reflects the floorplan of the IC. Each data flip-flop (DFF) consists of two latches, together realising 1 bit delay. The IC provides both single-ended (SE) and differential clock inputs. Selection between the two clock inputs is made via a 2:1 multiplexer, controlled via the 'sel'-input. The clock distribution inside the PRBS core is fully differential and uses a GSSG transmission line on top of a ground shield. The transmission line is terminated at both ends and in the middle, at the point at which the clock signal is driven onto the line. At the ends of the transmission line, the line is terminated differentially with its effective differential-mode characteristic impedance $Z_{0dm,eff}$. The termination resistors provide a common-mode termination impedance of $Z_{0dm,eff}/4$. The series impedance of the supply line adds to the common-mode termination impedance. Since there is some coupling between the signal lines of the GSSG transmission line, the common-mode termination impedance (assuming an ideal supply) is somewhat lower than the common-mode characteristic impedance of the transmission line. Some common-mode signal reflections may therefore occur. To avoid common-mode to differential-mode conversion, it is important to optimise the layouts of the circuits with respect to symmetry and matching of the differential half-circuits.

The combination of a data path placed in a ring with a clock path according to the floorplan shown in Fig. 6.5 allows simple alignment between data and clock at

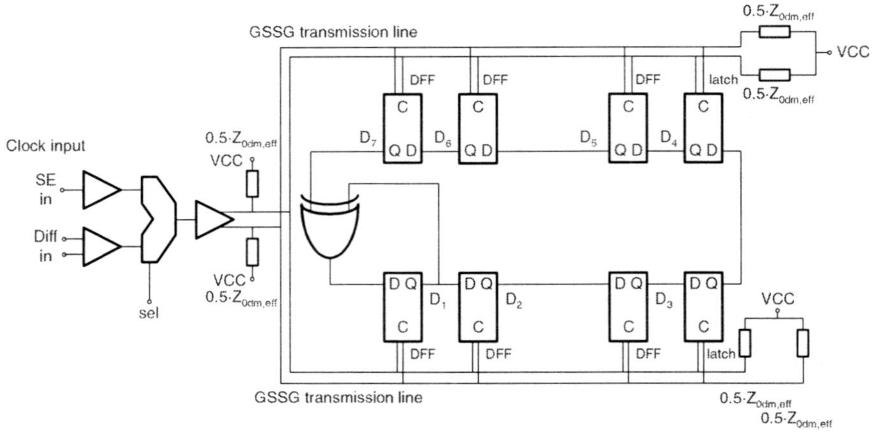

Fig. 6.5 PRBS generator concept, showing the basic PRBS core with clock distribution

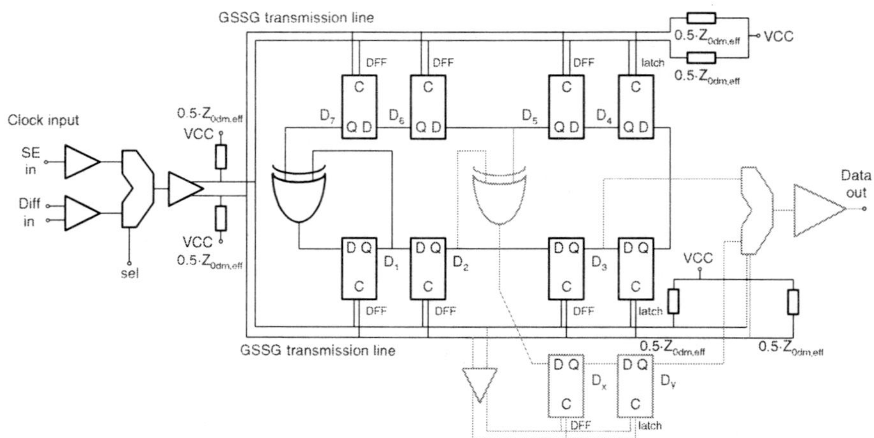

Fig. 6.6 Half-rate PRBS generator block diagram

the inputs of all the latches. So, the clock/data alignment at the first latch of D_1 is not affected by the clock delay in the loop. The maximum speed of the generator is limited by the data path between two consecutive latches with the longest delay. In the above design, the longest delay occurs between the output of DFF D_7 and the input of DFF D_1 due to the delay of the exclusive-OR gate. To transform the full-rate concept to a half-rate architecture, additional circuitry is needed, as shown in Fig. 6.6.

An exclusive-OR gate, a 2:1 multiplexer and a 1.5 bit shift register are added to the full-rate design. Again the block diagram reflects the floorplan. The location of the additional circuitry is chosen so that the alignment of the two data signals and the clock signal at the inputs of the 2:1 data multiplexer is optimum, while the impact of the additional circuitry on the PRBS core is minimal.

The clock input frequency in the half-rate design is 20 GHz for 40 Gb/s data, or one-half of the original design. The 2:1 output multiplexer is controlled by the on-chip clock, as can be seen in the block diagram. So, the half-rate PRBS generator shown in Fig. 6.6 requires only a single external clock signal.

A trigger output can be implemented on the basis of different concepts. If the bit clock is derived from a lower frequency reference clock signal via a CMU, the trigger signal may be derived from the low-frequency reference clock, possibly using standard CMOS logic. However, when the PRBS generator is implemented as a self-contained system, as is the case when the PRBS generator serves as a technology demonstrator, the trigger signal needs to be derived from the clock that drives the PRBS core.

Since the PRBS generator cycles through each of its 127 states for every PRBS sequence, a trigger signal may be derived by detecting any of the 127 states. Conceptually, any state can be detected using a 7-input logic gate. In [8], state 0000001 is detected. The detection logic for the derivation of the trigger signal can then largely be combined with detection of the all-zero state. A drawback of the detection of a single state is that all 7 DFF outputs in the generator core are loaded by the inputs of the trigger signal generator. Although the trigger output signal has a relatively low frequency ($f_{clock}/127$ or lower), a single state must be detected within a single bit-time (of the half-rate core). This requires the use of gates that operate at bit speed, thereby approximately doubling the capacitive load at each DFF output.

Another approach for generating a trigger signal is based on the property that there are exactly 32 rising edges within each $2^7 - 1$ PRBS sequence. The trigger signal can thus also be derived from the serial PRBS data using a frequency divide-by-32 circuit. The input of the trigger signal generator can be taken from any data signal inside the PRBS core. A position is chosen at which the loading is relatively low, so that the maximum speed of the PRBS core is not degraded. The block diagram of the half-rate PRBS core extended with trigger signal generator is shown in Fig. 6.7. The trigger signal generation circuit is implemented as a ripple counter: the

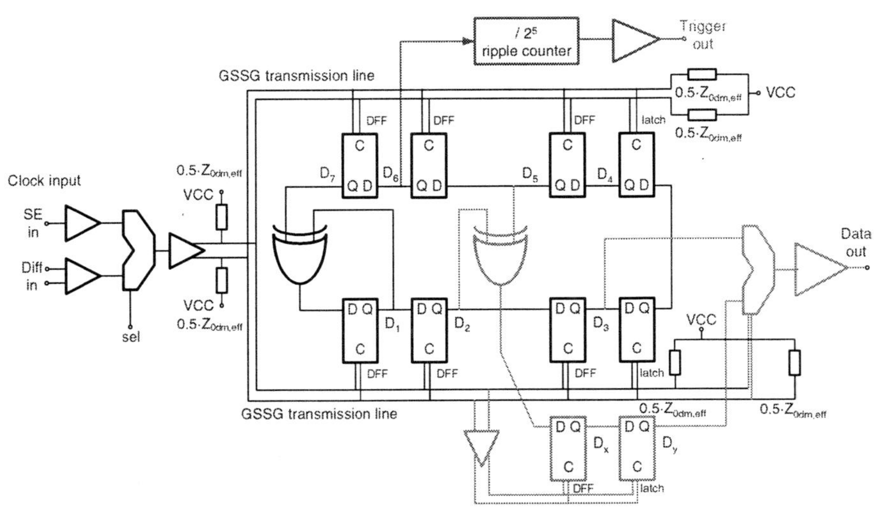

Fig. 6.7 Half-rate PRBS generator, extended with a trigger signal generator

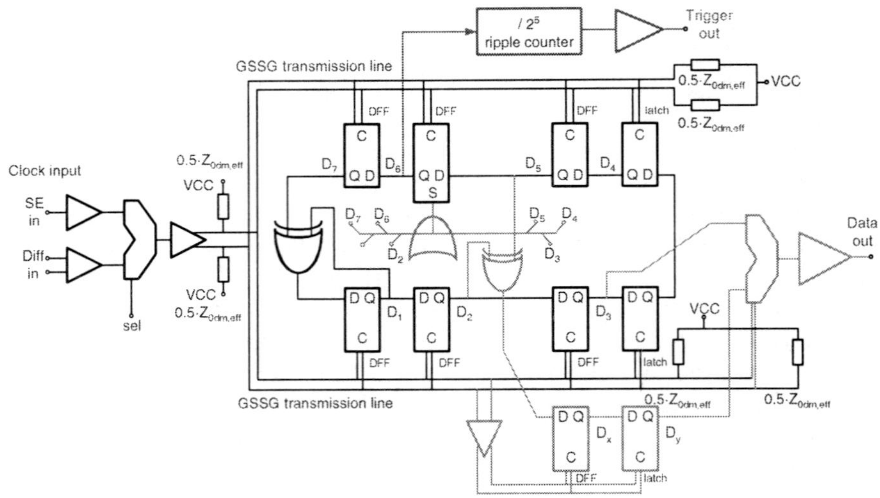

Fig. 6.8 Complete PRBS generator block diagram

output of each divide-by-2 circuit is used as a clock for the following divide-by-2 circuit. The first divide-by-2 circuit needs to operate at up to half the maximum bit-rate of the PRBS generator. For each subsequent divide-by-2 circuit, the speed requirements are relaxed by a factor of 2. It is possible to increase the impedance level and reduce bias currents in the circuits with relaxed speed requirements.

In this implementation, a 7-input wired-OR gate is used for all-zero detection, as shown in the complete generator block diagram of Fig. 6.8. To correct the all-zero state, it is sufficient to set one DFF output signal. DFF D_6 is for this purpose extended with an asynchronous set input.

6.4.2 All-Zero Detection and Correction

Since the all-zero detection function can be implemented using relatively slow logic, the bias current for the all-zero detector is lower than that for the fast logic gates inside the PRBS core. A low bias current results in a relatively low input capacitance for each data input of the 7-input wired-OR gate, thereby causing only a small increase to the load of the DFF outputs in the PRBS core. The circuit diagram of the all-zero detection and correction circuit is shown in Fig. 6.9. The input differential pair amplifiers operate from a relatively low bias current (e.g., $I_b = 1\,\text{mA}$, biasing the transistors a factor of 3 below the maximum allowed current density).

When all the differential data inputs are in the logic zero state, the current through the load resistors R and R/2 will be zero and the voltage at node V_a will equal VCC. Resistor R has a value so that $I_b \cdot R > V_{be}$. When n logic inputs are high, n integer and $n \in [1,\ 7]$, a current $n \cdot I_b$ will flow through the load network. The current through

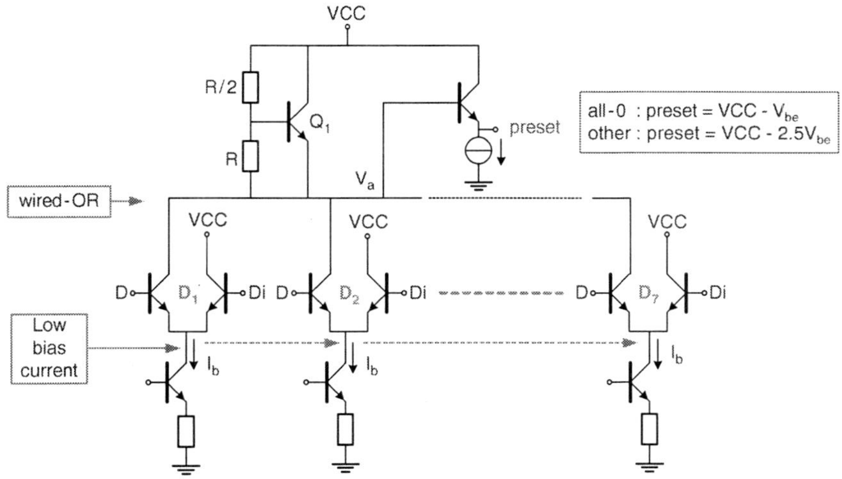

Fig. 6.9 Wired-OR CML circuit for all-zero detection and correction

resistor R is clamped to V_{be}/R, the remainder $n \cdot I_b - V_{be}/R$ flows through transistor Q_1. The resulting voltage at node V_a is $V_a = VCC - 1.5 \cdot V_{be}$.

The preset output activates the asynchronous set-input of a latch. Inside the latch with preset input, the preset signal actives a bypass path from the bias current to the load resistor of the inverted data output that is independent of the clock and data input signals.

The wiring from the DFF outputs to the detector inputs may provide a significant increase in the capacitive loading of each DFF. In this design, the amplifier interface circuit per DFF has been placed physically close to each DFF. The bias reference per amplifier is also generated locally per DFF. The resulting long wiring at the amplifier outputs is part of the wired-OR function for which no high-speed requirements have to be met.

6.4.3 Clock Distribution and Latch Design

The differential clock signal is distributed via a coplanar differential transmission line implemented in the top metal layer (Metal3) over a first metal ground shield (Metal1). The configuration conforms to the preferred transmission line configuration described in Chapter 2. The clock transmission line is indicated as a GSSG line in the block diagram of Fig. 6.8. The layout and equivalent lumped element model for the clock line are shown in Fig. 6.10.

The unloaded transmission line is designed for a differential-mode characteristic impedance of $Z_{0dm} = 100\ \Omega$ and has a common-mode characteristic impedance $Z_{0cm} = 50\ \Omega$. The signal delay of the line for common and differential modes is

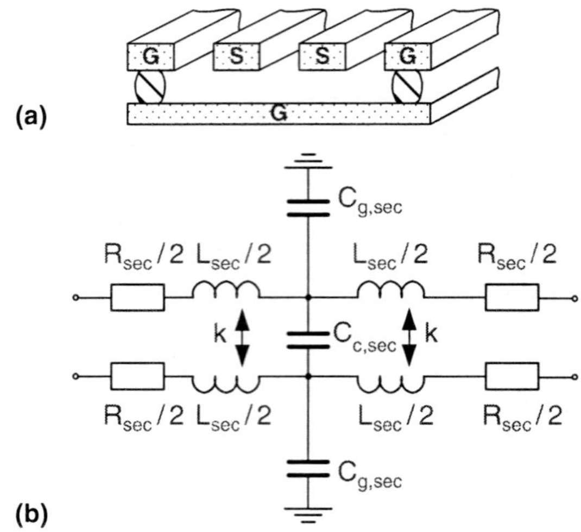

Fig. 6.10 Differential GSSG clock transmission line physical layout (a) and equivalent electrical model (b)

$t_{dm} \approx t_{cm} \approx 6\,\text{ps/mm}$. The physical length of the clock line between two consecutive latches is 0.125 mm. The transmission line model between two consecutive latches consists of one section as shown in Fig. 6.10b. One section thus represents a delay of $0.125\,\text{mm} \times 6\,\text{ps/mm} = 0.75\,\text{ps}$. At an output bit-rate of 50 Gb/s, the half-rate core operates at 25 Gb/s, so there are 53 sections per bit: sufficient to accurately model the signal distribution across the clock interconnect.

The relationship between the four line properties (Z_{0dm}, Z_{0cm}, t_{dm} and t_{cm}) and the equivalent lumped element model was described in Chapter 2 and [9] and is here repeated in equations (6.2)–(6.5). The losses are ignored in these equations (e.g., the equations are valid for $R_{sec} = 0$). In the following equations, the element values L, C_c and C_g represent the lumped sum across all the sections so that t_{dm} and t_{cm} are delay values across the total line length.

$$Z_{0dm} \approx \sqrt{\frac{2L(1-k)}{C_c + C_g/2}} \tag{6.2}$$

$$t_{dm} \approx \sqrt{2L(1-k)(C_c + C_g/2)} \tag{6.3}$$

$$Z_{0cm} \approx \sqrt{\frac{L(1+k)}{4C_g}} \tag{6.4}$$

$$t_{cm} \approx \sqrt{L(1+k) \cdot C_g} \tag{6.5}$$

The clock line is loaded by in total 14 latches from the PRBS core, one clock buffer and one data multiplexer, all distributed across the total line length. The differential input impedance of each latch can be mapped onto a parallel equivalent network

$R_l//C_l$. In the following analysis, it will be assumed that the clock buffer and data multiplexer provide the same loading to the clock line as a latch.

The concept of distributive capacitive loading is applied to the clock distribution. This concept is also applied to the distribution of signals inside the matrix of the cross-connect switch, as explained in Section 4.2. To minimise the clock line delay while applying distributed capacitive loading, the latch input capacitance C_l needs to be minimised while the latch input resistance R_l needs to be high, that is, $R_l \gg Z_{0dm}$. In addition, the line length between two consecutive clock inputs must be equal for all latches. The latch differential input capacitance C_l may then be added to the lumped differential capacitance of a single line section $C_{sec} = C_{c,sec} + C_{g,sec}/2$. The concept of clock distribution via distributed capacitive loading is shown in Fig. 6.11.

The distributed loading of the transmission line with the latches results in a lower characteristic impedance and increased time delay for the clock signal. The effective differential-mode characteristic impedance $Z_{0dm,eff}$ of the line loaded with the clock inputs of the latches equals

$$Z_{0dm,eff} = Z_{0dm}\sqrt{\frac{C_{sec}}{C_{sec}+C_l}} \tag{6.6}$$

Signal reflections across the clock transmission line can be minimized by terminating the line at the ends differentially with resistors of value $Z_{0dm,eff}$, as shown in Fig. 6.8. The need for reduced termination resistor values results in an increase in power dissipation in the clock driver of a factor of $(Z_{0dm}/Z_{0dm,eff})$ to maintain an ECL swing of $0.2\,V_{p,diff}$ at the clock line.

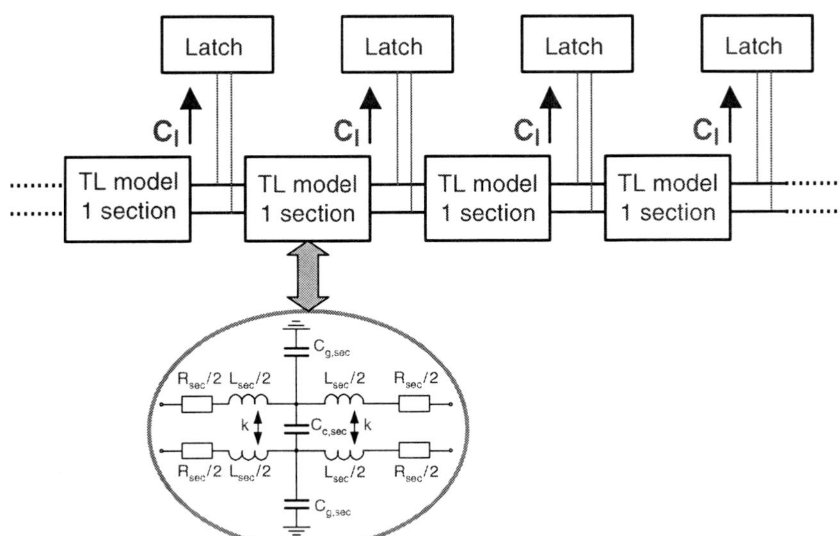

Fig. 6.11 Clock distribution based on the concept of distributed capacitive loading

The effective delay $t_{dm,eff}$ across the total clock transmission line loaded with the clock inputs of the latches equals

$$t_{dm,eff} = t_{dm}\sqrt{\frac{C_{sec}+C_l}{C_{sec}}} \tag{6.7}$$

As can be seen from equations (6.6) and (6.7), it is important to design the latch for minimum input capacitance C_l, both for minimum power dissipation and for minimum clock line delay. Obviously, a small physical size of the latch is as important for realising a low clock line delay. To determine whether the input capacitance C_l is low enough, the value of C_l can be compared with the equivalent lumped capacitance of a single line section between two latches, C_{sec}. At a typical section length of 0.125 mm and $Z_{0dm} = 100\,\Omega$, the lumped line capacitance of a single section is $C_{sec} = 7.5\,\text{fF}$.

The circuit diagram for the latch is shown in Fig. 6.12. The latch design is based on standard current-mode logic. Each latch generates its own bias currents. Emitter followers Q_1 and Q_2 are used to minimize the clock input capacitance. Since single emitter followers Q_3 and Q_4 are used at the differential data output, this type of logic is also often referred to as emitter-coupled logic (ECL).

The physical size of the latch (including a supply decoupling network) is $125\,\mu\text{m} \times 125\,\mu\text{m}$. The length of the clock line between two latches is $125\,\mu\text{m}$, giving $t_{dm,sec} = 0.75\,\text{ps}$ and $C_{sec} = 7.5\,\text{fF}$. Emitter followers Q_1 and Q_2 use minimum emitter area ($1\,\mu\text{m} \times 3\,\mu\text{m}$) transistors. The input capacitance $C_{i,dp}$ of the

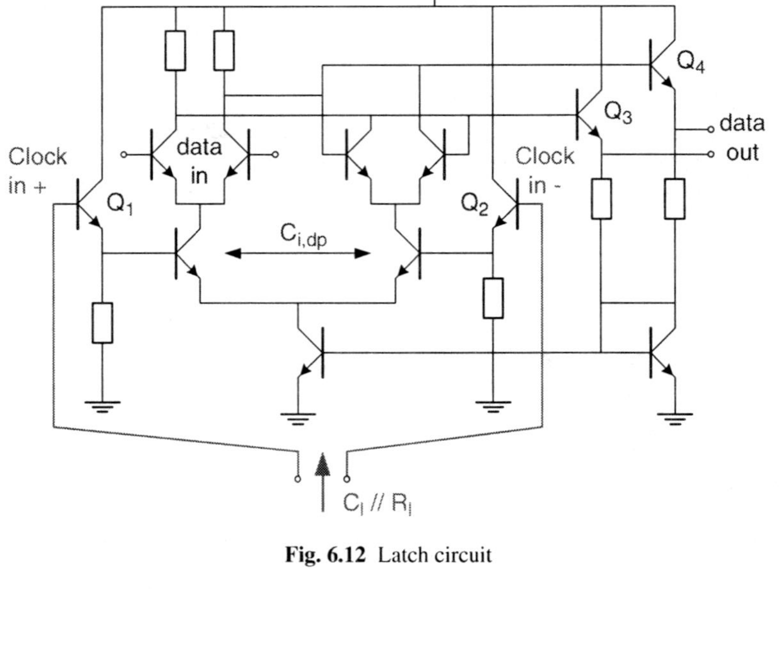

Fig. 6.12 Latch circuit

clock differential pair (see Fig. 6.12) is larger than C_{sec}. The value of $C_{i,dp}$ can be approximated via the f_T and dc bias of the transistors:

$$C_{i,dp} \approx 0.5 \cdot \left(\frac{1}{2\pi f_T \cdot \frac{1}{gm}} + C_{bc} \right) \tag{6.8}$$

The differential pair operates at $f_T \approx 170\,\text{GHz}$ at a tail current of 3 mA. With $C_{bc} = 8\,\text{fF}$, the result is $C_{i,dp} \approx 32\,\text{fF}$.

The input admittance Y_i from an emitter follower loaded with capacitance C_x can be derived as follows. When the collector-base capacitance C_{bc} is ignored, the input admittance can be approximated for $\omega < \omega_T$ by (see also equations (4.9) and (4.10))

$$Y_i = \frac{i_b}{v} = \frac{i_e}{v(\beta+1)} \approx \frac{j\omega C_x}{\beta+1} = j\left(\frac{\omega C_x}{\beta_0+1}\right) - \left(\frac{\omega^2 C_x}{\omega_T}\right) \tag{6.9}$$

The real part $\text{Re}(Y_i)$ corresponds to a frequency-dependent negative resistance. The imaginary part $\text{Im}(Y_i)$ corresponds to a capacitance of approximately C_x/β_0. The input capacitance of the emitter follower equals $C_x/\beta_0 + C_{bc}$.

With emitter followers Q_1 and Q_2 present, the differential clock input capacitance reduces to $C_l = C_{i,dp}/\beta_0 + C_{bc}/2 \approx C_{bc}/2 = 4\,\text{fF}$. Using equation (6.6), the effective differential-mode characteristic impedance becomes $Z_{0dm,eff} = 81\,\Omega$, and via equation (6.7) the delay of a section of the clock line between two latches is found to be $t_{dm,sec,eff} = 0.93\,\text{ps}$. If the clock is distributed in a ring, the total delay in the ring will be approximately (14 latches $\cdot$ 0.93 ps) or 13 ps, considerably reducing the allowable set-up time of the latch at the point where the clock enters the ring. Note that the bit-time in the half-rate PRBS core when operating at an output bit-rate of 40 Gb/s equals 50 ps. A better alternative is to use the fork-shaped clock distribution as shown in Fig. 6.8.

A simulation example showing the clock distribution in the lower half of the PRBS core is given in Fig. 6.13. Between the arrows, the clock line is loaded by seven latches, so the expected time delay Δt between the points indicated by the arrows is $6 \cdot 0.93$ ps or 5.58 ps. In the simulation, extra delay occurs due to the additional physical line length needed to fit the exclusive-OR function for half-rate functionality, and due to additional line loading for tapping the clock towards the 1.5 bit shift register for half-rate functionality.

From the simulation result it can be seen that the clock signal amplitude is increased at points further away from the clock driver. This is due to the negative input resistance R_l of the latches. The negative differential shunt resistance follows from the real part of equation (6.9) and equals

$$R_l = \frac{-\omega_T}{\omega^2 C_{i,dp}} \tag{6.10}$$

With $f_T = 170\,\text{GHz}$ and $C_{i,dp} = 32\,\text{fF}$ follows (at 25 GHz) $R_l = -1.35\,\text{k}\Omega$. Since $|R_l| \gg Z_{0dm,eff}$, a slight increase in signal amplitude occurs towards the line end. The increase in amplitude has no significant impact on the clock distribution.

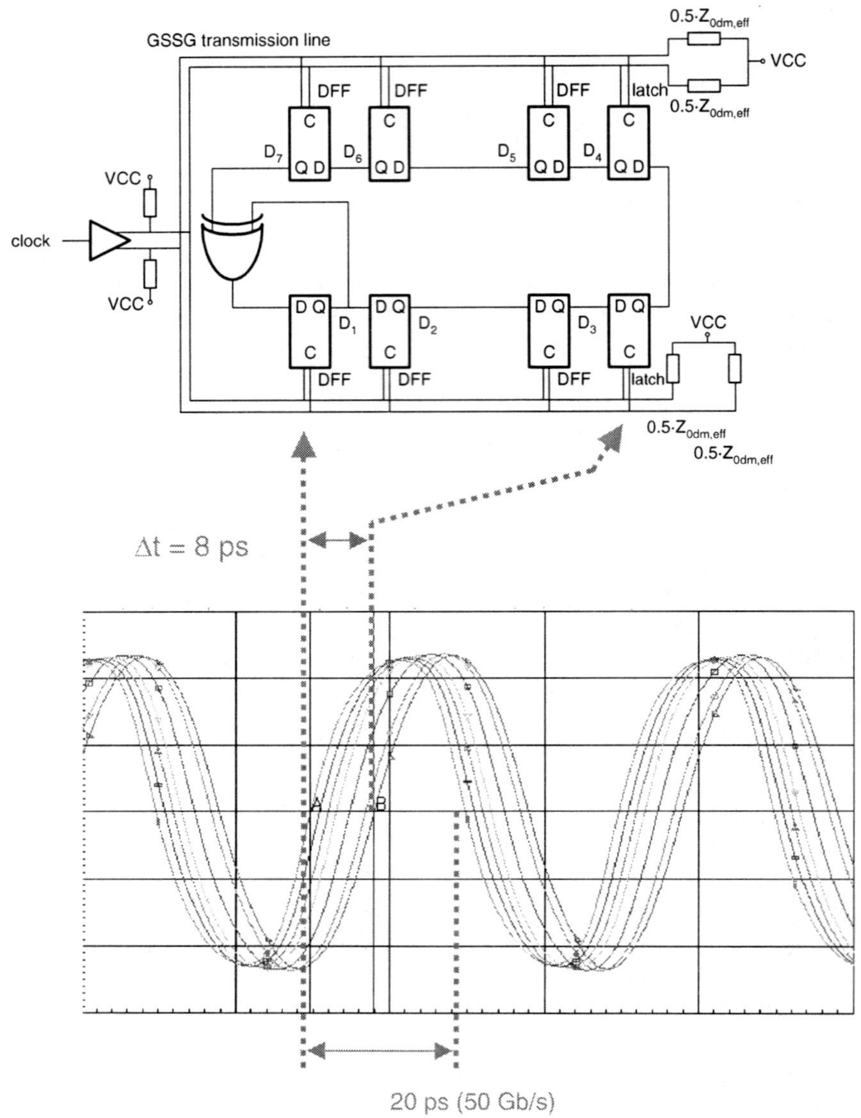

Fig. 6.13 Clock distribution simulation. The clock frequency is 25 GHz, corresponding to a 50 Gb/s bit-rate

The short transmission line for delay matching of the second PRBS core data output (providing the clock signal for flip-flop D_x and latch D_y in Fig. 6.8) does not require line termination at both ends. Due to the relatively short length (0.3 mm) required to cover the three latches, the total line delay is only 2 ps. The buffer, driving the signal onto the short line, provides a matched output impedance. Reflections from the open line end are insignificant in this case.

6.5 Experimental Results

A chip photomicrograph of the entire PRBS generator is shown in Fig. 6.14. The layout was designed according to the floorplan of the block diagram shown in Fig. 6.8. The clock transmission line is indicated.

The IC has been evaluated by means of on-wafer probing. Using the trigger output from the IC, the output data pattern was monitored and compared with results of simulations of a behavioural model. The measured output data at 20 Gb/s is shown in Fig. 6.15. The output data remains correct across an input frequency range of 0.5–29 GHz, which corresponds to an output bit-rate from 1 Gb/s to 58 Gb/s.

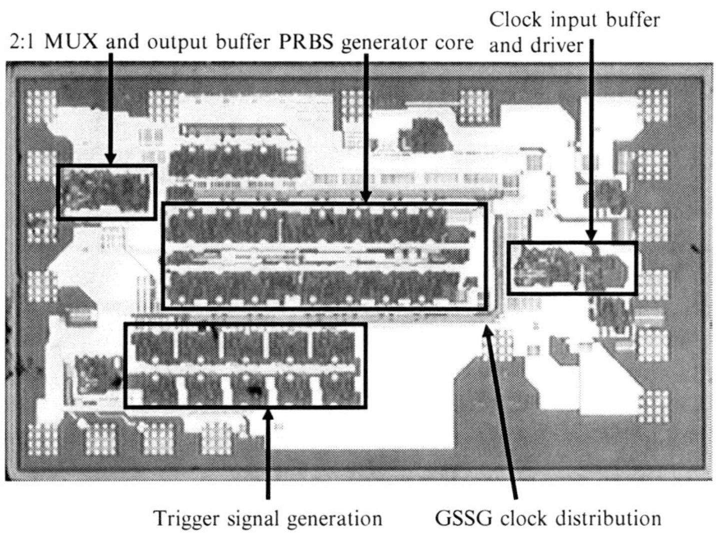

Fig. 6.14 PRBS generator chip photomicrograph. The IC measures $1.2 \times 2.2\,mm^2$

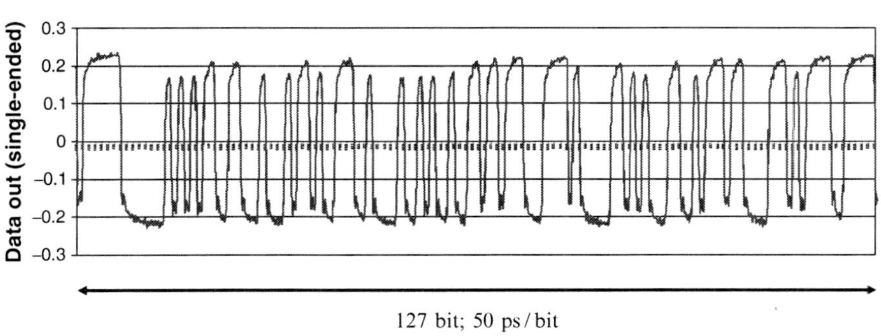

Fig. 6.15 Measured output data pattern at 20 Gb/s (using a 10 GHz clock), one sequence length. The result was obtained using the trigger output provided by the IC

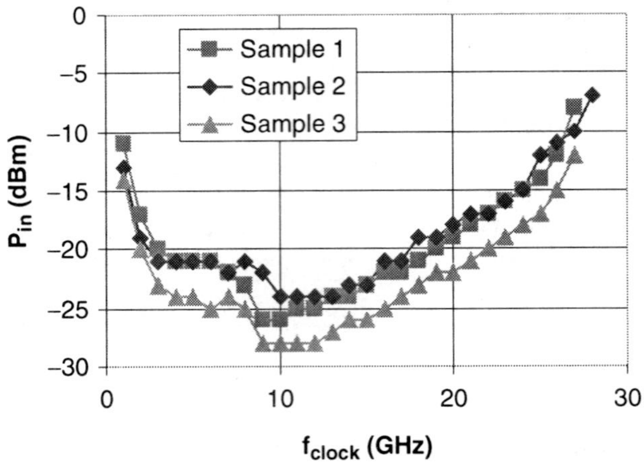

Fig. 6.16 Clock input sensitivity measured for three samples when driving the clock input with a sinusoidal signal. The single-ended clock input was used

The measured clock input signal sensitivity is shown in Fig. 6.16. The input sensitivity is below -10 dBm (corresponding to 0.1 V_p into 50 Ω) at frequencies between 1 and 26 GHz. The clock input selection circuit provides a high small-signal gain, resulting in high clock input sensitivity. At frequencies below 2 GHz, the input sensitivity drops due to the on-chip ac-coupling at the clock input and the reduced slew-rate of the sinusoidal input signal.

The results shown in Fig. 6.15 can only be used to evaluate the correctness of the output pattern, and should not be used to evaluate the quality of the output signal (rise-time, jitter, etc.) because the signal was obtained using a relatively low sampling rate for the communications analyser. The low sampling rate is necessary to capture at least one full PRBS sequence length. For evaluation of the quality of the data output signal, eye patterns of the output data are generated at the maximum (50 Gs/s) sampling rate of the communications analyser. An example eye diagram obtained at 58 Gb/s, the highest bit-rate at which the generator is still functional under nominal operating conditions, is shown in Fig. 6.17. Only one signal polarity of the differential data output was measured. Use was made of the single-ended clock input driven by a low-noise microwave signal generator at 29 GHz.

The output jitter remains below 1.1 ps RMS at output bit-rates up to 58 Gb/s. Table 6.3 provides a summary of results obtained at a nominal supply VCC = 3.5 V and room temperature.

A benchmark with previously published PRBS generators, comparing power dissipation and maximum output bit-rate, is shown in Fig. 6.18. Although it does not operate at the lowest power dissipation, the PRBS generator described in this chapter achieves the highest bit-rate to date.

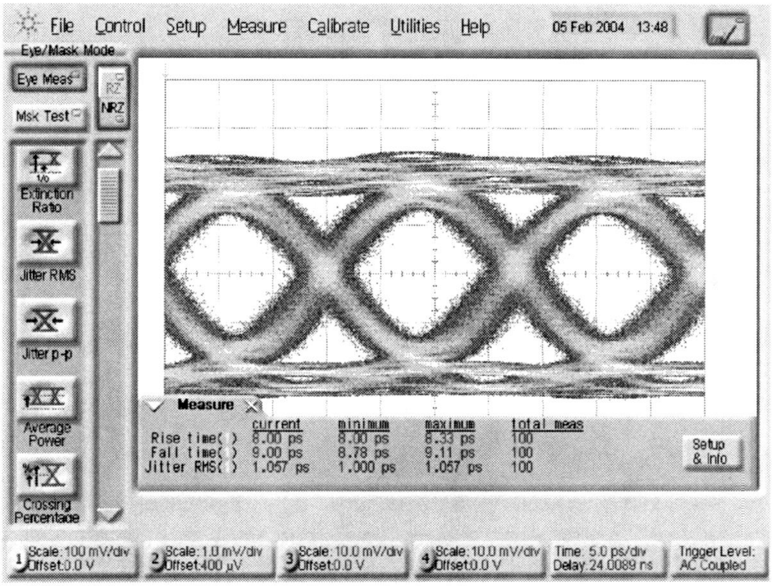

Fig. 6.17 Single-ended output eye diagram measured at 58 Gb/s

Table 6.3 InP PRBS generator performance summary

Parameter	Value
Output bit-rate	1–58 Gb/s
Core bit-rate	Half-rate (0.5–29 Gb/s)
Output jitter	<1.1 ps RMS
Sequence	$2^7 - 1$
Auto-start	Yes
Trigger output	Yes
# clock inputs required	1
Clock input sensitivity	<-15 dBm for 2–48 Gb/s; <-6 dBm for 1–58 Gb/s
IC technology	InP HBT; $f_T = f_{max} = 170$ GHz; 3 metal layers
Bit-rate/f_T	0.34
Chip area	$1.2 \times 2.2\,\text{mm}^2$
Power dissipation	1.75 W

6.6 Distributed Capacitive Loading Reviewed

The joint optimisation of circuits and interconnect for optimum RF signal transfer across relatively long distances on a chip is based on the concept of distributed capacitive loading, as explained in Section 4.2. This concept can only be applied when the total capacitance loading the transmission line can be sub-divided into smaller (but equal) parts. This concept considers the input capacitance of each circuit connected to the transmission line to be part of the transmission line, and consequently results in a reduced characteristic impedance (from Z_0 to $Z_{0,eff}$, see

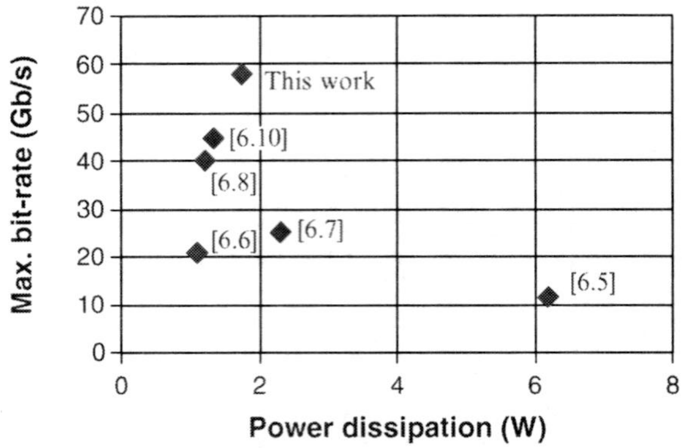

Fig. 6.18 Benchmark

equation (4.4)) and increased delay (from t_d to $t_{d,eff}$, see equation (4.5)). The distributed capacitive loading concept is widely used in distributed (broadband) amplifier design. In this book, distributed capacitive loading was applied to the RF signal distribution inside the cross-connect switch IC (in Chapter 4) and to the clock distribution inside the PRBS generator IC (in Chapter 6). The concept can be summarised as follows:

1. Initial design of the IC floorplan. The floorplan should result in an equal distribution of circuits and parasitics (e.g., from crossing interconnects) across the length of the transmission line
2. Initial design and modelling of the RF interconnect (Chapter 2)
3. Design of the circuits connecting to the transmission line with minimum input (or output) capacitance (not significantly larger than the lumped capacitance of the line section between two consecutive circuits) and high input (or output) resistance ($\gg Z_0$). Examples in this book are matrix node circuit design (Section 4.2) and latch design (Section 6.4.3)
4. Terminate the loaded transmission line with $Z_{0,eff}$

This procedure has been successfully applied to the design of the cross-connect switch IC and PRBS generator. Both ICs were first-time-right.

6.7 Conclusions and Outlook

High-speed digital circuits typically use current-mode logic (CML). To support Gb/s bit-rates, it is not sufficient to design the gates for minimum delay. In the case of digital circuits with more than a few gates, the delay of the interconnect for clock and data paths may be a significant portion of a bit-time and hence contribute to the critical path for speed.

From this study, it can be concluded that the maximum speed of operation depends not only on the gate delays, but also on the floorplan and the clock and data signal distribution. In addition to the maximum achievable speed of operation, the clock and data signal distribution also has an important impact on the quality of the data signal at high bit-rates. Signal reflections may adversely affect the jitter performance or introduce bit errors. So, for high bit-rate digital functions, it is important to optimise the circuits, interconnect and floorplan interactively. Accurate interconnect models are required for this purpose. Since high-speed CML is differential, there is a need for differential on-chip interconnect, very similar to the requirements of the cross-connect switch described in Chapter 4.

In this chapter, a fully integrated PRBS generator implemented in an InP HBT technology has been described. The fork-shaped clock routing and minimum (clock) input capacitance latch design in combination with a clock distribution based on the concept of distributed capacitive loading, enables the PRBS generator core to operate at up to 29 Gb/s, resulting in a 58 Gb/s output bit-rate.

The PRBS generator is an excellent example of a CML circuit of reasonable complexity (28 latches), demonstrating the design flow of Fig. 2.42:

- A rudimentary layout of a CML latch is used to design the initial floorplan. This information is needed to predict the clock interconnect delay. One important aspect of the floorplan is that the latches that are connected to the clock line are equally distributed along the line.
- A coplanar GSSG transmission line according to the preferred configuration shown in Fig. 2.20 is used for clock distribution. Knowledge of the back-end of the IC process (e.g., metal layer thickness, layer permittivities and design rules) is used to design an initial transmission line of a typical impedance level.
- A lumped RLMCG model according to Fig. 2.6 is used to capture the estimated characteristic impedance, delay and loss of the clock transmission line (for differential and common modes).
- The concept of distributed capacitive loading is applied to the clock signal. For minimum clock line delay, the latches (loading the clock line) are designed for minimum clock input capacitance. Simultaneously, the GSSG transmission line is designed using an EM simulator tool. The line impedance and delay for differential- and common-modes are analysed.
- The effective characteristic impedance and delay of the loaded clock transmission line are calculated using equations (4.4) and (4.5). The clock line is terminated with its effective characteristic impedance for differential mode and (preferably also for) common mode. The impact of a (potentially) wrong termination for common mode is analysed using Spectre simulations.
- The full PRBS generator core is simulated with refined information for latch circuit and transmission line parameters. The critical path for speed can now be analysed in detail and optimised. Optimisation may require several iterations with updates to the floorplan, transmission line and/or CML circuits. At any point in time, the simple manual calculations for effective interconnect delay $t_{d,eff}$ and impedance $Z_{0,eff}$ provide an essential verification tool for obtaining in-depth understanding of the simulation results.

- Other circuitry is designed (e.g., all-zero detection and correction, trigger pulse generation) and after each extension, the impact on the PRBS generator core performance is analysed.

Using this procedure, a first-time-right PRBS generator that supports output bit-rates of up to 58 Gb/s while requiring only a single clock input signal has been realised in a InP HBT technology.

The target bit-rate of at least 40 Gb/s while using only a single clock input has been achieved. As demonstrated, high bit-rate circuit and interconnect design in SiGe and InP HBT technologies are very similar, and the same techniques and approaches may be applied. The PRBS generator also shows that a complex digital function operating at a bit-rate of up to $2 \cdot f_A$ is feasible, while simulations have shown that too low an f_A in the SiGe technology is the main bottleneck (i.e., with respect to achieving the target bit-rate).

It is only a matter of time before SiGe production technologies will become available that show a performance similar, or even superior to the performance of the InP technology used for the PRBS generator. In the meantime it is likely that progress will be made in InP technology, too. With InP technology it is also feasible to combine photonic components such as waveguides, gratings and photodiodes with electronics on the same chip. So InP technology will continue to attract interest, for example for building early prototype circuits and supporting low-volume professional markets.

References

1. N.X. Nguyen, J. Fierro, G. Peng, A. Ly and C. Nguyen, "Manufacturable commercial 4-inch InP HBT device technology," in *Proc. GaAs MANTECH*, 2002.
2. P. Deixler, R. Colclaser, et al., "QUBiC4G: a $f_T/f_{max} = 70/100$GHz 0.25 μm low power SiGe-BiCMOS production technology with high quality passives for 12.5Gb/s optical networking and emerging wireless applications up to 20GHz," in *Proc. BCTM*, 2002, pp. 201–204.
3. H. Taub, D. Schilling, *Digital Integrated Electronics*, New York, McGraw-Hill, 1977, pp. 349–355.
4. F. Sinnesbichler, A. Ebberg, A. Felder, R. Weigel, "Generation of high-speed pseudorandom sequences using multiplex-techniques," *IEEE Trans. Microwave Theory Tech.*, vol. 44, No. 12, December 1996, pp. 2738–2742.
5. O. Kromat, U Langmann, G. Hanke, W.J. Hillery, "A 10-Gb/s silicon bipolar IC for PRBS testing," *IEEE J. Solid-State Circuits*, vol. 33, No. 1, January 1998, pp. 76–85.
6. M.G. Chen, J.K. Notthoff, "A 3.3-V 21-Gb/s PRBS generator in AlGaAs/GaAs HBT technology," *IEEE J. Solid-State Circuits*, vol. 35, No. 9, September 2000, pp. 1266–1270.
7. F. Schumann, J. Bock, "Silicon bipolar IC for PRBS testing generates adjustable bit rates up to 25Gbit/s," *Electronics letters*, November 1997, pp. 2022–2023.
8. H. Knapp, M. Wurzer, T. Meister, J. Bock, K. Aufinger, "40 Gbit/s 2^7–1 PRBS generator IC in SiGe bipolar technology," in *Proc. IEEE BCTM*, 2002, pp. 124–127.
9. H. Veenstra, E. van der Heijden, D. van Goor, "Optimising broadband signal transfer across long on-chip interconnect," in *Proc. ESSCIRC*, 2002, pp. 763–766.

Chapter 7
Analysis and Design of High-Frequency LC-VCOs

7.1 Introduction

High-frequency sources based on LC topologies will be described in this chapter. Oscillators are needed to synchronise circuits in digital systems and as a clock source for digital signal generators such as the PRBS generator described in Chapter 6 of this book. In the clock conversion function of optical networking systems, LC oscillators are preferred due to their low jitter generation, or low phase noise when viewed in the frequency domain. RC oscillators are often used in the data and clock recovery function, because they can provide a wide tuning range and typically occupy only a small chip area. In this book, only LC oscillators are considered since these are the most attractive for high-frequency applications. As will be shown, LC oscillators can be used to generate output signals with a predictable oscillation frequency and low phase noise up to very high frequencies (e.g., even beyond f_{cross} in bipolar circuit implementations).

Many tuneable LC oscillators apply a cross-coupled differential pair to undamp the LC-tank circuit, leading to the basic configuration shown in Fig. 7.1.

The maximum oscillation frequency that can be achieved with the oscillator topology of Fig. 7.1 in a given IC process depends on the active negative resistance R_x realised by the cross-coupled differential pair in relation to the shunt resistance of the tank R_t. In order to sustain oscillation, R_x must be negative and $|R_x| < R_t$. The cross-coupled differential pair provides a negative parallel equivalent input resistance R_x up to frequency f_{cross}. In Chapter 3 f_{cross} was analysed on the basis of the transistor y-parameters and evaluated for a simplified transistor model. The emitter series resistance R_e was ignored in the transistor model used in Section 3.3.5. In modern SiGe and SiGe:C technologies, however, the emitter series resistance may be comparable to $1/gm$ when the transistor is biased at peak-f_T and thus plays an important role in its effective transconductance.

In Section 7.2, the relationship between the input impedance of the cross-coupled differential pair and transistor parameters will be analysed for a transistor model including emitter series resistance. Expressions will be derived for the parallel equivalent input resistance R_x and capacitance C_x. The result of this analysis will indicate

H. Veenstra and J.R. Long, *Circuit and Interconnect Design for RF and High Bit-Rate Applications,* 197
doi: 10.1007/978-1-4020-6884-3_7, © Springer Science + Business Media B.V. 2008

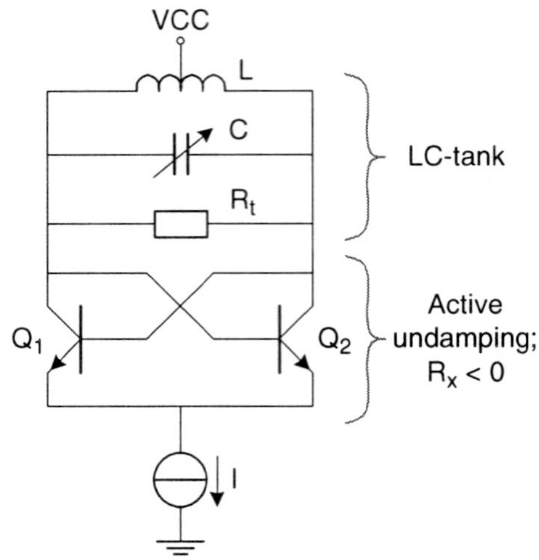

Fig. 7.1 LC oscillator using a cross-coupled differential pair to compensate for the losses of the tank. LC-tank losses are represented by resistance R_t

whether an IC technology ensures adequate performance for reaching a certain target oscillation frequency f_0, and for identifying the main limiting factors. Understanding the interplay between these parameters and circuit behaviour and limitations is necessary when refining a technology for the development of new products or new applications.

In addition to the cross-coupled differential pair, a negative resistance can also be implemented using a capacitively loaded emitter follower. The capacitive load for the emitter follower may result from the input capacitance of circuits connected to the output of the emitter follower in parallel to an (optional) capacitor. The parallel equivalent input impedance of a capacitively loaded emitter follower will be analysed in Section 7.3. With a capacitively loaded emitter follower, an input resistance is realised that may be negative at frequencies beyond f_{cross}. It will be shown in this chapter that this implementation is particularly interesting for oscillators operating at microwave frequencies.

In high bit-rate circuits, the oscillator typically needs to drive multiple latches, multiplexers and/or demultiplexers. Therefore, the VCO should be able to drive an on-chip transmission line, with characteristic impedance levels of 40–70 Ω (single-ended). An impedance of 50 Ω is often required if the VCO signal has to be driven off-chip. Such a low impedance load cannot be directly connected to the tank since it would severely affect the Q-factor of the loaded tank, causing the oscillator amplitude to drop, or even fail to oscillate. Therefore, buffering of the VCO signal (or signals in the case of multiple outputs) is needed to drive the relatively low impedance load and isolate the tank (e.g., to reduce frequency pulling or de-Qing of the tank). Usually, an oscillator output buffer is designed as a separate building block. The input impedance of the output buffer however loads the tank, and should be taken into account during the design of the oscillator. When the negative resistance is implemented using a capacitively loaded emitter follower, it is possible to combine the negative resistance and output buffer functions, as will be explained in Section 7.4.

Example realisations of LC-VCOs will be described in Sections 7.5 and 7.6. The oscillator described in Section 7.5 uses a cross-coupled differential pair and shows an oscillation frequency close to f_{cross}. The oscillator described in Section 7.6 uses a capacitively loaded emitter follower and shows an oscillation frequency above f_{cross}.

I/Q VCOs are used in many transceiver systems. I/Q sources are needed in image-reject mixers and half-rate clock- and data recovery circuits. Different techniques exist for the generation of I/Q signals. One method that is widely applied at microwave frequencies involves coupling two identical LC-VCO cores [1]. Typically, each of the LC-VCO cores is a standard LC-VCO, using a cross-coupled differential pair as a negative resistance cell, extended with a coupling circuit. The new LC-VCO topology based on the capacitively loaded emitter follower (as discussed in Sections 7.3 and 7.6) can also be used as a core for an I/Q VCO. This will be analysed in Section 7.7, and an example implementation will be described.

A discussion of the results and conclusions will be given in Section 7.8. All VCO implementations described in this chapter are implemented in a 0.25 μm BiCMOS SiGe technology [2].

7.2 Input Impedance of a Cross-Coupled Differential Pair

A cross-coupled differential pair (Q_1, Q_2) is shown in Fig. 7.1, where the transistors are assumed to be identical. The simplified small-signal model for the cross-coupled differential pair is shown in Fig. 7.2. Note that the transistor model used

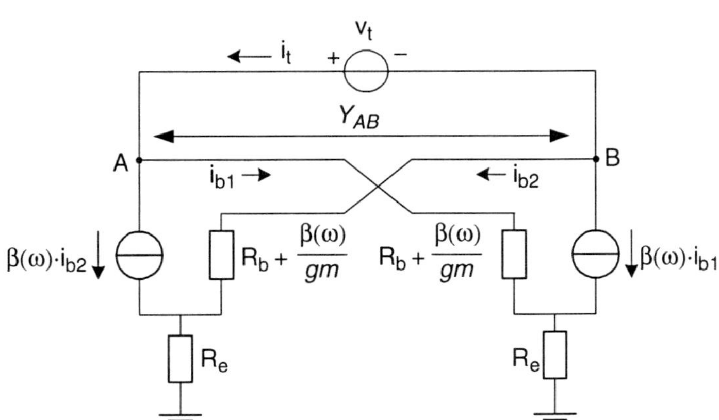

Fig. 7.2 Small-signal model used to calculate the input admittance Y_{AB} of the cross-coupled differential pair

in Fig. 7.2 is further simplified with respect to the model shown in Fig. 3.16, since the collector-base capacitance C_{bc} and collector-substrate capacitance C_{cs} are ignored in the transistor model. However, loading by the collector-base and collector-substrate parasitics is easily accounted for by adding a lumped capacitor of value $(2 \cdot C_{bc} + C_{cs}/2)$ in shunt with the collector terminals. As will be shown, the resulting approximation for the input admittance Y_{AB} is in good agreement with more accurate Spectre circuit simulations using the MEXTRAM transistor model. The input admittance across the collector terminals Y_{AB} can be represented by a frequency-dependent resistance R_x in parallel with a frequency-dependent capacitance C_x. The negative resistance provides the energy needed to sustain oscillation, while the capacitive parasitic is part of the LC-tank.

The transistor current gain $\beta(\omega)$ is assumed to be frequency-dependent with a first-order roll-off for frequencies above ω_T/β_0:

$$\beta(\omega) = \frac{\beta_0}{1 + \frac{j\omega\beta_0}{\omega_T}} \tag{7.1}$$

Since the target oscillation frequency is assumed to be well above the ω_T/β_0 cut-off frequency, the current gain can be approximated by

$$\beta(\omega) \approx \frac{-j\omega_T}{\omega} \tag{7.2}$$

In the case of matched transistors, the single-ended input impedances at each of the base terminals of transistors Q_1 and Q_2 (Z_{in1} and Z_{in2}, respectively) are equal, so that:

$$Z_{in1} = Z_{in2} = Z_{in} \tag{7.3}$$

Since transistors Q_1 and Q_2 operate in a common-emitter configuration, the input impedance at each of the base terminals of the model in Fig. 7.2 is

$$Z_{in} = R_b + \frac{\beta(\omega)}{gm} + (\beta(\omega)+1)R_e \approx R_b + R_e - j\frac{\omega_T}{\omega}\left(\frac{1}{gm} + R_e\right) \tag{7.4}$$

In the differential configuration the base currents are equal in magnitude but in anti-phase, so that:

$$i_{b1} = -i_{b2} \tag{7.5}$$

Admittance Y_{AB} can now be expressed in terms of the terminal current (i_t) and terminal voltage (v_t) as

$$Y_{AB} = \frac{i_t}{v_t} = \frac{i_{b1} + \beta(\omega) \cdot i_{b2}}{i_{b1} \cdot Z_{in1} - i_{b2} \cdot Z_{in2}} \tag{7.6}$$

Substituting equations (7.3) and (7.5) into (7.6) yields

$$Y_{AB} = \frac{1 - \beta(\omega)}{2 \cdot Z_{in}} \tag{7.7}$$

and combining equations (7.2), (7.4) and (7.7) gives the following result for the input admittance:

$$Y_{AB} = \frac{\frac{1}{2}\left(1 + \frac{j\omega_T}{\omega}\right)}{R_b + R_e - \frac{j\omega_T}{\omega}\left(\frac{1}{gm} + R_e\right)} \tag{7.8}$$

The parallel equivalent input resistance R_x and capacitance C_x derived from the real and imaginary parts of equation (7.8) are:

$$R_X = \frac{2\left(R_b + R_e\right)^2 + 2\left(\frac{\omega_T}{\omega}\right)^2\left(\frac{1}{gm} + R_e\right)^2}{\left(R_b + R_e\right) - \left(\frac{\omega_T}{\omega}\right)^2\left(\frac{1}{gm} + R_e\right)} \tag{7.9}$$

$$C_X = \frac{\left(\frac{\omega_T}{\omega^2}\right)\left(R_b + R_e + \frac{1}{gm} + R_e\right)}{2\left(R_b + R_e\right)^2 + 2\left(\frac{\omega_T}{\omega}\right)^2\left(\frac{1}{gm} + R_e\right)^2} \tag{7.10}$$

The input resistance crossover frequency, f_{cross}, is defined as the frequency at which the denominator of equation (7.9) is zero:

$$f_{cross} = f_T\sqrt{\frac{\left(1/g_m + R_e\right)}{\left(R_b + R_e\right)}} \tag{7.11}$$

As can be seen from equation (7.9), resistance R_x is negative at $f < f_{cross}$ and positive at $f > f_{cross}$. Figure 7.3 compares calculated and simulated results obtained for the shunt resistance R_x on the basis of the predictions of equation (7.9) and circuit simulations in the SiGe technology described in [2] using the MEXTRAM 504 transistor model. A good fit was obtained between Spectre circuit simulations and calculations using the small-signal model of Fig. 7.2.

Equation (7.9) predicts an f_{cross} of 39 GHz, while the circuit simulation reports that f_{cross} is 35 GHz, representing an error of about 12%. In this example, the cross-coupled transistors have emitter areas that are five times the minimum size and operate at a tail current of 2 mA with transistor parameters $f_T = 50$ GHz, $R_b = 50\ \Omega$,

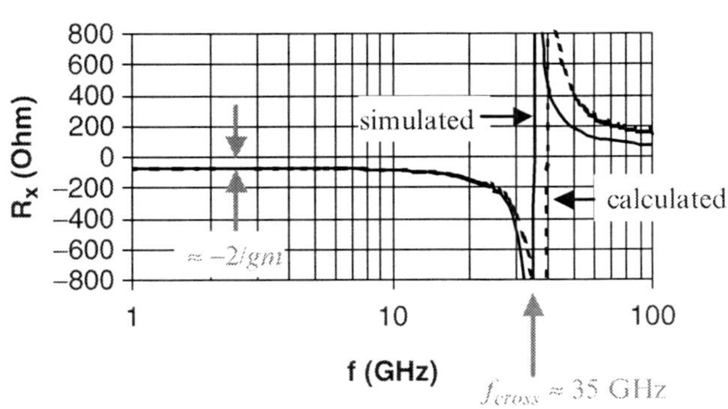

Fig. 7.3 Calculated and simulated parallel equivalent input resistance R_x of the cross-coupled differential pair

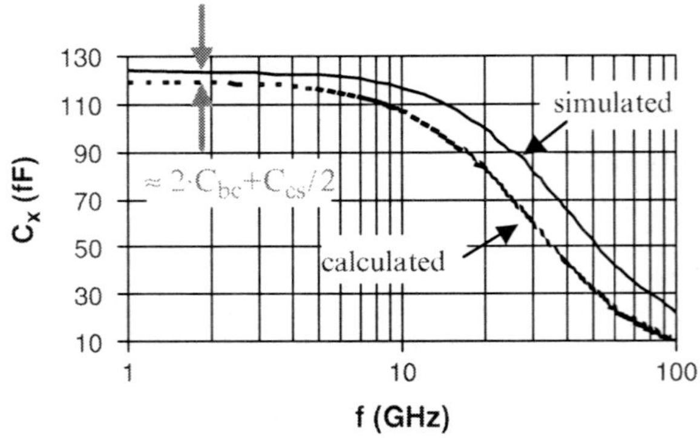

Fig. 7.4 Calculated and simulated parallel equivalent small-signal input capacitances C_x of the cross-coupled differential pair

$R_e = 10\ \Omega$ and $1/gm = 26\ \Omega$ from the simulated operating point summary. It should be noted that it is necessary to bias the transistors below peak-f_T (70 GHz) because the collector currents swing up to the full bias current I when operating in an oscillator. Also, the maximum-f_T for the SiGe transistor is obtained with 2 V reverse bias across the collector-base junction, whereas in the oscillator using a cross-coupled pair the collector-base junction operates close to zero bias (as assumed in the preceding analysis).

As previously mentioned, f_{cross} indicates the maximum frequency for an oscillator designed around the cross-coupled pair. From equation (7.11) it can be seen that it is essential to minimise the base and emitter series resistances of the transistor in order to obtain a high f_{cross}. In Chapter 3 it was shown that f_{cross} remains relatively constant across bias variations of an order of magnitude. So, the size of the transistors of the cross-coupled differential pair has little effect on f_{cross}.

Figure 7.4 compares the calculated and simulated results obtained for the shunt capacitance C_x. Again, a good fit is obtained between simulations and measurements. The offset between the simulated and calculated C_x is mainly due to the term $2 \cdot C_{bc} + C_{cs}/2$ that is seen in parallel to the differential input but was ignored in the calculations. It should be noted that the capacitance shown in Fig. 7.4 is valid for operation in the small-signal regime, and is therefore most relevant to the oscillation frequency during start-up of the VCO. Due to non-linearity of the transistor parasitic junction capacitors contributing to C_x, the input capacitance shrinks with increasing oscillation amplitude, resulting in a frequency that is amplitude-dependent, as will be demonstrated in Section 7.5.2.

7.3 Input Impedance of a Capacitively Loaded Emitter Follower

The frequency-dependent negative resistance needed to implement an oscillator can also be realised using a capacitively loaded emitter follower, as shown in Fig. 7.5. A small-signal equivalent circuit for calculation of the input admittance Y_i is shown

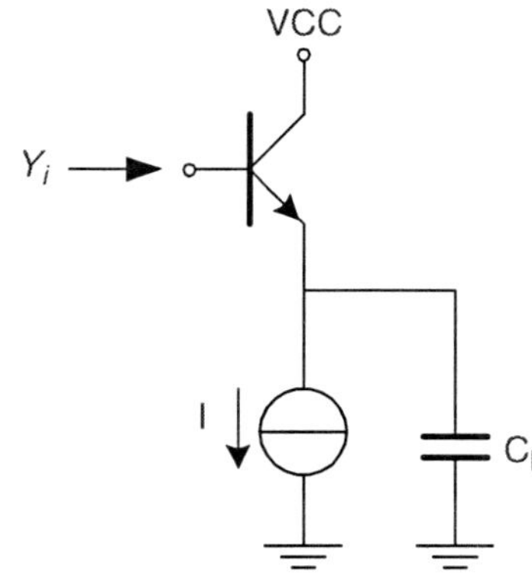

Fig. 7.5 Capacitively loaded emitter follower

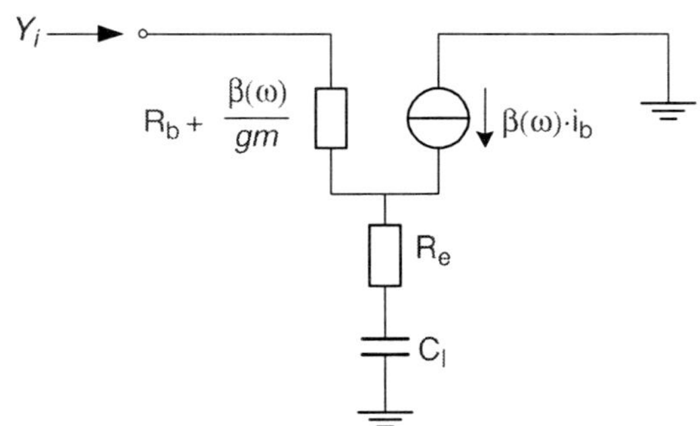

Fig. 7.6 Small-signal model for the capacitively loaded emitter follower

in Fig. 7.6. The transistor model includes base (R_b) and emitter (R_e) series resistances. The transistor current gain $\beta(\omega)$ is frequency-dependent and assumed to be complex according to equation (7.1). The transistor model used in Fig. 7.6 is further simplified with respect to the model shown in Fig. 3.16, since the collector-base and collector-substrate capacitances are ignored. The collector-substrate capacitance is connected between the power supply and ground and therefore plays no role in defining the input admittance Y_i. Also, as the collector is a small-signal ground, the collector-base capacitance shunts the input and can simply be added to the parallel equivalent input capacitance. As will be shown, the resulting approximation for the input admittance Y_i is in good agreement with more accurate Spectre circuit simulations using the MEXTRAM transitor model.

The input admittance of the circuit of Fig. 7.6 is

$$Y_i = \frac{1}{R_b + \frac{\beta(\omega)}{gm} + (\beta(\omega)+1)\left(R_e + \frac{1}{j\omega C_l}\right)} \tag{7.12}$$

Since the oscillation frequency is assumed to be well above ω_T/β_0, the current gain $\beta(\omega)$ is again approximated according to equation (7.2). The combination of equations (7.2) and (7.12) gives

$$Y_i = \frac{1}{\left(R_b + R_e - \frac{\omega_T}{\omega^2 C_l}\right) - j\left(\frac{\omega_T}{\omega g_m} + \frac{\omega_T R_e}{\omega} + \frac{1}{\omega C_l}\right)} \tag{7.13}$$

From the real part of equation (7.13) ($\mathrm{Re}(Y_i)$) it follows that the highest frequency, f_{LIMIT}, at which $\mathrm{Re}(Y_i) < 0$ and the circuit provides undamping in an oscillator is:

$$f_{LIMIT} = \sqrt{\frac{f_T}{2\pi C_l (R_b + R_e)}} \tag{7.14}$$

To maximise f_{LIMIT}, the sum of the base and emitter series resistances must be minimised (this is also necessary for a high f_{cross}).

The parallel equivalent differential input resistance R_x is obtained from the $\mathrm{Re}(Y_i)$ (i.e., $R_x = 2/\mathrm{Re}(Y_i)$), while the imaginary part $\mathrm{Im}(Y_i)$ yields the parallel equivalent input capacitance (i.e., $C_x = \mathrm{Im}(Y_i)/2\omega$).

$$R_x = \frac{2\left(R_b + R_e - \frac{\omega_T}{\omega^2 C_l}\right)^2 + 2\left(\frac{\omega_T}{\omega \cdot gm} + \frac{\omega_T R_e}{\omega} + \frac{1}{\omega C_l}\right)^2}{R_b + R_e - \frac{\omega_T}{\omega^2 C_l}} \tag{7.15}$$

$$C_x = \frac{\frac{\omega_T}{\omega \cdot gm} + \frac{\omega_T R_e}{\omega} + \frac{1}{\omega C_l}}{2\omega\left(\left(R_b + R_e - \frac{\omega_T}{\omega^2 C_l}\right)^2 + \left(\frac{\omega_T}{\omega \cdot gm} + \frac{\omega_T R_e}{\omega} + \frac{1}{\omega C_l}\right)^2\right)} \tag{7.16}$$

Equation (7.14) predicts that a high f_{LIMIT} can also be obtained by choosing a low value for the load capacitance C_l of the emitter follower. However, reducing C_l results adds damping, because the absolute value of the negative resistance $|R_x|$ increases. The influence of the load capacitance on the input resistance is illustrated in Fig. 7.7 for an emitter follower biased at 3 mA with $R_b = 30\,\Omega$, $R_e = 5\,\Omega$ and $f_T = 64\,\mathrm{GHz}$. The differential input resistance plotted in the figure was obtained by applying equation (7.15) to three different load capacitance values. Note that the input resistance for a frequency-independent load capacitance will drop by 40 dB/decade (for $f \ll f_{LIMIT}$).

The actual maximum attainable oscillation frequency using a capacitively loaded emitter follower will depend on the tank loss resistance (e.g., R_t in Fig. 7.1), but may be well above the crossover frequency for a cross-coupled pair (i.e., f_{cross}) in a given technology. This motivates the study of the emitter follower topology for millimeter-wave frequency LC oscillators.

7.4 Combining Negative Resistance and Output Buffer Functions

The load capacitance of the emitter follower (C_l) may be realised by the input capacitance of a second (resistively loaded) emitter follower in parallel with the output capacitance of the current source biasing the first emitter follower (see Fig. 7.8). The resulting topology combines the negative resistance function with the output buffer

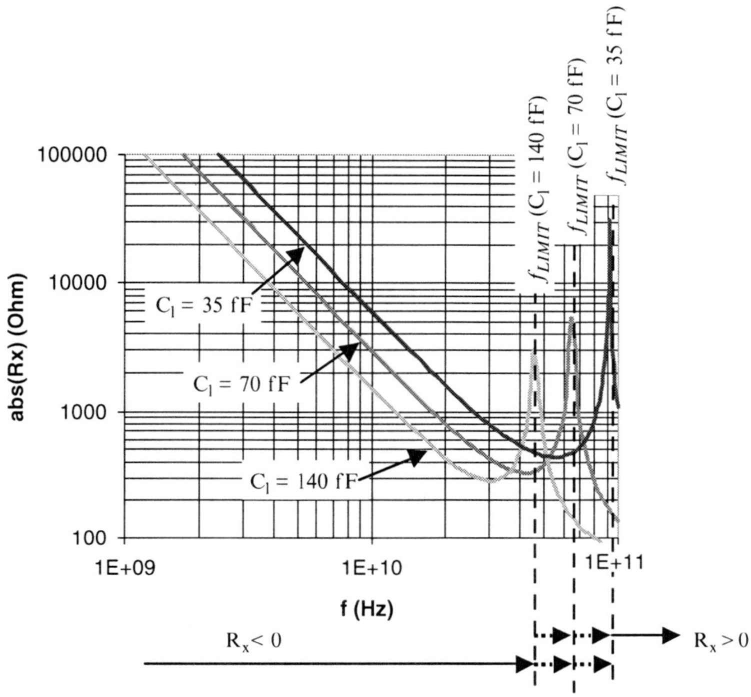

Fig. 7.7 Effect of the load capacitance C_l on the shunt input resistance R_x

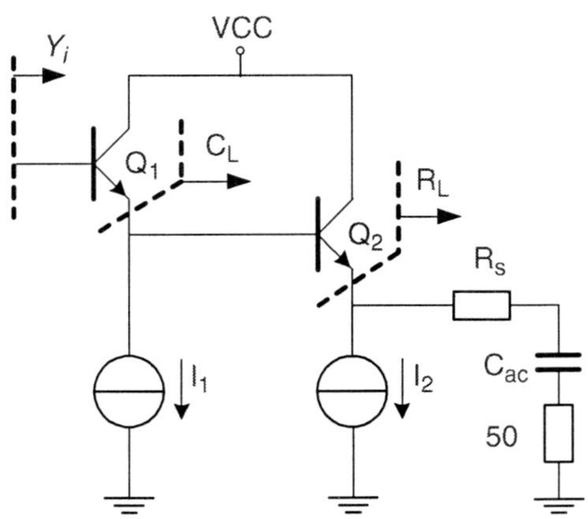

Fig. 7.8 Two cascaded emitter followers, the output emitter follower with resistive load

function. The resistive load of emitter follower Q_2 forces the emitter current of Q_2 to be in phase with the emitter voltage. The resistive emitter current translates via the phase shift in the current gain $\beta(\omega)$ of Q_2 into capacitive behaviour at the base terminal of Q_2, thereby implementing the load capacitance C_l for emitter follower Q_1. Therefore, the input impedance of Q_1 has a negative real component (i.e., negative resistance) that is defined by the total load resistance R_L at the output emitter follower, which can be well defined and is not a parasitic.

In the following example, emitter followers Q_1 and Q_2 are biased at 3 mA and 4.5 mA, respectively. Capacitance C_1 represents the total capacitive load at the output of Q_1 and originates from three contributions: input capacitance from the resistively loaded output emitter follower Q_2 ($C_{in,Q2} \approx 27$ fF), base-collector capacitance of the output emitter follower Q_2 ($C_{bc,Q2} \approx 13$ fF) and output capacitance of the bias current source I_1 ($C_{out,I1} \approx 16$ fF).

The values used for the calculations for transistor Q_1 are $R_b = 30\,\Omega$, $R_e = 5\,\Omega$, $gm = 115.8$ mS, $C_1 = C_{in,Q2} + C_{bc,Q2} + C_{out,I1} = 27 + 13 + 16 = 56$ fF and $f_T = 64$ GHz (parameters for a transistor with an emitter area of $0.5\,\mu\text{m} \times 10.2\,\mu\text{m}$, biased at 3 mA). The calculated input resistance and capacitance are shown in Figs. 7.9 and 7.10, respectively, together with the more accurate values obtained in Spectre circuit simulations using MEXTRAM 504 transistor models. Resistance R_x undamps the LC-tank, while capacitance C_x is seen in parallel to the tank and shifts the resonant frequency somewhat. A comparison of Figs. 7.10 and 7.4 shows that the value of C_x in this new topology is significantly smaller than that of a cross-coupled differential pair. This enables a larger tuning range when a voltage variable tuning capacitor and resonant tank are added to implement a VCO. The input capacitance may be almost identical in both small-signal and large-signal regimes, which is not the case with the cross-coupled differential pair. This reduces the sensitivity of oscillation frequency to oscillation amplitude. The input capacitance is only weakly dependent on the signal amplitude if the base-collector junction of Q_1 is reverse biased, and bias currents I_1 and I_2 are designed to be large with respect to the signal currents of Q_1 and Q_2.

As shown in Fig. 7.9, the f_{LIMIT} resulting from equation (7.14) is approximately 70 GHz (assuming $f_T = 64$ GHz, $R_b = 30\,\Omega$, $R_e = 5\,\Omega$), which is somewhat higher

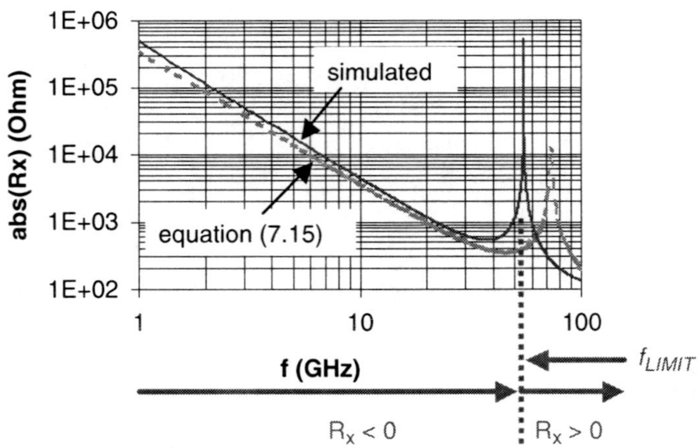

Fig. 7.9 Calculated and simulated absolute values of the differential parallel equivalent input resistance for a 50 Ω single-ended output load resistance

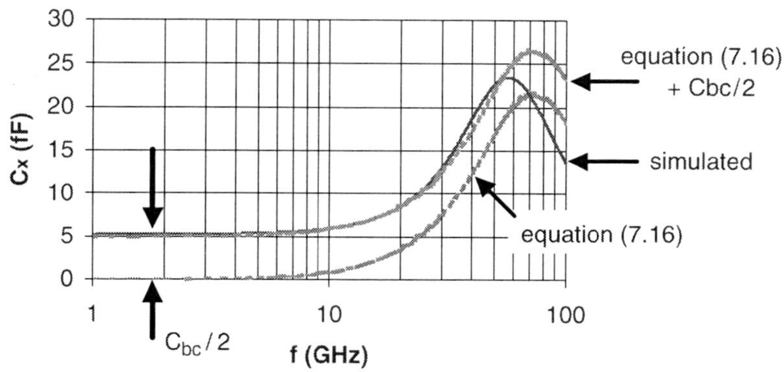

Fig. 7.10 Calculated and simulated differential parallel equivalent input capacitances for a 50 Ω single-ended output load resistance

than the 55 GHz obtained in the computer simulation. The frequency limit is significantly higher than f_{cross} (approximately 35 GHz), which defines the equivalent operating bandwidth for a cross-coupled pair topology in the same technology. In particular, the capacitively loaded emitter follower topology is suited to the frequency range in which $|R_x|$ is close to its minimum, which is the 30–45 GHz range for the example shown.

The offset in C_x for simulation versus calculation according to equation (7.16) in Fig. 7.10 is mainly due to the base-collector capacitance of transistor Q_1 (e.g., $C_{bc,Q1}/2 \approx 5\,fF$) that is not taken into account in the small-signal model shown in Fig. 7.6.

7.5 LC-VCO Operating at a Frequency Close to f_{cross}

The maximum frequency for an oscillator using a cross-coupled differential pair equals f_{cross}, which is approximately 35 GHz in this technology. A lossless LC-tank is needed to reach f_{cross}. With a practical tank circuit, the maximum attainable oscillation frequency occurs at the point at which the negative resistance R_x exactly counteracts the positive resistance from the tank R_t. To account for temperature and process variations, a safety factor of 2–3 is often chosen as the ratio of R_t and $-R_x$. In the following, an oscillator will be designed using a cross-coupled differential pair and targeting an oscillation frequency close to f_{cross}. The on-chip tank inductor and varactor will be described in Section 7.5.1. As will be shown for a practical on-chip tank circuit, 20 GHz is a realistic oscillation frequency target for the technology used. The complete VCO circuit will be described in Section 7.5.2. Evaluation results will be presented in Section 7.5.3. The VCO described in this section is an improved version of the LC-VCO presented in [4]; it uses an improved output buffer to obtain a higher output signal amplitude.

7.5.1 Inductor and Varactor

A 0.5 nH single-turn inductor with center-tap is implemented in the 3 μm thick top-metal layer above a deep trench isolation grid. The inductor is kept free of tiling, because metal tiles reduce the quality factor at frequencies above 10 GHz [3]. Also, it is important to avoid the formation of a closed loop at a short distance from the inductor, e.g., within a radius corresponding to the inductor diameter. Such a loop may easily be formed accidentally, for example in implementing a fully symmetrical layout of a differential circuit, via the supply network (note that the supply and ground nets are usually shorted on-chip via supply decoupling capacitors) or in contacting a patterned shield underneath the inductor. The inductor in this design has no shield, because measurements have shown that a shield does not improve the quality factor at 20 GHz whereas it does lower the self-resonant frequency of the inductor. Measurements have been performed for frequencies up to 50 GHz. The stand-alone inductor achieves a measured Q of 20 at 20 GHz (see Fig. 7.11), while the self-resonant frequency is well above 50 GHz. The measurement results are corrected for the probe-pad impedance using the open-short de-embedding technique.

In order to obtain the highest possible differential Q-factor, the varactor is implemented as a differential configuration as shown in Fig. 7.12. To obtain the lowest possible differential series resistance, interdigitated p+ diffusion stripes of minimum width (constituting the anodes of the differential varactor) are placed as close

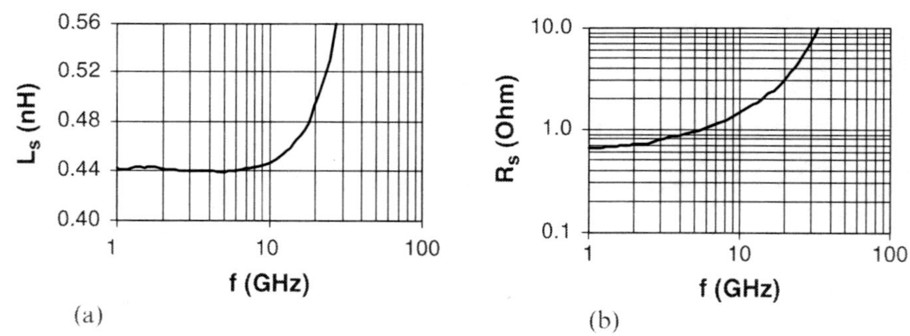

Fig. 7.11 Measured series inductance L_s (a) and resistance R_s (b) of the on-chip inductor

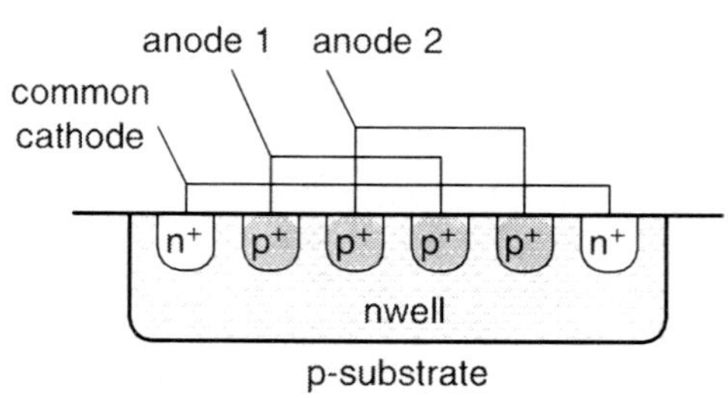

Fig. 7.12 Differential varactor layout for maximum quality factor

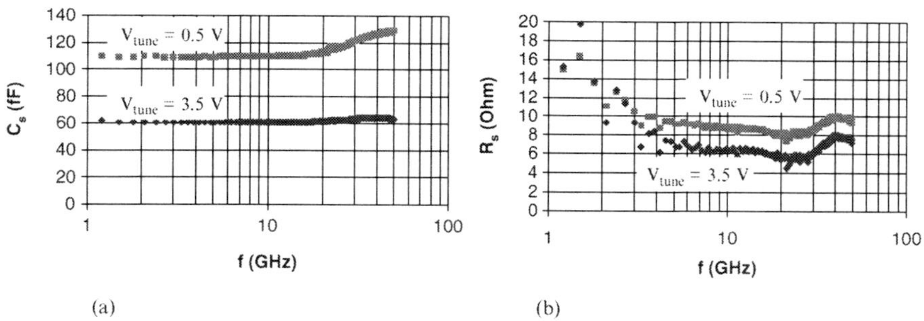

Fig. 7.13 Differential series capacitance C_s (a) and resistance R_s (b) of the varactor measured at $V_{tune} = 0.5\,V$ and $V_{tune} = 3.5\,V$

together as possible within a common nwell. The nwell constitutes the common cathode of the varactor and has a large associated parasitic capacitance between nwell and the substrate. In the application, this parasitic capacitance is connected between the varactor tuning voltage V_{tune} and ground and plays no role in the differential tank impedance. The resistance in series with the common cathode is not minimised in this differential configuration, but that is irrelevant for the differential behaviour of the varactor. The measured differential series capacitance and resistance are shown in Fig. 7.13.

At 20 GHz, the differential varactor obtains a measured worst-case quality factor $Q = 9$ (at $V_{tune} = 0.5\,V$), a best-case $Q = 21$ (at $V_{tune} = 3.5\,V$). The measurement results shown in Fig. 7.13 are corrected for the probe-pad impedance using the open-short de-embedding technique.

The measured series resistance of the inductor and varactor can be translated into equivalent parallel resistances using the relation

$$R_p = (Q^2 + 1) \cdot R_s \tag{7.17}$$

The parallel loss resistances resulting for the inductor $R_{p,L}$ and the capacitor $R_{p,C}$ are shown in Fig. 7.14. The tank loss resistance R_t is equal to the parallel resistance of $R_{p,L}$ with $R_{p,C}$.

As can be seen in Fig. 7.14, the varactor is dominant in the tank loss at frequencies above approximately 15 GHz; the inductor is dominant at frequencies below approximately 15 GHz.

Oscillation is only possible if the active negative resistance is strong enough to undamp the tank, e.g., $-R_x < R_t$. In the example shown in Fig. 7.14, the frequency at which $R_t = -R_x$ is 26 GHz. So, the highest frequency at which oscillation is possible is 26 GHz. Note that oscillation at 26 GHz is only possible if the oscillator output buffer provides an infinite parallel equivalent input resistance, because the output buffer has not yet been taken into consideration. Besides, the inductor in the oscillator must behave in the same way as the inductor as measured in stand-alone mode. To allow for processing and temperature variations, a finite input resistance of the output buffer and some degradation of the inductor Q-factor in the oscillator layout,

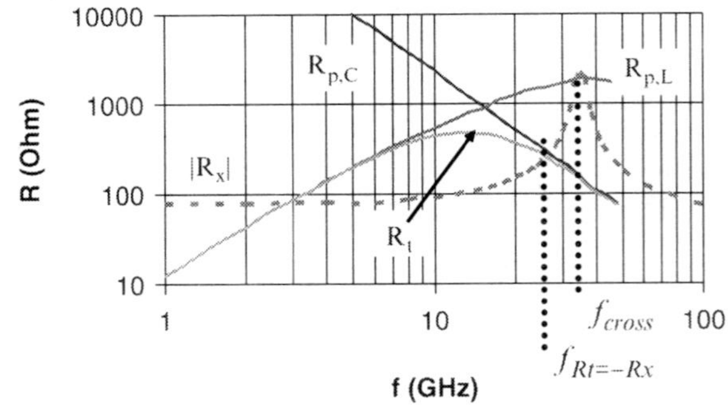

Fig. 7.14 Tank parallel loss resistance R_t measured at $V_{tune} = 0.5\,V$ with contributions of the varactor $R_{p,C}$ and inductor $R_{p,L}$. The simulated active negative resistance $|R_x|$ of Fig. 7.3 is also shown

an oscillation frequency of approximately 20 GHz is a realistic target to demonstrate reliable oscillation at a frequency close to f_{cross}.

7.5.2 VCO and Output Buffer Circuits

The detailed circuit of the VCO core is shown in Fig. 7.15. The circuit is designed for a nominal supply voltage of 4 V. The diode D_1 in series with the center-tap of the inductor prevents clamping of the signal at the tank by the base-collector junctions of the emitter followers Q_3 and Q_4. The emitter followers in turn provide a level-shift to the bases of the cross-coupled differential pair Q_1, Q_2. Without emitter followers Q_3, Q_4, there will always be one base-collector junction of the cross-coupled pair operating in the forward-bias region, introducing an extra loss-resistance in parallel to the tank.

With the circuit shown in Fig. 7.15 a differential peak voltage swing across the tank up to $2 \cdot V_{be}$ (with $V_{be} \approx 0.8\,V$ a forward biased diode voltage) is possible, although the base-collector junctions of Q_1 and Q_2 become forward biased for a differential peak voltage swing above V_{be}.

The output buffer circuit is based on two pairs of cascaded emitter followers driving the external 50 Ω loads. The circuit is shown in Fig. 7.16. Nodes V_{in+} and V_{in-} are connected to the LC-tank. This configuration is identical to the circuit analysed in Section 7.4. At 20 GHz, the input resistance is negative and relatively high in absolute terms (in Fig. 7.9: $R_{i,diff} \approx -1\,k\Omega$, e.g., $|R_{i,diff}| > R_t$); it does not provide sufficient undamping to the tank to sustain oscillation. So, the cross-coupled differential pair is needed for undamping. Furthermore, the output buffer provides a low input capacitance (at 20 GHz in Fig. 7.10: $C_{i,diff} \approx 9\,fF$, i.e., an order of magnitude lower than the varactor capacitance).

Series resistors ($R_s = 42\ \Omega$) implement a 50 Ω output resistance (single-ended). The overall gain of the output buffer in loaded condition equals −6 dB; the simulated small-signal bandwidth of the buffer equals 67 GHz. Given the high swing

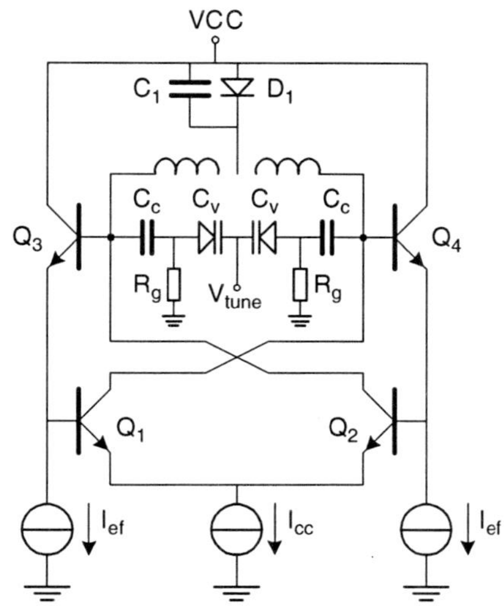

Fig. 7.15 Detailed circuit diagram of the LC-VCO core

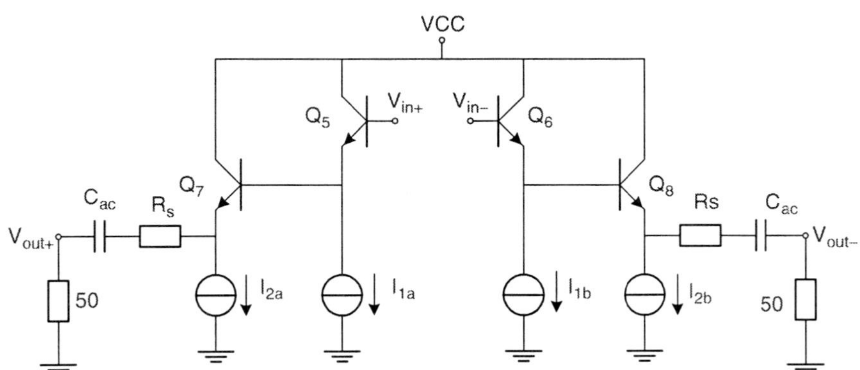

Fig. 7.16 Differential output buffer with 50 Ω single-ended output impedance

across the tank, the low gain is sufficient to allow accurate measurements. During on-wafer evaluation, a differential probe connects to the differential output. One single-ended signal drives a 50 Ω passive probe plus cable to the 50 Ω input of a spectrum analyser. The other single-ended output is terminated into a 50 Ω resistor.

It is interesting to observe that the oscillation frequency changes between start-up and steady state. This effect is demonstrated by the simulation results presented in Fig. 7.17. The oscillation frequency is derived from the time difference between the zero crossings, and increases from $f_0 = 16.2\,\mathrm{GHz}$ at start-up to $f_0 = 19.2\,\mathrm{GHz}$ in steady state.

The reason for the lower oscillation frequency at start-up is the voltage dependence of the differential input capacitance of the cross-coupled differential pair. At the differential input of a cross-coupled differential pair, there are two base-emitter

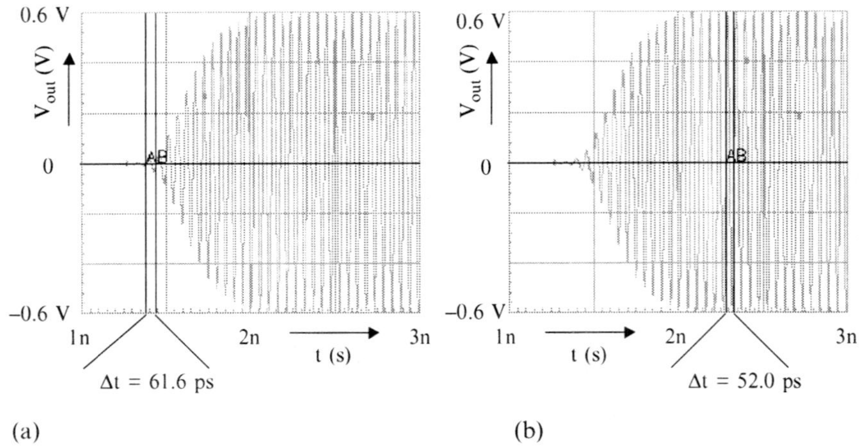

Fig. 7.17 Simulation result of the entire VCO at $V_{tune} = 0.5\,V$, showing the oscillation period at start-up (a) and in steady state (b). The signal shown is at the buffer output

capacitances connected in anti-series. Since the sum of the two emitter currents is constant (fixed by the bias current source of the cross-coupled pair), if the base-emitter capacitance of one transistor increases, the base-emitter capacitance of the other will decrease. The series capacitance of the two base-emitter junctions thus depends on the signal level at the tank, and shows a maximum when the voltage across the tank is zero.

7.5.3 Experimental Results

A photomicrograph of the IC placed in the wafer probing station is shown in Fig. 7.18. The VCO excluding bondpads measures $0.30 \times 0.30\,mm^2$; including bondpads the size is $0.68 \times 0.54\,mm^2$. At the bottom side, a Cascade differential GSSG probe connects to the differential VCO output. The two probe pads at the top are for power supply (VCC) and ground (GND). The probe pads in the middle row connect to the tuning voltage V_{tune} (left) and the bias circuitry (right). If the V_{tune} pad is not connected, an on-chip network provides a default tuning voltage V_{tune} equal to VCC/2. If the bias pad is not connected, an on-chip resistor sets the default bias current level for all the current sources.

The difference between the measured and simulated oscillation frequencies is very small, as demonstrated in Fig. 7.19. The measured VCO frequency is approximately 4% higher than would be expected on the basis of Spectre circuit simulations. The simulated curve represents the steady state oscillation frequency.

Table 7.1 compares the oscillation frequencies obtained in simulations, calculations and measurements. It should be noted that the measured oscillation frequency must be compared with the simulated steady state results of the VCO, while the

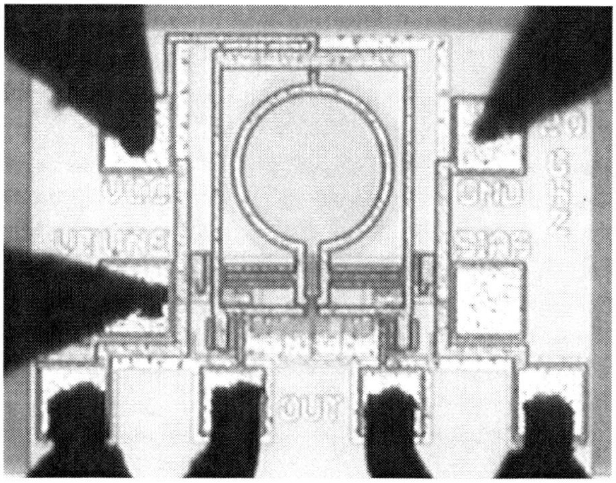

Fig. 7.18 LC-VCO chip photomicrograph

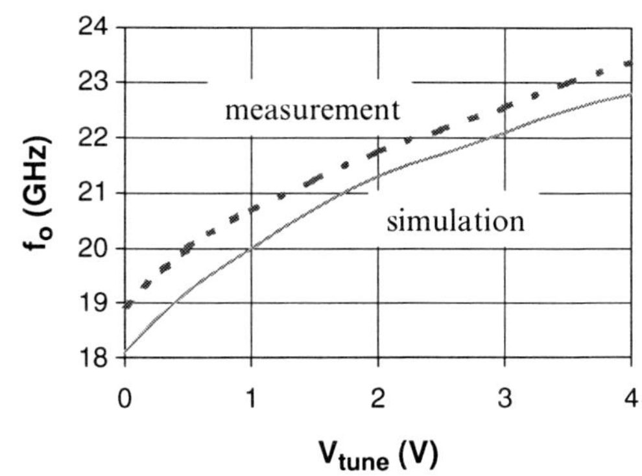

Fig. 7.19 Measured and simulated oscillation frequency versus tuning voltage

Table 7.1 Analysis of measured, simulated and calculated oscillation frequencies

Simulation	$\mathbf{V_{tune} = 0.5\,V}$	$\mathbf{V_{tune} = 3.5\,V}$
Varactor	$C_p = 117\,fF$ at 20 GHz	$C_p = 70\,fF$ at 20 GHz
Inductor	$L_p = 0.45\,nH$ at 20 GHz	
Tank	$f_{res} = 21.9\,GHz$	$f_{res} = 28.2\,GHz$
Buffer	$C_p = 9\,fF$ at 20 GHz (see Fig. 7.10)	
Active part	$C_p = 100\,fF$ at 20 GHz (see Fig. 7.4)	
VCO, start-up (Fig. 7.17)	$f_o = 16.2\,GHz$	
VCO, steady state (Fig. 7.17)	$f_o = 19.2\,GHz$	$f_o = 22.4\,GHz$
Calculation	$\mathbf{f_o = 15.8\,GHz}$	$\mathbf{f_o = 17.7\,GHz}$
Measurement	$\mathbf{V_{tune} = 0.5\,V}$	$\mathbf{V_{tune} = 3.5\,V}$
Varactor (Fig. 7.13)	$C_p = 111\,fF$ at 20 GHz	$C_p = 62\,fF$ at 20 GHz
Inductor (Fig. 7.11)	$L_p = 0.50\,nH$ at 20 GHz	
Entire VCO	$f_o = 20.0\,GHz$	$f_o = 23.0\,GHz$

(small-signal) calculated oscillation frequency must be compared with the simulated start-up results of the VCO. The frequency f_{res} represents the resonance frequency of the LC-tank circuit including ac coupling capacitors C_c and dc bias resistors R_g as shown in Fig. 7.15.

Good agreement is found between the results of the simulations, calculations and measurements. A small difference is observable between the simulated and measured results obtained for the varactor and the inductor. Many small errors may contribute to the difference between the measured and simulated oscillation frequencies, such as process and temperature variations, layout parasitic capacitance and inductance, model imperfections, etc. The results however demonstrate that all these effects are within acceptable limits to allow a prediction of the oscillation frequency to within 5% accuracy.

An example measurement result of the single-ended output spectrum is shown in Fig. 7.20.

Some signal loss occurs due to probes, connectors and cables. To find the combined losses of the cables, connectors and RF probe, a 'through-connect' calibration structure was measured using two identical sets of GSSG probes, cables and connectors. The resulting loss (7.2 dB at 20 GHz) must be divided by 2 to find the losses of the cable, connector and probe between the oscillator circuit output and the spectrum analyser input. The fundamental signal power found with the spectrum analyser is −5.8 dBm. Ignoring impedance mismatch effects at the output, the signal level at the single-ended output on-wafer is −2.2 dBm. Since the buffer output is matched to 50 Ω using a series resistor (R_s = 42 Ω in Fig. 7.16), the output impedance is well controlled, and mismatch at the output may be ignored. The −2.2 dBm output

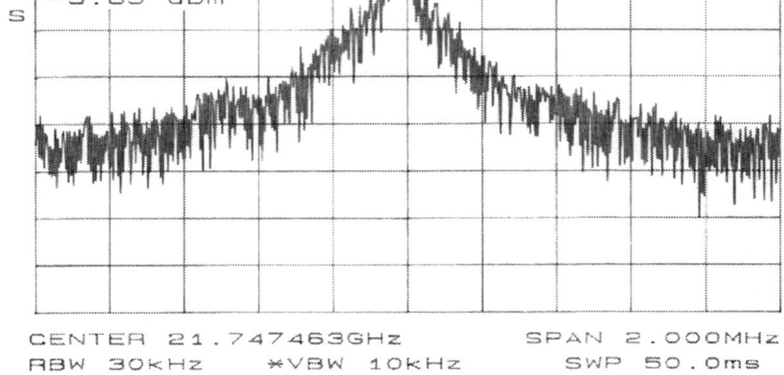

Fig. 7.20 Measured single-ended output spectrum of the VCO at $V_{tune} = 2\,V$

power corresponds to a signal amplitude at the differential output of 0.5 $V_{p,diff}$. The simulated output signal amplitude equals 0.6 $V_{p,diff}$ (see Fig. 7.17). The (simulated) output buffer voltage gain equals −6 dB, so the signal amplitude on the tank estimated on the basis of measurement results equals 1.0 $V_{p,diff}$.

The phase noise measured under the same conditions is shown in Fig. 7.21. The −112 dBc/Hz phase noise at 2 MHz from the carrier is 5 dB better than simulated. The discrepancy between the measured and the simulated phase noise is not understood. It is unlikely that the noise of the LC-tank plays a role in this discrepancy since the measured quality factors of the L and C are in agreement with the simulations. Also, the measured 0.5 $V_{p,diff}$ oscillator output signal amplitude is close to the simulated 0.6 $V_{p,diff}$. So, the difference cannot be explained by a different signal swing on the tank. The difference may have something to do with the transistor model or model parameters. For example, $1/f$-noise parameters are usually not verified during IC process development. The noise discrepancy is an interesting area for future research. An unexplained 8 dB discrepancy between measured and simulated phase noise was previously reported for a 40 GHz LC-VCO in [5].

The measured performance of the VCO is summarised in Table 7.2.

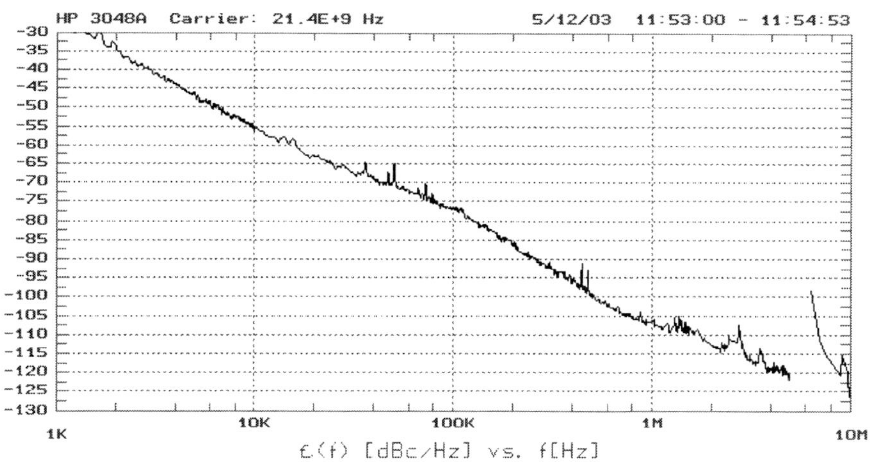

Fig. 7.21 Measured phase noise of the single-ended output signal at $V_{tune} = 2\,V$

Table 7.2 Summary of measurement results

Parameter	Value
Oscillation freq. ($V_{tune} = 0.5$–$3.5\,V$)	20.0–23.0 GHz
Phase noise ($V_{tune} = 2\,V$)	−112 dBc/Hz @ 2MHz
Chip area	$0.68 \times 0.54\,mm^2$
Supply pushing	35 MHz/V
Supply current at VCC = 4 V:	
VCO core only	6 mA
including biasing, output buffers	23 mA
Output power into 50 Ω load	−2.2 dBm (single-ended)
Signal amplitude on tank	2.0 $V_{pp,diff}$

7.6 LC-VCO Operating at a Frequency above f_{cross}

In this section, the design of an oscillator operating above f_{cross} and using a capacitively loaded emitter follower will be demonstrated. Experimental characterisation of the inductor and varactor will be described in Section 7.6.1, followed by a description of the test circuit in Section 7.6.2. VCO measurements will be presented in Section 7.6.3.

7.6.1 Inductor and Varactor

To enable a high oscillation frequency, the values of the tank inductance and capacitance are reduced with respect to the 20 GHz oscillator described in Section 7.5. Use is made of a single-turn inductor with a center-tap, without tiling or ground shield. The inductor is placed above a grid of deep-trench isolation to increase its self-resonant frequency. The inductor is designed with the LSIM3.1 tool [7]. The measured series-equivalent network parameters are given in Fig. 7.22.

At 40 GHz, the inductive reactance corresponds to an inductance of 0.36 nH and it achieves a quality factor Q of approximately 25 without tiling. As can be seen from the equivalent series resistance (see Fig. 7.22b), tiling increases the losses but has virtually no effect on the inductance. As a result, the inductor Q is reduced by tiling.

A differential varactor was implemented as shown in Fig. 7.12. A test structure with 10 varactors in parallel was designed in order to accurately characterise the varactor at 40 GHz. The capacitance of a single varactor cannot be measured accurately because it is of the order of only a few tens of femto-Farads. The de-embedding of the bondpad impedance would introduce a significant measurement uncertainty if only a single varactor were to be evaluated. The measured results after de-embedding of the bondpads and interconnect of the test structure are shown in Fig. 7.23. The measured worst-case quality factor is 5.2 at 40 GHz (at a reverse bias voltage of $V_{tune} = 0.5\,V$), and the best-case Q is 12 (at $V_{tune} = 3.5\,V$).

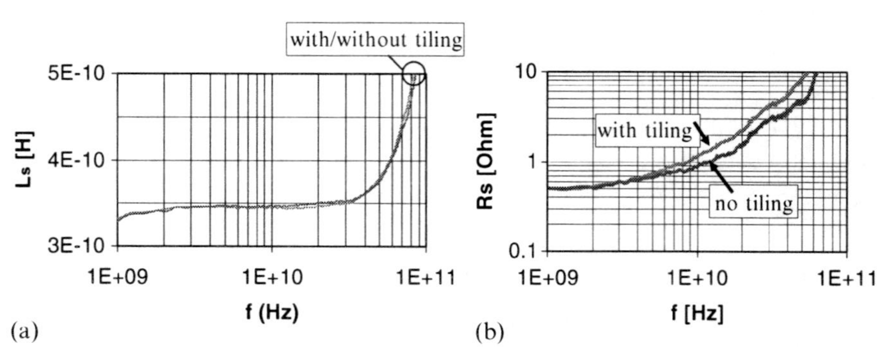

Fig. 7.22 Measured series inductance L_s (a) and resistance R_s (b) of the on-chip inductor

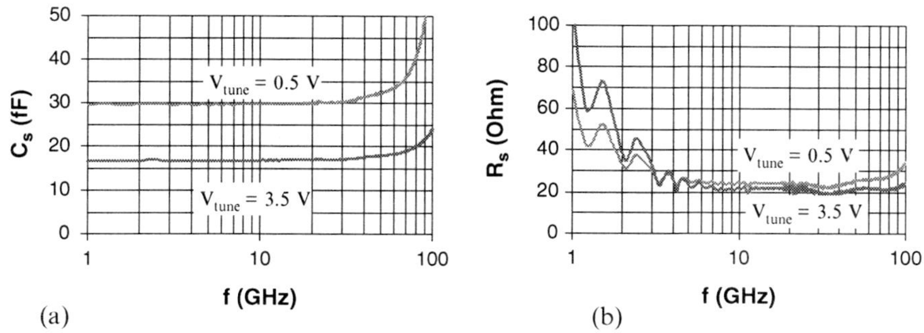

Fig. 7.23 Measured differential series equivalent capacitance C_s (a) and resistance R_s (b) of the on-chip varactor

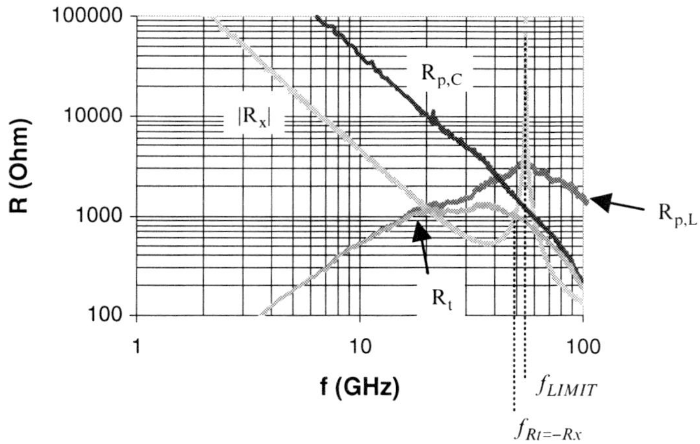

Fig. 7.24 Measured tank parallel loss resistance R_t at $V_{tune} = 3.5$ V with contributions of the varactor $R_{p,C}$ and inductor $R_{p,L}$. The simulated active resistance $|R_x|$ of Fig. 7.9 is also shown

The varactor dominates the quality factor of the LC-tank at frequencies above 30 GHz (at $V_{tune} = 0.5$ V, see Fig. 7.25) to 40 GHz (at $V_{tune} = 3.5$ V, see Fig. 7.24). However, the relatively high quality factor of the varactor at a V_{tune} of 3.5 V results in sufficient margin between the tank resistance R_t and the negative resistance R_x of the active circuit between 30 and 40 GHz as shown in Fig. 7.24. At V_{tune} at 0.5 V, the ratio $R_t/-R_x$ is only slightly greater than unity, as shown in Fig. 7.25, so sustained oscillation may not be guaranteed under all conditions at minimum bias (e.g., with variations in process and temperature).

The lowest margin in the oscillation condition occurs at the lowest tuning voltage. It is interesting to observe that the inductor is dominant in the quality factor up to relatively high frequencies (30–40 GHz). In the case of the 20–23 GHz oscillator in the same IC technology, the inductor is dominant up to only 15 GHz (see Fig. 7.14).

The total tank capacitance results from three contributions: varactor, input capacitance of the active negative resistor circuit and interconnect parasitic capacitance.

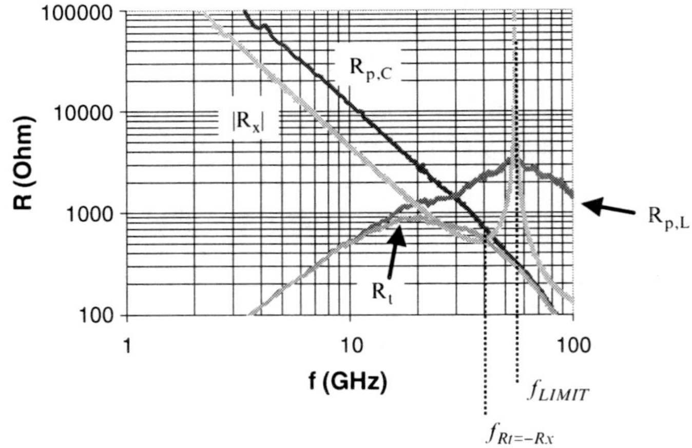

Fig. 7.25 Measured tank parallel loss resistance R_t at $V_{tune} = 0.5\,V$ with contributions of the varactor $R_{p,C}$ and inductor $R_{p,L}$. The simulated active resistance $|R_x|$ of Fig. 7.9 is also shown

In the oscillator circuit, the input capacitance of the active circuit has no associated series resistance since it represents the imaginary part of the input impedance. Also, the quality factor of the differential interconnect parasitic capacitance is typically higher than the quality factor of the varactor. In the oscillator circuit described here, the varactor contributes only about 30–50% (depending on the tuning voltage) to the tank capacitance. So, the quality factor of the varactor does not dominate the total tank Q. As a result, there is more margin on top of the oscillation condition for the entire VCO circuit than suggested by Figs. 7.24 and 7.25.

7.6.2 VCO and Output Buffer Circuits

The entire VCO schematic is shown in Fig. 7.26. The circuit operates from a nominal supply voltage of 4 V.

Current source I_D (with $I_D \ll I_1, I_2$) biases diode D_1 to create a low-ohmic path between the center-tap of the inductor and the ac ground. Capacitor C_1 across diode D_1 further reduces the ac impedance between the center-tap of the inductor and ground. Diode D_1 is connected in series with the center-tap of the inductor to avoid forward biasing of the base-collector junctions of followers Q_1 and Q_3 and subsequent de-*Q*ing of the tank. A differential peak voltage swing across the tank of up to $2 \cdot V_{be}$ ($V_{be} \approx 0.8\,V$) is possible without any junction operating in forward-bias across the tank. However, the simulated signal swing at the tank equals $0.64\,V_{p,diff}$, which is below this maximum limit, because the oscillator is current rather than voltage-limited. Diode D_1 is still needed in the current-limited regime to prevent forward biasing of the base-collector junctions of Q_1 and Q_3.

Series resistors R_s realise a 50 Ω single-ended output impedance. With 50 Ω load resistors, the simulated small-signal gain of the buffer devices is −6dB, and the

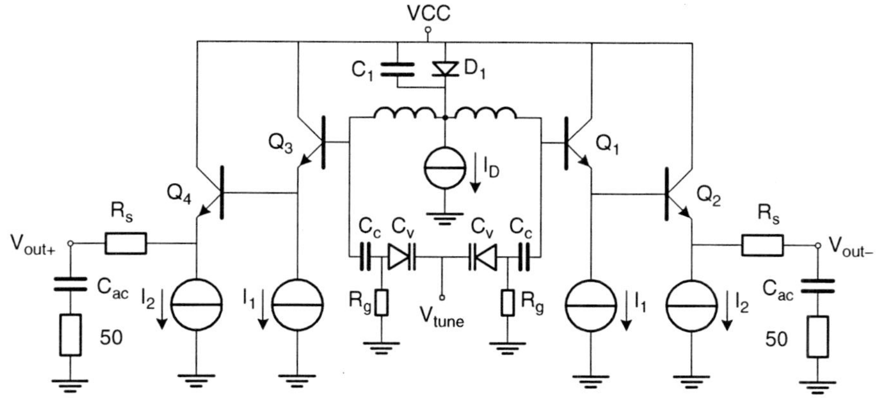

Fig. 7.26 Detailed schematic for the LC-VCO using two cascaded emitter followers with resistive load per single-ended output. The 50 Ω resistors represent the off-chip output loads

single-ended output power is −6 dBm. In Fig. 7.26, the resistive load to the emitter followers Q_2, Q_4 is off-chip. When the VCO is part of a larger IC, it is also possible to integrate the load resistors. With on-chip load resistors, frequency pulling due to potential load impedance variations is minimized.

7.6.3 Experimental Results

A photomicrograph of the IC and on-wafer probes is shown in Fig. 7.27. The VCO excluding bondpads measures $0.30 \times 0.30\,\mathrm{mm}^2$; including bondpads the testchip size is $0.68 \times 0.40\,\mathrm{mm}^2$. At the bottom, a differential GSSG probe connects to the differential VCO output. The two probe pads at the top supply power (VCC) and ground (GND). The probe pads in the middle (shown unconnected in Fig. 7.27) connect to the tuning voltage V_{tune} (left) and the bias circuitry (right). An on-chip bias network provides V_{tune} of VCC/2 when the V_{tune} pad is not connected. Similarly, an on-chip resistor sets the default bias current level for all current sources when the bias pad is not connected.

The measured and simulated oscillation frequencies are shown in Fig. 7.28.

At a tuning voltage of 1 V, the measured and simulated oscillation frequencies are equal. Table 7.3 compares the oscillation frequencies obtained in measurements, simulations and calculations. In the table, frequency f_{res} represents the resonant frequency of the LC-tank circuit including ac-coupling capacitors C_c and bias resistors R_g as shown in Fig. 7.26. It should be noted that, contrary to the VCO topology using a cross-coupled differential pair as an active negative resistance, the oscillation frequency for the topology used here is almost independent of the VCO amplitude. For example, in simulations the frequency at $V_{tune} = 0.5\,\mathrm{V}$ was found to decrease from 35.8 GHz at start-up to 35.1 GHz in steady state.

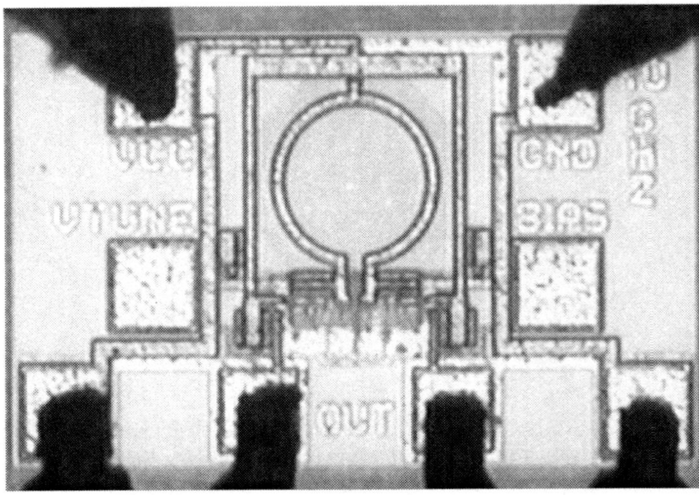

Fig. 7.27 Chip photomicrograph of the LC-VCO

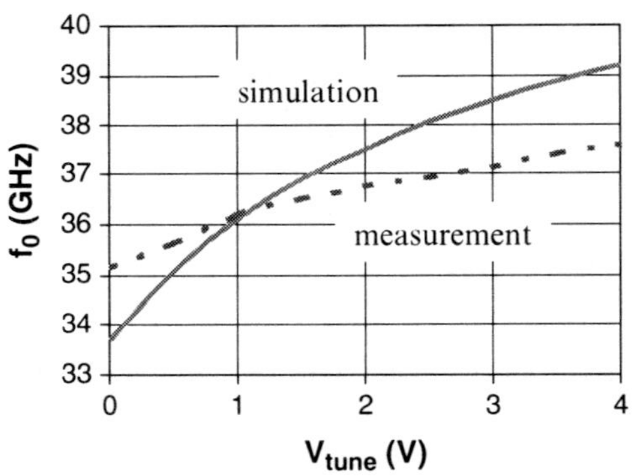

Fig. 7.28 Measured and simulated oscillation frequency versus tuning voltage

Table 7.3 Analyses of measured, simulated and calculated oscillation frequencies

Simulation	$\mathbf{V_{tune} = 0.5\,V}$	$\mathbf{V_{tune} = 3.5\,V}$
Varactor	$C_p = 32.2\,fF$ at 40 GHz	$C_p = 20.0\,fF$ at 40 GHz
Inductor	$L_p = 0.37\,nH$ at 40 GHz	
Tank	$f_{res} = 45.6\,GHz$	$f_{res} = 56.4\,GHz$
Buffer	$C_p = 19\,fF$ at 40 GHz (see Fig. 7.10)	
Current sources	$C_p = 4\,fF$ at 40 GHz	
VCO	$f_o = 35.1\,GHz$	$f_o = 38.9\,GHz$
Calculation	$\mathbf{f_o = 35.2\,GHz}$	$\mathbf{f_o = 39.9\,GHz}$
Measurement	$\mathbf{V_{tune} = 0.5\,V}$	$\mathbf{V_{tune} = 3.5\,V}$
Varactor (Fig. 7.23)	$C_p = 30.5\,fF$ at 40 GHz	$C_p = 17.4\,fF$ at 40 GHz
Inductor (Fig. 7.22)	$L_p = 0.36\,nH$ at 40 GHz	
Entire VCO	$f_o = 35.7\,GHz$	$f_o = 37.4\,GHz$

The measured tuning range is narrower than the simulated tuning range. This is not due to the tuning ratio of the varactor, since the stand-alone varactor measurements are in agreement with the results of the simulations. A plausible explanation is provided by the layout. The supply interconnect was modelled as ideal short-circuits in the simulations. However, the supply interconnect between the center-tap of the inductor and the collectors of the emitter followers is of a significant length. One line of approximately 0.3 mm length is observable between the center-tap of the inductor and the collectors of Q_1, Q_2 in Fig. 7.26; a second line of the same length is observable between the center-tap of the inductor and the collectors of Q_3, Q_4. To avoid a shorted loop around the inductor, the two lines are open ended near the active area. The supply routing is indicated in Fig. 7.29.

Since there is no nearby return path for the current in these interconnects, each supply line has a significant associated inductance L_s. Seen from the LC-tank, the collector-base capacitance C_{bc} of the first emitter followers (Q_1 and Q_3, respectively in Fig. 7.26) is in series with each L_s. The network shown in Fig. 7.30 is used to analyse the effect of L_s.

Capacitor C_2 represents the collector capacitance of the second emitter follower. A typical value of C_2 is $C_2 \approx C_{bc} + C_{cs} \approx 18\,\text{fF}$. Seen from the LC-tank, the network shown in Fig. 7.30 behaves as a frequency-dependent capacitance C_{bc}' with

$$C'_{bc} = C_{bc}\frac{1-\omega^2 L_s C_2}{1-\omega^2 L_s (C_2 + C_{bc})} \tag{7.18}$$

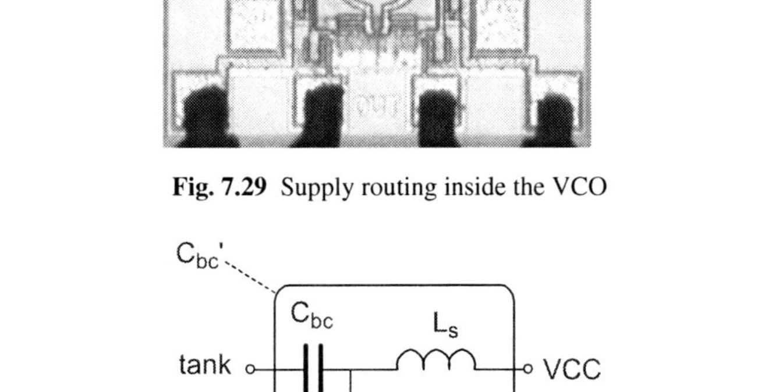

Fig. 7.29 Supply routing inside the VCO

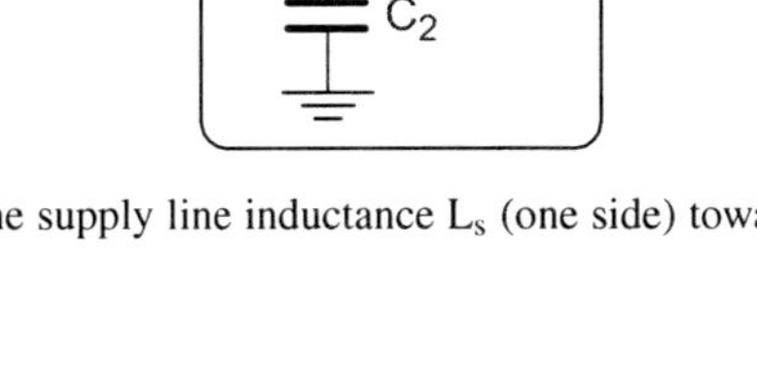

Fig. 7.30 Network with the supply line inductance L_s (one side) towards the collector of the first emitter follower

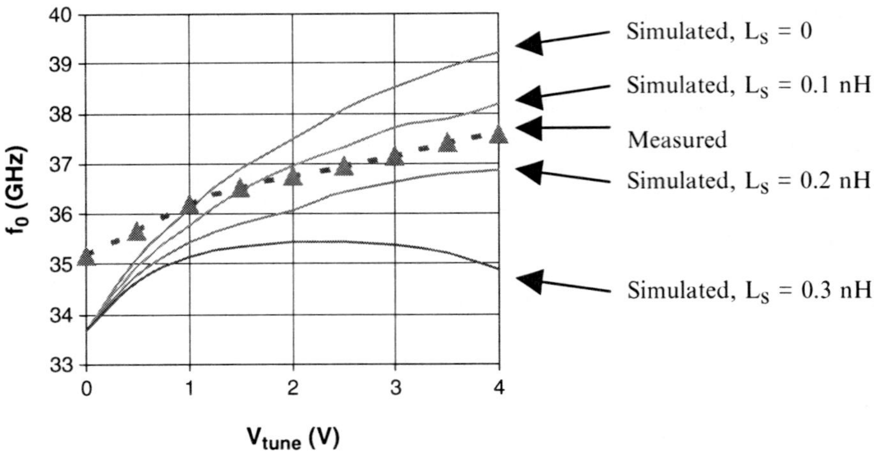

Fig. 7.31 Effect of the supply line series inductance L_s on the tuning range

The self-resonance from L_s with C_2 introduces a Miller gain to the base-collector capacitance of the first emitter followers. The increase in C_{bc}' with frequency results in a reduced tuning range of the VCO. The simulated effect of the supply line inductance L_s on the VCO tuning curve is shown for four different values of L_s in Fig. 7.31. For a supply line inductance of $L_s \approx 0.2\,\text{nH}$, the measured and simulated tuning ranges are in close agreement. A lumped inductance of 0.2 nH can be expected for a 0.3-mm-long supply line.

In order to improve the tuning range of the VCO, the layout must be improved. For example, a ground path can be placed underneath the supply line and the supply interconnect can be made wider in order to reduce the supply line inductance to less than 0.1 nH.

Figure 7.32 shows an example measurement result of the single-ended output frequency spectrum at $V_{tune} = 2\,\text{V}$ (for VCC = 4 V; V_{tune} terminal disconnected).

The measured single-ended output power is −14 dBm into 50 Ω, which corresponds to −9 dBm (versus −6 dBm in simulations) after correction for probe and cable losses of 5 dB at 37 GHz. The phase noise measured at $V_{tune} = 2.0\,\text{V}$ is shown in Fig. 7.33.

The −105 dBc/Hz phase noise at 2 MHz from the carrier is 8 dB better than the value obtained in simulations. The discrepancy between the measured and the simulated phase noise cannot be explained. It is unlikely that the noise of the LC-tank plays a role in this discrepancy since the measured quality factors of the L and C are in agreement with those obtained in simulations. Also, the measured $0.45\,V_{p,diff}$ oscillation signal amplitude is within 3 dB from the simulated $0.64\,V_{p,diff}$. So, the difference cannot be explained by a different signal swing on the tank. The difference could be due to the transistor model or model parameters, or due to a deviation in the load impedance from 50 Ω. This is an interesting area for future research. An unexplained 8 dB discrepancy between measured and simulated phase noise

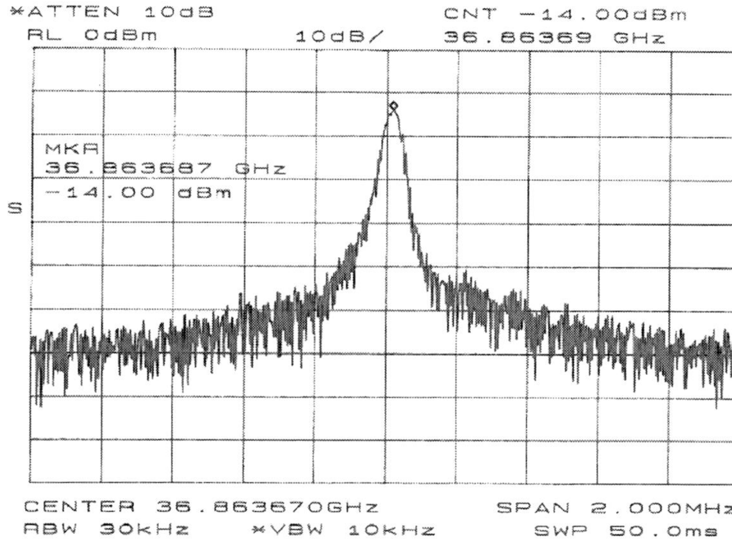

Fig. 7.32 Measured output spectrum at $V_{tune} = 2\,V$

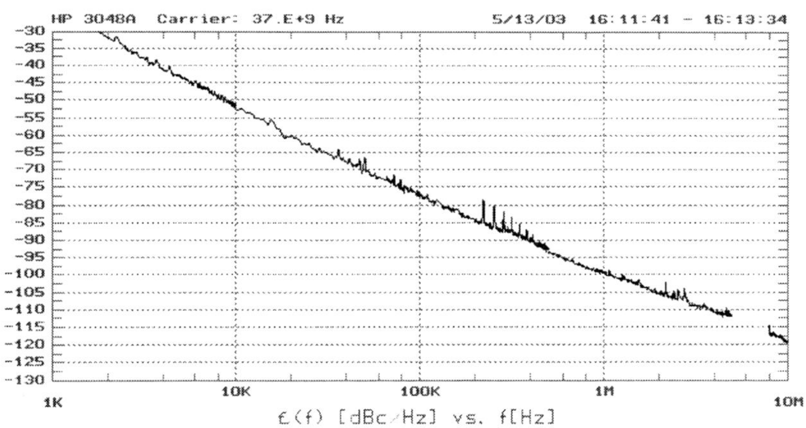

Fig. 7.33 Phase noise of the single-ended output signal measured at $V_{tune} = 2\,V$

was reported for a 40 GHz LC-VCO in [5]. In more recent publications from the same authors [10], the discrepancy in phase noise was only 2 dB. The oscillator from [10] has improved buffering of the oscillator output signal, which reduces the load pulling. The contribution of the off-chip load impedance, seen from the probe tips, to the tank impedance may explain the difference between measured and simulated phase noise for the oscillator presented in this section. The quality factor of a typical 1-m-long coaxial cable can easily exceed 100 when the input reflection coefficient of the receiver is worse than −20 dB [11]. This effect on the oscillator phase noise requires further study, for example by integrating the load resistors and

Table 7.4 Summary of measurement results

Parameter	Value
Oscillation freq. ($V_{tune} = 0.5$–$3.5\,V$)	35.7–37.4 GHz
Phase noise ($V_{tune} = 2\,V$)	-105 dBc/Hz @ 2MHz
Chip area	$0.68 \times 0.40\,mm^2$
Supply pushing	100 MHz/V
Supply current at VCC = 4 V including biasing, output buffers	21 mA
Output power into 50 Ω load	-9.0 dBm (single-ended)
Signal amplitude on tank	$0.45\,V_{p,diff}$

including additional buffering to improve the isolation between the LC-tank and the off-chip load impedance.

The measured performance of the VCO is summarised in Table 7.4.

7.7 I/Q Signal Generation

Different techniques exist for the generation of I/Q signals. A widely used technique is based on the coupling of two identical VCO cores in a loop as shown in Fig. 7.34. Both VCO cores are tuned to the same frequency and are assumed to provide a differential output. One signal inversion inside the loop creates a 180° phase shift. The oscillation condition requires a phase shift of in total 360° inside the loop and thus forces a phase difference of 90° between the differential outputs (I, nI) and (Q, nQ) of the two identical VCO cores.

Several implementations for LC-type VCO cores are known. The core uses an inductor, a capacitor and an active negative resistance. A widely used implementation for the negative resistance is the cross-coupled differential pair. So, a possible implementation for the VCO core (without interface for coupling) is as shown in Fig. 7.1.

Several implementations are known for the coupling interface, for example those published by Andreani [1]. These implementations are shown in Fig. 7.35 (parallel coupling), Fig. 7.36 (top-series coupling) and Fig. 7.37 (bottom-series coupling).

The parallel coupling is implemented by differential pairs (Q_3, Q_4) and (Q_7, Q_8), biased at currents I_c. The differential pairs are directly connected to the LC-tank circuits by both their inputs (i.e., base terminals) and outputs (i.e., collector terminals). The additional capacitance connected to each LC-tank reduces the tuning range. Besides, the finite Q-factor of the input impedance of the differential pairs reduces the margin for the oscillation condition and reduces the maximum achievable oscillation frequency with respect to the single-phase VCO core. Therefore, the parallel coupling topology is not attractive for oscillators targeting frequencies close to or above f_{cross}.

The parallel-coupled topology shows a trade-off between phase noise and quadrature phase error. At a given mismatch between the LC-tanks of the two resonator cores, a stronger coupling, implemented by an increased I_c/I_t ratio, results in better phase accuracy but higher phase noise. Phase shifters have been proposed in series with the coupling stages to achieve operation of each resonator at its peak-Q [6]. The quality factor Q_n of an n-stage LC oscillator may be higher than the quality factor of a single resonator. In [6] an approximation for Q_n is given:

$$Q_n = n \cdot Q \cdot \cos(\varphi_r) \tag{7.19}$$

with Q being the quality factor of the LC-tank, n the number of stages (e.g., for a quadrature oscillator, $n = 2$) and φ_r the phase shift at which each resonator is forced to operate within the n-stage oscillator. Reduction of Q_n leads to phase noise degradation and should therefore be avoided. In [6] it is also shown that the phase noise for an n-stage oscillator may be $10 \cdot \log(n)$ dB better than the phase noise of the resonator, provided that the phase shifters and coupling circuits are noiseless. In practice, this is typically not the case.

The series coupling circuits proposed in [1], shown in Figs. 7.36 and 7.37, typically use NMOS transistors. Some transistors operate at drain-source voltages close to zero, which is not a problem for MOS transistors. However, additional level shifts are needed if the series-coupled topologies are implemented with bipolar transistors. The series-coupled implementations have been demonstrated to produce lower phase noise than the parallel-coupled topology [1]. Of the two series-coupled topologies, the bottom-series topology shows the best phase noise performance; the top-series topology shows the best phase accuracy. In the case of neither topology is there any trade-off between phase noise and quadrature phase error; the phase error acts approximately as a design constant [1].

On the basis of the new LC-VCO core described in Section 7.6 (e.g., Fig. 7.26), I/Q VCOs can be realized using parallel or series coupling circuits. The proposed parallel coupling variant is shown in Fig. 7.38. Coupling is implemented with differential pairs (Q_9, Q_{10}) and (Q_{11}, Q_{12}), each biased at a current I_c. The input impedance of transistors Q_9–Q_{12} adds to the load capacitance of the first emitter followers Q_1, Q_3, Q_5 and Q_7. The effect of a change in the load capacitance of the first emitter followers was visualized in Fig. 7.7. At bias currents $I_c < I_2$, the impact of the coupling transistors on the maximum achievable oscillation frequency f_{LIMIT} is expected to be relatively small. The collector terminals of Q_9–Q_{12} are connected to the LC-tank circuits. The output capacitance C_{out} at each collector terminal of Q_9–Q_{12} adds to the tank capacitance. If C_{out} is large with respect to the varactor capacitance C_v (i.e., $C_{out} > C_v$), the tuning range will be significantly narrower.

An important advantage of the series-coupled topology over the parallel-coupled topology is that the loading to the LC-tank is minimised. Besides, as demonstrated in [1], the series-coupled topology shows lower phase noise and has no trade-off between phase noise and quadrature phase error. Therefore, a series-coupled topology was selected for IC implementation. The series-coupled topology is shown in Fig. 7.39. The total bias current of the I/Q VCO does not need to be increased with respect to the sum of the bias currents of the two VCO cores, due to the series connection of the coupling transistors Q_9–Q_{12}. These coupling transistors share the bias currents I_2 with the output emitter followers Q_2, Q_4, Q_6 and Q_8. A supply voltage

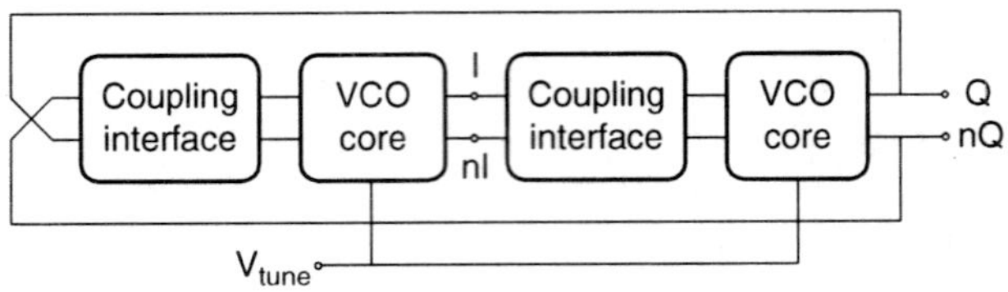

Fig. 7.34 I/Q signal generation using two identical VCO cores with coupling interfaces

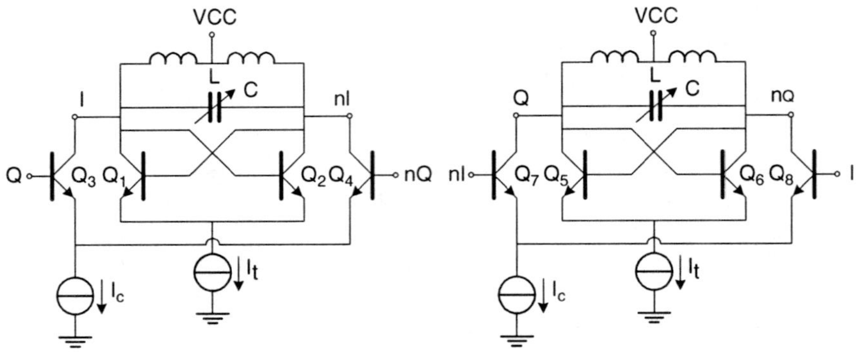

Fig. 7.35 I/Q VCO based on a cross-coupled differential pair with parallel coupling

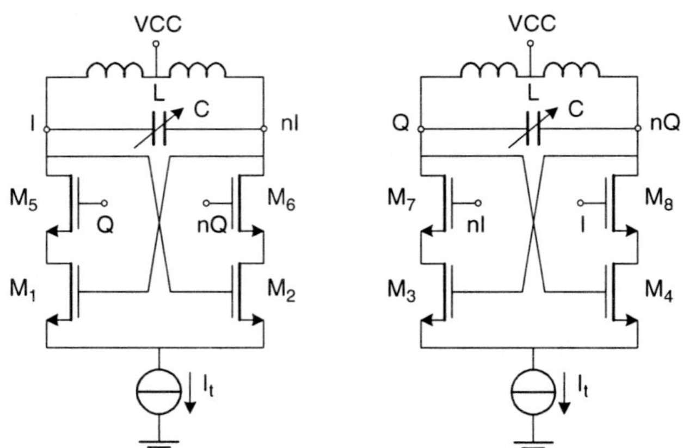

Fig. 7.36 I/Q VCO based on a cross-coupled differential pair with top-series coupling

higher than that of the non-quadrature oscillator shown in Fig. 7.26 is needed due to the extra level shift introduced by the coupling transistors. The circuit shown in Fig. 7.39 operates at a typical supply voltage of 5 V. The circuit has been implemented in the QUBiC4G technology [2].

A tuning range of 33.3–39.1 GHz was obtained in simulation. To allow accurate evaluation of the I/Q accuracy, a single side-band (SSB) mixer has been included on the I/Q VCO chip, as shown in the block diagram of Fig. 7.40.

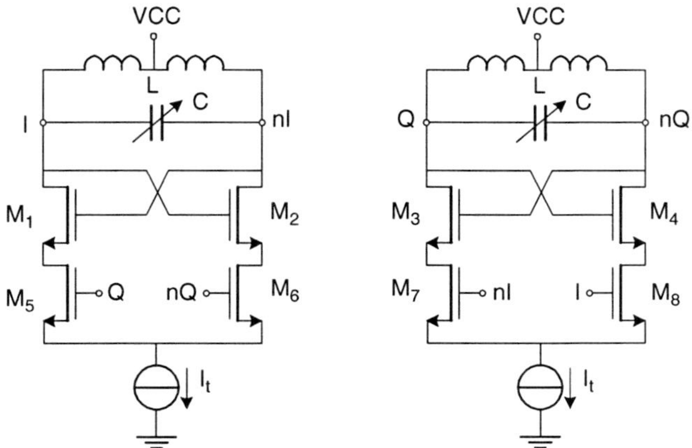

Fig. 7.37 I/Q VCO based on a cross-coupled differential pair with bottom-series coupling

The two mixers are implemented using Gilbert cells. The external input signal f_{RF} drives both mixers via a passive power splitter, implemented as shown in Fig. 7.41. The two GSG transmission lines from the f_{RF} input to the mixer input are designed for a characteristic impedance Z_0 of $Z_0 = \sqrt{(Z_i \cdot Z_l)} \approx 71\ \Omega$ and an electrical length of $\lambda/4$ for $f_{RF} = 40\,\text{GHz}$. Each 50 Ω termination resistor at the end of the transmission line then transforms into a 100 Ω input resistance at the f_{RF} input, as follows from equation (2.33). The parallel combination of the two transmission lines provides a correct 50 Ω termination of the f_{RF} source at $f_{RF} = 40\,\text{GHz}$.

The input frequency f_{RF} must be close to the VCO frequency f_0. The wanted mixer output signals IF_I and IF_Q are both at an intermediate frequency $f_{IF} = f_0 \pm f_{RF}$. The low-pass filters (LPF) are used to suppress the sideband at frequency $f_0 + f_{RF}$. Using a mixer for evaluation of the I/Q accuracy has been demonstrated before, see for example [1,8]. The relationship between the image rejection ratio (IRR), I/Q phase error $\Delta\varphi$ (with $\Delta\varphi$ in radian) and I/Q relative amplitude difference $\Delta A/A$ is [9]:

$$IRR = \frac{\left(\frac{\Delta A}{A}\right)^2 + (\Delta\varphi)^2}{4} \tag{7.20}$$

To obtain a good IRR, the amplitude accuracy is as important as the phase accuracy. A lot of attention has been paid to the layout of the IC, in order to obtain a good amplitude and phase accuracy between the I and Q signals. Ground shields have been used to avoid direct crosstalk between the I and Q signal wires at the centre of the IC, where the loop of the I/Q VCO is closed. Dummy interconnect is used to balance the parasitic capacitances loading the I and Q signals. To obtain a good phase accuracy, wires of equal length are used for the I and Q signals. To ensure a safe margin on top of the oscillation condition, care must be taken to avoid closed loops near the inductors, because they would reduce the Q-factor of the inductors.

A chip photomicrograph is shown in Fig. 7.42. The IC measures $1.7 \times 1.2\,\text{mm}^2$, of which the I/Q VCO occupies $0.7 \times 0.5\,\text{mm}^2$.

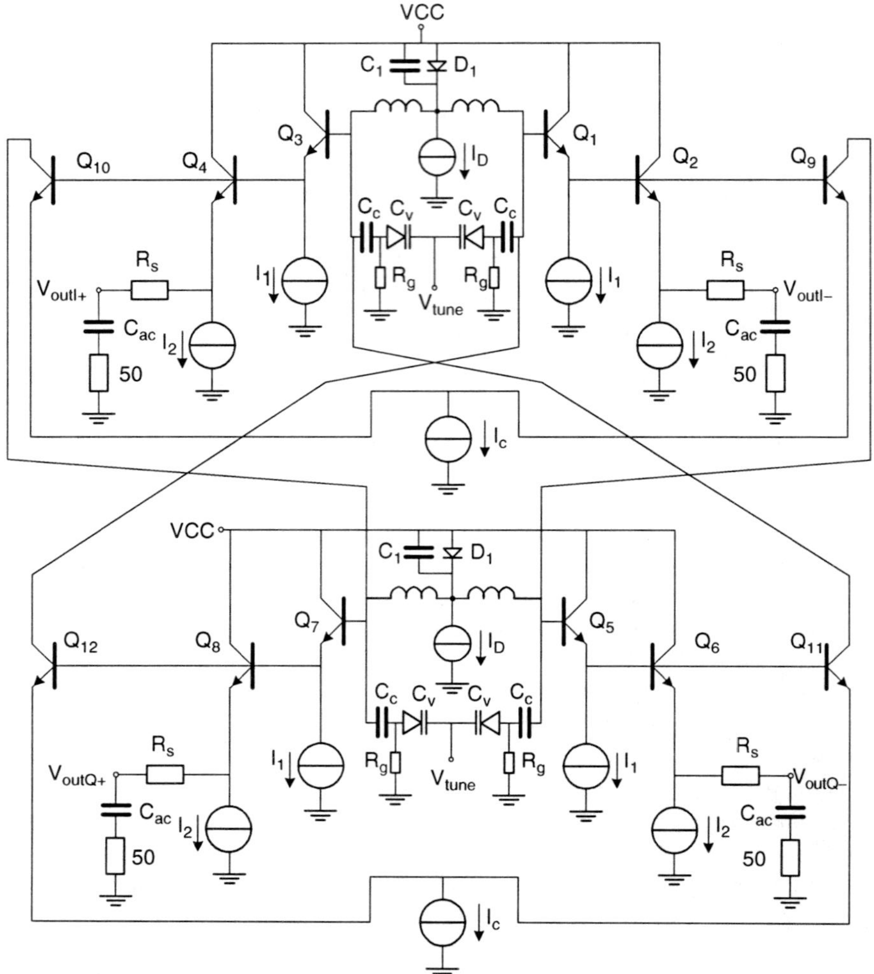

Fig. 7.38 Proposed I/Q VCO using parallel coupling, based on the new LC-VCO topology presented in Section 7.6

The measured oscillation frequency is approximately 10% lower than simulated. Besides, the measured tuning range (30.6–32.6 GHz) is narrower than expected (33.3–39.1 GHz). Similar problems were encountered when a single-phase VCO was present on the same wafers. A possible cause is the varactor capacitance density, which may be off-target. This will have to be verified by a varactor characterisation for this specific batch of wafers. Another possible explanation could be imperfections in the layout parasitic extraction routine. This requires further study.

An example oscillator output signal spectrum obtained from measurements on a single-ended I or Q output signal is shown in Fig. 7.43. The SSB mixer was loading the VCO outputs during the measurement.

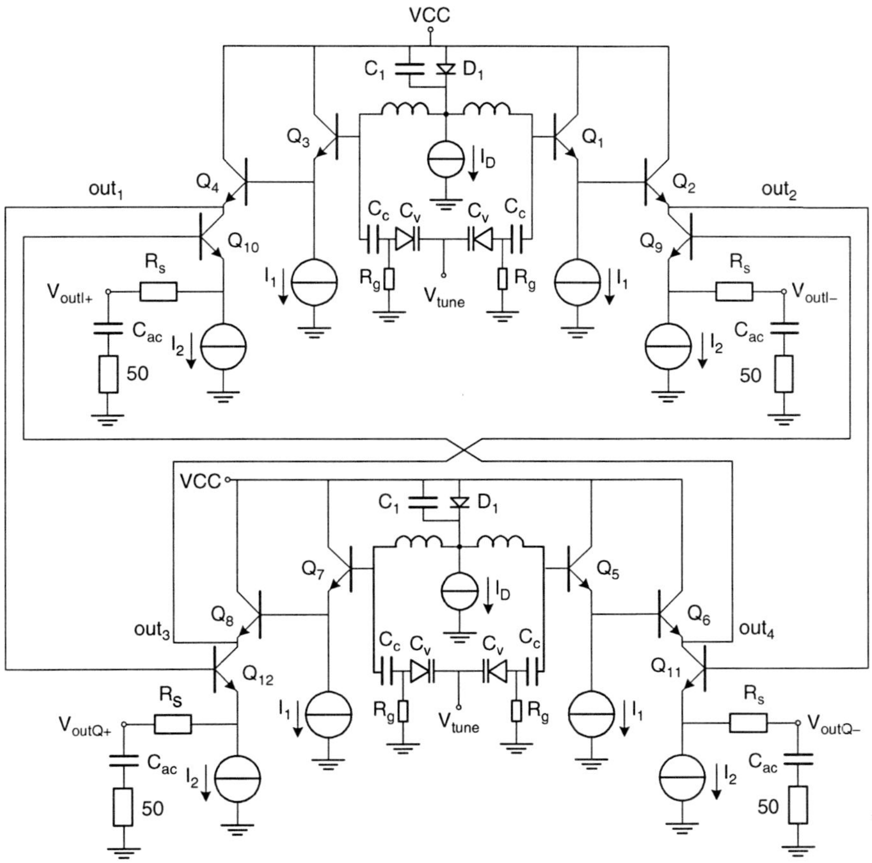

Fig. 7.39 Realised I/Q LC-VCO using series coupling

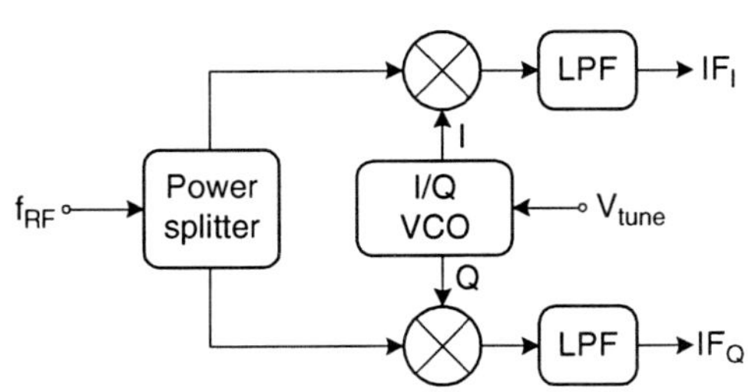

Fig. 7.40 I/Q VCO with SSB mixer

The image rejection ratio of the down-converted output signal was analysed. This measurement was performed using a Rohde & Schwarz SMIQ system. The down-converted IF_I and IF_Q signals from the I/Q VCO/down-converter IC were applied to the inputs of the SMIQ system. The SMIQ system uses the two input signals to drive an up-conversion mixer. An example measurement result obtained for the

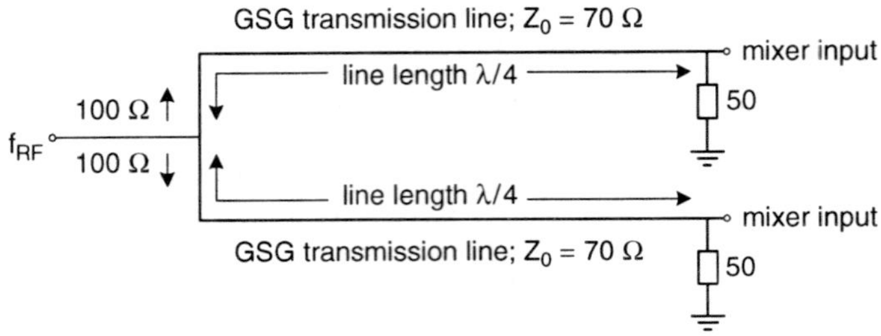

Fig. 7.41 Passive power splitter for the test signal at f_{RF}

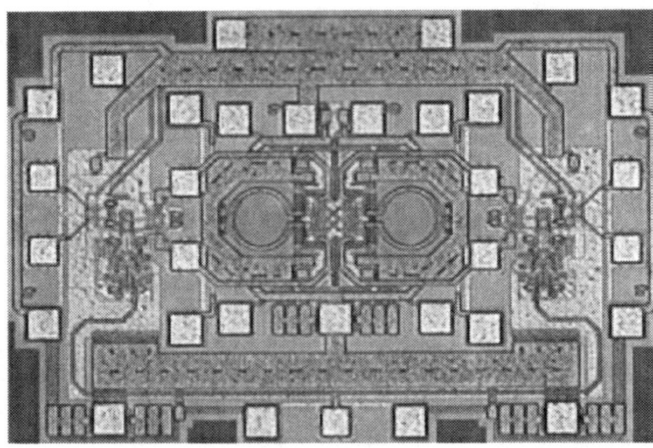

Fig. 7.42 Chip photomicrograph of the I/Q VCO with quadrature downconverter

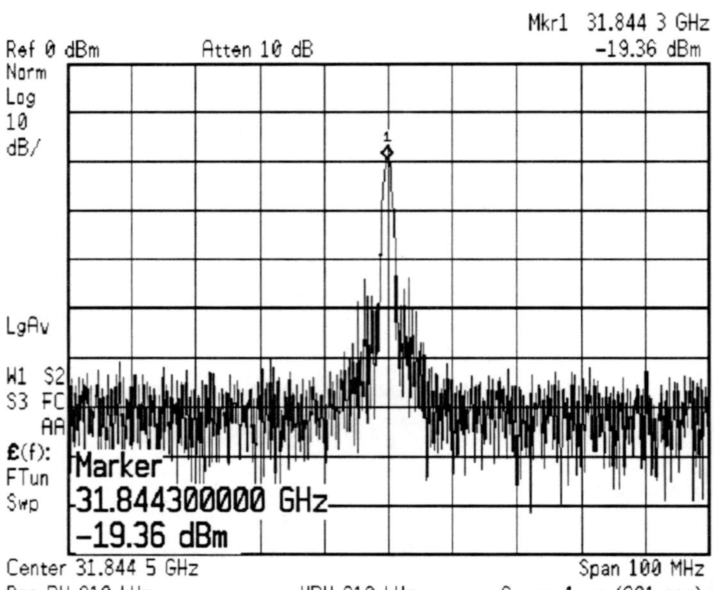

Fig. 7.43 Example output spectrum, measured single-ended at $V_{tune} = 2\,V$

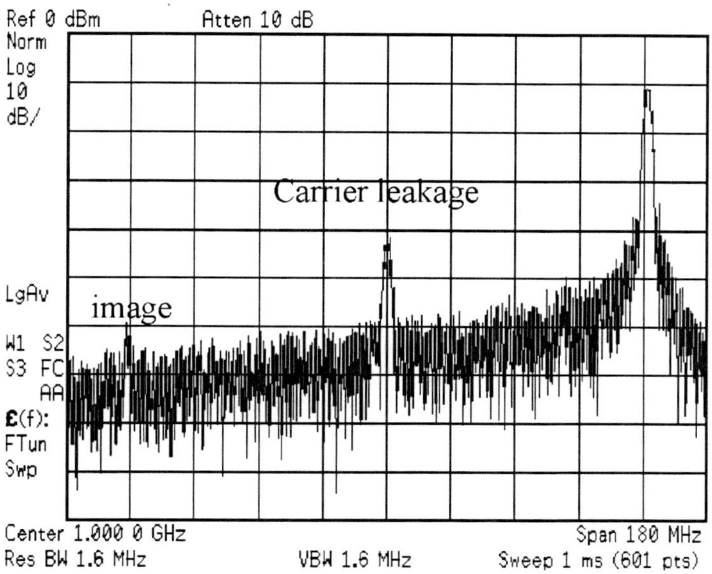

Fig. 7.44 Example IRR measurement result

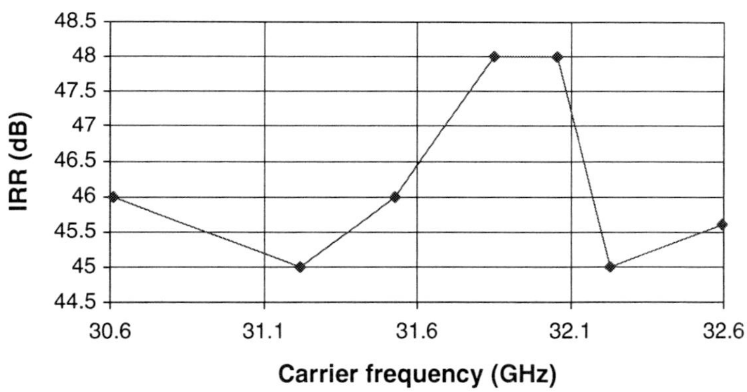

Fig. 7.45 Measured IRR versus oscillation frequency at $f_{IF} = 75\,\text{MHz}$

down-converted (at $f_{IF} = 75\,\text{MHz}$) and in the SMIQ system to 1 GHz up-converted spectrum is shown in Fig. 7.44.

The measurement results obtained for the IRR across the tuning range of the VCO are shown in Fig. 7.45. The RF signal f_{RF} applied to the down-conversion mixers was varied with V_{tune} to obtain a fixed intermediate frequency f_{IF} of approximately 75 MHz.

If an ideal matching between the amplitudes of the I and Q outputs of the VCO is assumed, the measured 45 dB IRR corresponds to a phase error of 0.6° (as follows from equation (7.20)).

The measurement results are summarised in Table 7.5.

Table 7.5 Summary of measurement results

Parameter	Value
Oscillation freq. ($V_{tune} = 0.5$–$3.5\,V$)	30.6–32.6 GHz
Phase noise ($V_{tune} = 2\,V$)	$-103\,dBc/Hz$ @ 2MHz
Chip area I/Q VCO (excluding mixers)	$0.7 \times 0.5\,mm^2$
Supply current at VCC = 5 V including biasing, output buffers	43 mA
Image reject ratio	>45 dB
I/Q phase error	<0.6°

7.8 Conclusions and Outlook

This chapter discussed the question whether an IC technology provides adequate performance for ensuring a certain target oscillation frequency f_0 and identifying the main limiting factors. The device metric f_{cross} introduced for the first time in [4] was shown to provide a direct relation between the maximum attainable oscillation frequency of an LC-VCO based on a cross-coupled differential pair and the IC technology. Equation (7.11) provides f_{cross} as a function of the IC technology parameters f_T, R_b and R_e. Using this equation, the impact of potential IC technology improvements on the maximum attainable VCO frequency (when using a cross-coupled differential pair as the negative resistance) can be predicted. An LC-VCO that operates at an oscillation frequency close to the theoretical maximum f_{cross} was demonstrated using a cross-coupled differential pair as an active negative resistance in Section 7.5.

An alternative to the active negative resistance was introduced, enabling oscillator circuits that generate frequencies beyond the limitations of present topologies when using a cross-coupled differential pair. These alternative circuits implement the active negative resistance with a capacitively loaded emitter follower. In Section 7.3 it was shown that a capacitively loaded emitter follower provides a negative shunt input resistance up to a frequency f_{LIMIT}. The value of f_{LIMIT} depends on IC technology parameters (i.e., f_T, R_b and R_e) and on the load capacitance C_l, as follows from equation (7.14). To reach a high f_{LIMIT}, the same technology requirements hold as for f_{cross}; a low base and emitter series resistance are essential. At practical values of the load capacitance C_l it is feasible to achieve f_{LIMIT} higher than f_{cross}. The shunt input negative resistance of the capacitively loaded emitter follower shows a minimum at a frequency somewhat below f_{LIMIT}. In the technology used, this minimum occurs at approximately $0.65 \cdot f_{LIMIT}$, as shown in Fig. 7.7.

A negative resistance built from a capacitively loaded emitter follower has several advantages over the cross-coupled differential pair. In the first place, f_{LIMIT} is typically higher than f_{cross} (in the technology used $f_{LIMIT} \approx 55\,GHz$ and $f_{cross} \approx 35\,GHz$) and thus the new topology allows implementation of oscillators generating frequencies that cannot be reached with a cross-coupled differential pair. In the second place, the absolute value of the shunt input capacitance is considerably lower in the new topology (for example, for the technology used at $f = 20\,GHz$, a reduction by a factor of 9 is obtained). A reduced shunt input capacitance enables a larger tuning range for the VCO. In the third place, the negative resistance can simultaneously implement the output buffer function, avoiding the need for a separate output buffer. To do so, the output buffer is formed from a second pair of emitter followers that drive a resistive load and simultaneously provide a capacitive load to the first pair of emitter followers. In microwave applications, it is common practice to apply a resistive load to the output buffer (e.g., 50 Ω single-ended), either on-chip or off-chip. In the fourth place, when use is made of a cross-coupled differential pair, the oscillation frequency will depend on the oscillation signal amplitude due to the non-linearity of the input capacitance of the cross-coupled differential pair – an effect that is significantly less pronounced in the case of an oscillator using a capacitively loaded emitter follower. In the technology used, an 18.5% increase in frequency between start-up and steady state operation was found in simulations for the cross-coupled differential pair (see Fig. 7.17). This effect could be reduced to −2.2% by using the example oscillator implemented with a capacitively loaded emitter follower.

An oscillator based on a cross-coupled differential pair, achieving an oscillation frequency close to (but below) f_{cross}, was demonstrated in Section 7.5. In addition, an oscillator based on a capacitively loaded emitter follower achieving an oscillation frequency above f_{cross} was demonstrated in Section 7.6. In the case of both VCO implementations a good match was achieved between calculated, simulated and measured oscillation frequencies. The discrepancy in phase noise between measurements and simulations has not yet been explained. The difference could be due to the transistor model or model parameters, or due to a deviation in the load impedance from 50 Ω. This requires further study, for example by integrating the load resistors and including additional buffering to improve the isolation between the LC-tank and the off-chip load impedance in a redesign.

An I/Q LC-VCO based on the new topology with a capacitively loaded emitter follower was demonstrated in Section 7.7. A series coupling topology was used to couple the two VCO cores of the I/Q VCO because this allows a minimal impact on the maximum achievable oscillation frequency and tuning range.

Many layout aspects have to be considered to obtain a good match between simulation results and measurements of a VCO intended for use at microwave frequencies. An optimum quality factor of the inductor can only be achieved if tiling is not applied inside or close to the inductor. Also, a closed interconnect loop near the inductor should be avoided. To avoid a closed loop, the differential circuits are typically implemented as two individual circuit halves, each with its own supply path. This makes the supply line inductance not fully common-mode – an effect that was overlooked in the VCO demonstrated in Section 7.6 and is believed to be the reason why the tuning range was lower than expected.

In the literature it is often stated that the varactor is dominant in the quality factor of the tank for oscillators operating at frequencies of 10 GHz and beyond. In the technology used, the quality factors of the inductor and the varactor were found to be equally important for the quality factor of the tank near 15 GHz in the 20 GHz oscillator demonstrated in Section 7.5 (see Fig. 7.14). However, this does not mean that the varactor will be dominant in any VCO operating at frequencies above 15 GHz. For example, the quality factors of the inductor and varactor are equally important for the quality factor of the tank near 30–40 GHz (30 GHz at $V_{tune} = 0.5$ V, see Fig. 7.25; 40 GHz at $V_{tune} = 3.5$ V, see Fig. 7.24) in the 37 GHz oscillator demonstrated in Section 7.6. It is therefore expected that the quality factor of the inductor will remain important for the quality factor of the LC-tank in oscillators designed for even higher oscillation frequencies in next-generation IC processes.

In more complex transceiver ICs, the oscillator output signal may depend on other signals that are generated on-chip. For example, frequency pulling may be an issue for transmitter ICs operating at low or zero IF. The transmit signal (i.e., the modulated carrier) may pull the oscillation frequency away from its target, thereby generating spurious tones in the transmit signal. Such a coupling may occur via the substrate (capacitive coupling) or via the air (inductive coupling) between bondwires or other on-chip inductors and the VCO inductor.

The new theory and oscillator designs described in this chapter are of particular interest for existing and emerging microwave applications such as 40 Gb/s optical networking, broadband data networking (11–66 GHz, 802.16), local to multipoint distribution systems (28–30 GHz, LMDS) and automotive radar at 24 and 77 GHz. With the theory presented and some basic information on the available IC technology, it is possible to determine whether an oscillator covering the frequency range for the target application is feasible.

References

1. P. Andreani, "A 2 GHz, 17% tuning range quadrature CMOS VCO with high figure-of-merit and 0.6° phase error," in *Proc. ESSCIRC*, 2002, pp. 815–818.
2. P. Deixler, R. Colclaser, et al., "QUBiC4G: a $f_T/f_{max} = 70/100$GHz 0.25 µm low power SiGe-BiCMOS production technology with high quality passives for 12.5Gb/s optical networking and emerging wireless applications up to 20GHz," in *Proc. BCTM*, 2002, pp. 201–204.
3. W. de Cock and M. Steyaert, "A 2.5V, 10GHz fully integrated LC-VCO with integrated high-Q inductor and 30% tuning range", *Analog Integrated Circuits and Signal Processing*, vol. 33, No. 2, November 2002.
4. H. Veenstra, E. van der Heijden, "A 19–23 GHz integrated LC-VCO in a production 70 GHz f_T SiGe technology," in *Proc. ESSCIRC*, 2003, pp. 349–352.
5. H. Li, H.-M. Rein, "Millimeter-wave VCOs with wide tuning range and low phase noise, fully integrated in a SiGe bipolar production technology,"*IEEE J. Solid-State Circuits*, vol. 38, No. 2, February 2003, pp. 184–191.
6. J. van der Tang, P. van de Ven, D. Kasperkovitz, A. van Roermund, "Analyses and design of an optimally coupled 5-GHz quadrature LC oscillator," *IEEE J. Solid-State Circuits*, vol. 37, May 2002, pp. 657–661.
7. L.F. Tiemeijer, et al., "Predictive spiral inductor compact model for frequency and time domain," in *Proc. IEDM*, 2003, pp. 878–878.
8. S.L.J. Gierkink, et al., "A low-phase-noise 5-GHz CMOS quadrature VCO using superharmonic coupling, *IEEE J. Solid-State Circuits*, vol. 38, July 2003, pp. 1148–1154.
9. L. Der, B. Razavi, "A 2-GHz CMOS image-reject receiver with LMS calibration," *IEEE J. Solid-State Circuits*, vol. 38, February 2003, pp. 167–175.
10. H. Li, H.-M. Rein, T. Suttorp and J. Böck, "Fully integrated SiGe VCOs with powerful output buffer for 77-GHz automotive radar systems and applications around 100 GHz," *IEEE J. Solid-State Circuits*, vol. 39, October 2004, pp. 1650–1658.
11. M. Schott, F. Lenk and P. Heymann, "On the load-pull effect in MMIC oscillator measurements," in *Proc. EUMW*, 2003, pp. 367–370.

Glossary

Abbreviations

BERT	Bit-error rate tester
BGA	Ball grid array
BIST	Built-in self-test
BPI	Bits per inch (linear track density)
CML	Current-mode logic
CMU	Clock multiplier unit
CPW	Coplanar waveguide
DCR	Data and clock recovery
DFF	Data flip-flop, consisting of two cascaded latches with inverted clock inputs
DFT	Design for testability
DMUX	Demultiplexer
ECL	Emitter-coupled logic
EECL	Double emitter-coupled logic
EM	Electromagnetic
ESD	Electro-static discharge
ETDM	Electrical time division multiplexing
FOM	Figure of merit
GaAs	Gallium-arsenide
GSG	Ground-signal-ground configuration for a transmission line
GSSG	Ground-signal-signal-ground configuration for a transmission line
IC	Integrated circuit
InP	Indium phosphide
IRR	Image rejection ratio
LNA	Low-noise amplifier
LPF	Low-pass filter
MUX	Multiplexer
NRZ	Non-return to zero
OADM	Optical add drop multiplexer

OXC	Optical cross-connect switch
PA	Power amplifier
PLL	Phase-locked loop
PRBS	Pseudo-random binary sequence
RF	Radio frequency
SE	Single-ended
Si	Silicon
SiGe	Silicon-germanium
SiGe:C	Silicon-germanium:carbon
SOI	Silicon on insulator
SSB	Single side-band
TEM	Transversal electromagnetic mode
TIA	Transimpedance amplifier
TL	Transmission line
TPI	Tracks per inch (radial track density)
UWB	Ultra wide-band
VCO	Voltage controlled oscillator
WDM	Wavelength division multiplexing

Constants

k	$1.38 \cdot 10^{-23}$	J/molecule · K	Boltzmann constant
q	$1.6 \cdot 10^{-19}$	C	Charge of a single electron
c	$3 \cdot 10^{8}$	m/s	Speed of light in vacuum
ε_0	$8.854 \cdot 10^{-12}$	F/m	Permittivity in vacuum
μ_0	$1.257 \cdot 10^{-6}$	H/m	Permeability in vacuum
V_T	0.0259	V	Thermal voltage at 300 K

Symbols

A_{cm}		Common-mode voltage gain
A_{dm}		Differential-mode voltage gain
BV_{CBO}	V	Collector-base junction breakdown voltage
BV_{CED}	V	Collector-emitter breakdown voltage when a diode is connected between base and emitter terminals
BV_{CEO}	V	Breakdown voltage between collector and emitter in open base configuration
C	F	Capacitance
f	Hz	Frequency

f_A	Hz	Available bandwidth. Bandwidth of a single transistor amplifier with resistive load driven with a voltage source at a specified low-frequency voltage gain (e.g., usually 20 dB)
f_{cross}	Hz	Maximum frequency at which a cross-coupled differential pair provides a negative shunt input conductance
f_{max}	Hz	Maximum oscillation frequency of a transistor
f_{out}	Hz	Output bandwidth at grounded base when the collector is driven with a current source
f_T	Hz	Transition frequency of a transistor. This occurs where the (extrapolated) current gain equals 1
f_V	Hz	Input bandwidth at grounded collector when the base is driven with a voltage source
f_δ	Hz	Skin effect corner frequency
f_ε	Hz	Corner frequency of the substrate network (e.g., dielectric relaxation frequency)
G	S	Conductance
gm	A/V	Transconductance i_c/v_{be}
gm_p	A/V	Transconductance when biased at peak-f_T
J_c	A/m^2	Collector current density
$J_{c,fAp}$	A/m^2	Collector current density when biased at peak-f_A
$J_{c,fTp}$	A/m^2	Collector current density when biased at peak-f_T
k		Coupling coefficient
L	H	Inductance
L_{cm}	H	Common-mode inductance
$L_{cm,sec}$	H	Common-mode inductance of one section of the line
L_{dm}	H	Differential-mode inductance
$L_{dm,sec}$	H	Differential-mode inductance of one section of the line
M	H	Mutual inductance
M		Avalanche current multiplication factor
Q		Quality factor
R	Ω	Resistance
R_{TH}	K/W	Thermal resistance
T	K	Temperature
t_d	s	Delay
$t_{d,eff}$	s	Effective delay
t_{cm}	s	Common-mode delay
t_{dm}	s	Differential-mode delay
$t_{dm,sec,eff}$	s	Effective differential-mode delay of one section
t_r	s	Rise-time (20%–80% unless otherwise indicated)
Z_{even}	Ω	Even-mode characteristic impedance
Z_{odd}	Ω	Odd-mode characteristic impedance
Z_0	Ω	Characteristic impedance
$Z_{0,eff}$	Ω	Effective characteristic impedance
Z_{0cm}	Ω	Common-mode characteristic impedance

Z_{0dm}	Ω	Differential-mode characteristic impedance
$Z_{0dm,eff}$	Ω	Effective differential-mode characteristic impedance
α	Np/m	Attenuation constant
β	rad/m	Propagation constant
β		Current gain
β_0		dc current gain
δ	m	Skin depth
δ_x	m	Effective skin depth in the x-direction
ε_r		Relative permittivity
$\varepsilon_{r,eff}$		Effective relative permittivity
$\varepsilon_{r,eff,cm}$		Effective relative permittivity for common-mode signals
$\varepsilon_{r,eff,dm}$		Effective relative permittivity for differential-mode signals
$\varepsilon_{r,sub}$		Relative permittivity of the substrate
γ	1/m	Complex propagation constant
λ	m	Wavelength
μ_n	$m^2/V \cdot s$	Electron mobility
μ_p	$m^2/V \cdot s$	Hole mobility
μ_r		Relative permeability
ρ	$\Omega \cdot m$	Resistivity
ρ_{sub}	$\Omega \cdot m$	Substrate resistivity
σ	S/m	Conductivity

Appendix A
y-parameters for a Transistor Model with Arbitrary R_e, R_b and R_c

In Section 3.3, device metrics were defined on the basis of y-parameters. The results of Section 3.3 are most useful for calculating FOMs on the basis of measured y-parameters. In Section 3.4, approximate equations for the device metrics were derived on the basis of the simplified transistor model shown in Fig. 3.16. The model includes a base series resistance R_b, but ignores the emitter series resistance R_e and collector series resistance R_c. Although ignoring R_c and R_e leads to simple approximations for the device metrics that are useful in many practical situations, both R_c and R_e may be significant for the FOMs in certain bias condition ranges. For example, in modern SiGe and SiGe:C IC processes operating at high collector current densities, the f_T may be seriously degraded due to a series resistance R_c because a Miller effect of the collector-base capacitance C_{bc} occurs (introducing a Miller gain $(1+gm \cdot R_c)$ to C_{bc}).

In this appendix, the y-parameters will be derived for a transistor model including series resistances R_e, R_b and R_c as shown in Fig. A.1. The calculations will be performed in a 3-step procedure. In the first step, the y-parameters will be derived for $R_b = R_c = 0$. This will result in the y-parameters for the transistor model shown in the inner box in Fig. A.1. In the second step, the effect of a non-zero base resistance on the y-parameters obtained in the first step will be calculated. Finally, in the third step, the effect of a non-zero collector resistance R_c on the y-parameters obtained in the second step will be calculated.

Step 1: y-parameters for a transistor with $R_b = R_c = 0$ and an arbitrary R_e

The common-emitter y-parameters if $R_b = R_c = 0$ are calculated using the definitions of equation (3.5) for the circuit in the inner box of Fig. A.1. To calculate y_{11} and y_{21}, the base is driven by an ac voltage source v_b and the collector is grounded. The voltage $v_{e'}$ at node e′ then becomes

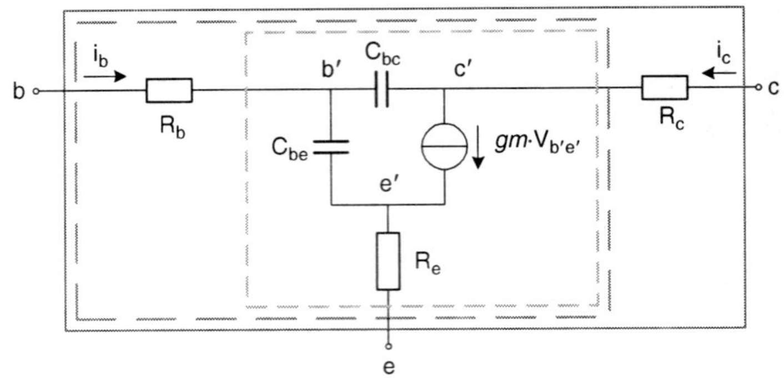

Fig. A.1 Transistor model including series resistances

$$v_{e'} = v_b \frac{(gm + j\omega C_{be})R_e}{1 + (gm + j\omega C_{be})R_e} = v_b \frac{F}{1+F} \tag{A.1}$$

Here, factor F has been introduced to simplify the equation; it equals

$$F = (gm + j\omega C_{be})R_e \tag{A.2}$$

Next, the base and collector currents are calculated, and the results are used to find y_{11} and y_{21}. This gives:

$$y_{11} = \left.\frac{i_b}{v_b}\right|_{v_c=0} = j\omega C_{bc} + \frac{j\omega C_{be}}{1+F} \tag{A.3}$$

$$y_{21} = \left.\frac{i_c}{v_b}\right|_{v_c=0} = -j\omega C_{bc} + \frac{gm}{1+F} \tag{A.4}$$

To calculate y_{12} and y_{22}, the collector is driven by an ac voltage source v_c and the base is grounded. The voltage at node e$'$ then becomes zero, resulting in:

$$y_{12} = \left.\frac{i_b}{v_c}\right|_{v_b=0} = -j\omega C_{bc} \tag{A.5}$$

$$y_{22} = \left.\frac{i_c}{v_c}\right|_{v_b=0} = j\omega C_{bc} \tag{A.6}$$

Step 2: *y*-parameters for a transistor with $R_c = 0$ and arbitrary R_b and R_e

In the next step, the effect of adding a base series resistance (or more general: a resistance in series with port 1) on the *y*-parameters of an arbitrary 2-port is calculated.

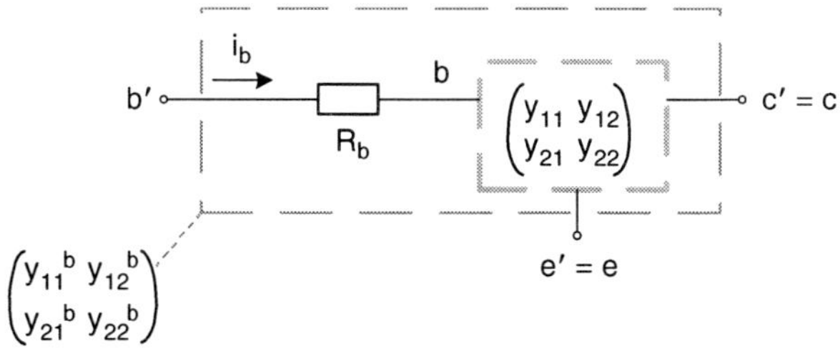

Fig. A.2 Adding a series resistance to port 1

The addition of the base resistance leads to the relation

$$v_b{}' = v_b + i_b \cdot R_b \tag{A.7}$$

The new common-emitter y-parameters $y_{ii}{}^{b}$, expressed in the common-emitter y-parameters y_{ii} of the 2-port excluding base series resistance, are calculated using the definitions of equation (3.5) for the circuit in the outer box of Fig. A.2. To calculate $y_{11}{}^{b}$ and $y_{21}{}^{b}$, the base is driven by an ac voltage source v_b and the collector is grounded. This leads to

$$y_{11}^b = \left.\frac{i_b{}'}{v_b{}'}\right|_{v_c{}'=0} = \frac{i_b}{v_b + i_b \cdot R_b} = \frac{y_{11} v_b}{v_b + y_{11} v_b R_b} = \frac{y_{11}}{1 + y_{11} R_b} \tag{A.8}$$

$$y_{21}^b = \left.\frac{i_c{}'}{v_b{}'}\right|_{v_c{}'=0} = \frac{i_c}{v_b + i_b \cdot R_b} = \frac{y_{21} v_b}{v_b + y_{11} v_b R_b} = \frac{y_{21}}{1 + y_{11} R_b} \tag{A.9}$$

To calculate $y_{12}{}^{b}$ and $y_{22}{}^{b}$, the collector is driven by an ac voltage source v_c and the base is grounded. Although the base terminal is grounded, a voltage equal to $v_b = v_b{}' - i_b \cdot \mathrm{R_b} = -i_b \cdot \mathrm{R_b}$ occurs at the internal node b. In this situation, the following relations exist:

$$i_b = y_{11} v_b + y_{12} v_c = y_{11}(-i_b R_b) + y_{12} v_c \tag{A.10}$$

$$i_c = y_{21} v_b + y_{22} v_c = y_{21}(-i_b R_b) + y_{22} v_c \tag{A.11}$$

From equation (A.10) follows:

$$y_{12}^b = \left.\frac{i_b{}'}{v_c{}'}\right|_{v_b{}'=0} = \left.\frac{i_b}{v_c}\right|_{v_b{}'=0} = \frac{y_{12}}{1 + y_{11} R_b} \tag{A.12}$$

Eliminating i_b from equations (A.10) and (A.11) yields:

$$y_{22}^b = \left.\frac{i_c{}'}{v_c{}'}\right|_{v_b{}'=0} = \left.\frac{i_c}{v_c}\right|_{v_b{}'=0} = y_{22} - \frac{y_{12} y_{21} R_b}{1 + y_{11} R_b} \tag{A.13}$$

So, the y-parameters for the two-port including a base series resistance, $y_{ii}{}^{b}$, can be expressed in terms of the y-parameters of the two-port excluding the base resistance, y_{ii}, and $\mathrm{R_b}$.

Step 3: *y*-parameters for a transistor with arbitrary R_c, R_b and R_e

In the final step, the effect of adding a collector series resistance (or more general: a resistance in series with port 2) on the *y*-parameters of an arbitrary 2-port is calculated.

The addition of the collector resistance leads to the relation

$$v_c' = v_c + i_c \cdot R_c \tag{A.14}$$

The new common-emitter *y*-parameters $y_{ii}{}^{c}$, expressed in the common-emitter *y*-parameters y_{ii} of the 2-port excluding a collector series resistance, are calculated using the definitions of equation (3.5) for the circuit in the outer box of Fig. A.3. To calculate $y_{12}{}^{c}$ and $y_{22}{}^{c}$, the collector is driven by an ac voltage source v_c and the base is grounded. This leads to

$$y_{12}^{c} = \left.\frac{i_b'}{v_c'}\right|_{v_b'=0} = \frac{i_b}{v_c + i_c \cdot R_c} = \frac{y_{12} v_c}{v_c + y_{22} v_c R_c} = \frac{y_{12}}{1 + y_{22} R_c} \tag{A.15}$$

$$y_{22}^{c} = \left.\frac{i_c'}{v_c'}\right|_{v_b'=0} = \frac{i_c}{v_c + i_c \cdot R_c} = \frac{y_{22} v_c}{v_c + y_{22} v_c R_c} = \frac{y_{22}}{1 + y_{22} R_c} \tag{A.16}$$

To calculate $y_{11}{}^{b}$ and $y_{21}{}^{b}$, the base is driven by an ac voltage source v_b and the collector is grounded. When the collector terminal c′ is grounded, a voltage equal to $v_c = v_c' - i_c \cdot R_c = -i_c \cdot R_c$ occurs at the internal node c. In this situation, the following relations exist:

$$i_b = y_{11} v_b + y_{12} v_c = y_{11} v_b + y_{12}(-i_c R_c) \tag{A.17}$$

$$i_c = y_{21} v_b + y_{22} v_c = y_{21} v_b + y_{22}(-i_c R_c) \tag{A.18}$$

From equation (A.18) follows:

$$y_{21}^{c} = \left.\frac{i_c'}{v_b'}\right|_{v_c'=0} = \left.\frac{i_c}{v_b}\right|_{v_c'=0} = \frac{y_{21}}{1 + y_{22} R_c} \tag{A.19}$$

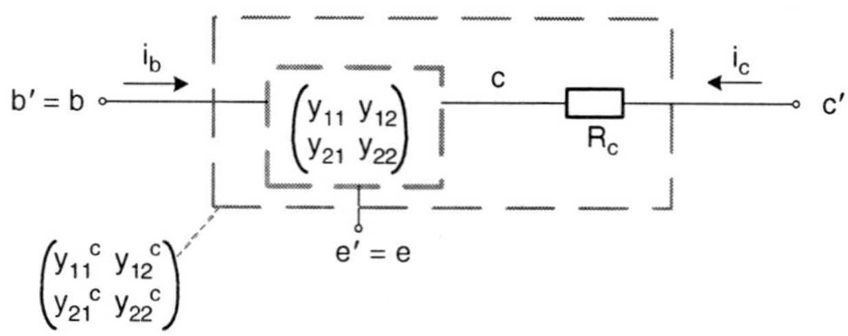

Fig. A.3 Adding a series resistance to port 2

Eliminating i_c from equations (A.17) and (A.18) yields:

$$y_{11}^{c} = \left.\frac{i_b'}{v_b'}\right|_{v_c'=0} = \left.\frac{i_b}{v_b}\right|_{v_c'=0} = y_{11} - \frac{y_{12}y_{21}R_c}{1+y_{22}R_c} \tag{A.20}$$

So, the y-parameters for the two-port including a collector series resistance, $y_{ii}{}^{c}$, can be expressed in terms of the y-parameters of the two-port excluding the collector resistance, y_{ii}, and R_c.

Index

Printed in the United Kingdom
by Lightning Source UK Ltd.
130908UK00001B/265-282/P

9 781402 068829